The Ecology of Wildlife Diseases

The Ecology of Wildlife Diseases

EDITED BY

Peter J. Hudson
Department of Biological Sciences, University of Stirling, Stirling, Scotland

Annapaola Rizzoli
Centro di Ecologia Alpina, Trento, Italy

Bryan T. Grenfell,
Department of Zoology, University of Cambridge, England

Hans Heesterbeek
Centre for Biometry, Wageningen, The Netherlands

Andy P. Dobson
Department of Ecology and Evolutionary Biology, Princeton University, USA

OXFORD
UNIVERSITY PRESS

Great Clarendon Street, Oxford OX2 6DP

Oxford University Press is a department of the University of Oxford.
It furthers the University's objective of excellence in research, scholarship, and education by publishing worldwide in

Oxford New York

Auckland Bangkok Buenos Aires Cape Town Chennai
Dar es Salaam Delhi Hong Kong Istanbul Karachi Kolkata
Kuala Lumpur Madrid Melbourne Mexico City Mumbai Nairobi
São Paulo Shanghai Singapore Taipei Tokyo Toronto

and an associated company in Berlin

Oxford is a registered trade mark of Oxford University Press in the UK and in certain other countries

Published in the United States
by Oxford University Press Inc., New York

First published 2002

British Library Cataloguing in Publication Data

Data available

Library of Congress Cataloging in Publication Data

Ecology of wildlife diseases/edited by Peter J. Hudson ... [et al.].
p. cm.
1. Animal ecology—Congresses. 2. Wildlife diseases—Congresses.
I. Hudson, Peter J.

QH540 .E32 2001 571.9′1—dc21 2001036223

ISBN 0 19 850620 1 hbk.
ISBN 0 19 850619 8 pbk.
10 9 8 7 6 5 4 3 2 1

Typeset by Integra Software Services Pvt. Ltd., Pondicherry, India
www.integra-india.com
Printed in Great Britain
on acid-free paper by Bookcraft

Foreword

As president of the Centro di Ecologia Alpina, I was extremely pleased to be part of this meeting and to present this volume. The CEA is a recently founded organization that is dedicated to the study and preservation of the alpine environment and this volume reflects many of the research themes that it has sponsored.

The meeting this book is based on was particularly important because it offered an occasion to bring together numerous colleagues from different fields with a common interest in wildlife biology. Furthermore, it was an opportunity for the host country, Italy, to consolidate years of experience in wildlife and domestic animal biology and diseases.

The Alpine environment is changing rapidly, global climate change and other pressures such as tourism have made the study and conservation of the Alps a growing priority. Local communities depend upon a stable and healthy wildlife fauna for their continued welfare. Indeed the Alps represent one of the most diverse environments that we know of with over 4500 species of plants and perhaps as many as 30 000 animal species. Human activity has changed the landscape dramatically and has modified its biodiversity, creating artificial or semi-natural habitats alongside the more traditional habitats. As a result, there are numerous vertebrate species that are declining in number and some are close to extinction. Clearly, in a delicate balance such as this, disease may well accelerate the process.

The importance of studying wildlife diseases in the alpine environment is fundamental. The increase in ecotonal areas, together with habitat fragmentation and shrinkage will tend to concentrate different species in limited areas and increase the risk of parasite transmission. At the same time this concentration of species increases the risk of new diseases emerging. For example the increase in species like ungulates has led to a dramatic rise in the number of ixodid ticks, resulting in a higher risk of transmission of tick-borne diseases to humans. There is a need for effective control strategies against ticks and their associated diseases. These can only be accomplished when we have studied the biology and behaviour of ticks in the environment.

In this ecological analysis and review of the role of disease in wildlife, the reader will find a wide range of new and exciting questions. The final chapter looks to the future and is particularly interesting. I feel personally that the integration of new techniques in molecular biology into the study of epidemiology and genetic interactions between the eternal triad of tick–disease–host particularly challenging. This volume will be an indispensable reference tool for all those interested in wildlife biology and disease, from students to professional researchers and teachers in biology and veterinary medicine.

Claudio Genchi
President of Centro di Ecologia Alpina

Preface

The initiation and production of this volume involved so many people that it is not easy to acknowledge them all fairly. In many respects the planning of this book and conference started on the completion of the Cambridge meeting, although in reality it was not until the five of us got together at a parasitological meeting in Parma that we decided that a follow-up meeting, with more empirical workers and active veterinary workers, was needed. With initial ideas and plans, two of us met in Bryan Grenfell's kitchen in Cambridge and sat down over a cup of coffee with Bryan's 'character-infested' cat known fondly as 'Fat Bastard'. We sketched out the preliminary plans for a Wildlife Epidemiology meeting that highlighted the questions but unified the discipline. Bryan has moved on to a much posher house and, sadly, the cat has moved on to another kitchen 'in the sky', but we wanted to record for posterity the cat's contribution. We know many epidemiologists remember Fat Bastard with fondness, and yet his name never appears in any scientific publication. Now it does.

The real trick to a good conference is charming people, excellent food, copious quantities of good wine in a setting that is both conducive to thought and yet sufficiently attractive to encourage people to travel half-way round the world. Getting the charming people is the easy bit—the five editors drew up a short-list of the 50 people we thought would make the meeting work. However, the real challenge was to provide the correct setting and we were particularly fortunate that Anna could provide the ideal location in Trento. Through the continuing and remarkable support of the Centre's President, Professor Claudio Genchi, the Director, Gianni Nicolini and Research Director, Claudio Chemini, Anna acquired generous support from the *Centro di Ecologia Alpina, The Autonomous Province of Trento* and from the newly formed *Societá Italiana di Ecopatologia della Fauna*, an Italian wildlife disease society. We would also like to thank *The Park Services, The Wildlife Service, The Hunting Association of The Province of Trento, The Istituto Agrario di S.Michele all'Adige, Informatica Trentina, Castello del Buonconsiglio* and Nerio Giovanazzi.

Throughout the conference, the staff at *CEA* were remarkably helpful and it would be a long list to mention all by name. Nevertheless, we must make two exceptions. First, we wish to provide special thanks to our friend and colleague Isabella Cattadori. Isa organized every little detail, made sure everything and everyone worked when they should have done, kept us sane, and yet made sure we enjoyed ourselves. Second, Roberto Rosà who was heroic in his assistance with the checking and double-checking of the references during the birth of the book.

The saying goes that one should never work with children and animals—at times we thought that could be extended to include 50 independently thinking scientists from various parts of the world who were asked to contribute to discussions and write paragraphs for various chapters, which must then be moulded into a book on diseases. We jest of course—they were all amazing and tolerant, although we do realize that this tolerance was an emergent property of the excellent location and good-quality Italian food and wine. We would never have achieved any of this without our lead authors. Particular thanks for their perseverance and tolerance were, in chapter order: Ken Wilson, Dan Tompkins, Mick Roberts, Jonathan Swinton,

George Hess, Sarah Randolph, and Sarah Cleaveland. We also wish to thank the people from related disciplines who helped to focus our attention during the meeting, in particular Ottar Bjørnstad, Bill Amos, Jon Greenman, Rachel Norman, Angela McLean, and Andrew Read. Also, special thanks to our colleagues and friends in SIEF who were so kind during our visit to Italy. Finally many thanks to Ian and Cathy and the people at Oxford University Press for helping us get this finished. We never believed any of your threats.

Peter Hudson, Annapaola Rizzoli, Bryan Grenfell, Hans Heesterbeek and Andy Dobson

Contents

Contributors

Bill Amos, Department of Zoology, Cambridge University, Downing Street, Cambridge, England, UK. CB2 3EJ. e-mal w.amos@zoo.cam.ac.uk

Per Arneberg, Department of Biology, University of Tromsø, 9037 Tromsø, Norway. e-mail pera@ibg.uit.no

Michael Begon, SOBS, Nicholson Building, University of Liverpool, L 69 3BX UK. e-mail mbegon@liv.ac.uk

Ottar Bjørnstad, National Centre for Ecological Analysis and Synthesis, University of Santa Barbara, California, USA. e-mail ottarnb@bio.uio.no

Isabella Cattadori, Centro di Ecologia Alpina, 38040 Viote del Monte Bondone, Trento, Italy. e-mail cattadori@cealp.it. Department of Biological Sciences, University of Stirling FK9 4LA Scotland, UK. e-mail imcc1@stir.ac.uk

Claudio Chemini, Centro di Ecologia Alpina, 38040 Viote del Monte Bondone, Trento, Italy. e-mail chemini@cealp.it

Sarah Cleaveland, Centre for Tropical Veterinary Medicine, University of Edinburgh, Easter Bush, Roslin, EH 25 9RG Midlothian, Scotland, UK. e-mail sarah.cleaveland@ed.ac.uk

Giulio de Leo, Dip. Elettronica, Politecnico di Milano, Via Ponzio 34, 20133 Milano, Italy. e-mail deleo@elet.polimi.it

Andy Dobson, EEB, Eno Hall, Princeton University, NJ 08540 New Jersey, USA. e-mail andy@eno.princeton.edu

Chris Dye, Global TB programme, World Health Organization, CH 121127, Geneva, Switzerland. e-mail dyec@who.ch

Nicola Ferrari, Centro di Ecologia Alpina, 38040 Viote del Monte Bondone, Trento, Italy. e-mail ferrari@cealp.it

Ezio Ferroglio, Departmento di Produzioni Animali Epidemiologia e Ecologia, Via Nizza 52, 10126, Torino, Italy. e-mail ferrogli@veter.unito.it

Cesare Furlanello, ITC – IRST, 38050 Povo, Trento, Italy. e-mail furlan@ite.it

Claudio Genchi, Istituto di Patologia Generale Veterinaria, Università di Milano, Via Celoria 10, 20100, Milano, Italy. e-mail claudio.genchi@unimi.it

Jon Greenman, Department of Computing Sciences and Mathematics, University of Stirling FK9 4LA Scotland, UK. e-mail j.v.greenman@stir.ac.uk

Bryan Grenfell, Department of Zoology, Cambridge University, Downing Street, Cambridge, England, UK. CB2 3EJ.e-mail bryan@zoo.ca.ac.uk

Vittorio Guberti, Istituto Nazionale per la Fauna Selvatica, Via Ca Fornacetta 9, 40064 Ozzano E., Bologna, Italy. e-mail infsvete@iperbole.bologna.it

Rosie Hails, CEH Oxford, Mansfield Road, Oxford OX1 3SR England, UK. e-mail rha@mail.nerc-oxford.ac.uk

John Harwood, Sea Mammal Research Unit, University of St. Andrews, KY16 9TS Scotland. e-mail jh17@st-andrews.ac.uk

Hans (J.A.P.) Heesterbeek, Centre for Biometry, Wageningen, P.O. Box 16, 6700 AA Wageningen, Netherlands. e-mail j.a.p.heesterbeek@plant.wag-ur.nl

George Hess, Forestry Department, North Carolina State University, Raleigh, NC 27695 – 8002 North Carolina, USA. e-mail grhess@ncsu.edu

Peter Hudson, Department of Biological Sciences, University of Stirling FK9 4LA Scotland, UK. e-mail p.j.hudson@stir.ac.uk

Linda Jones, CEH Oxford, Mansfield Road, Oxford OX1 3SR England, UK. e-mail ldj@mail.nerc-oxford.ac.uk

Rosina Claudia Tammi Krecek, Department of Veterinary Tropical Diseases, Faculty of Veterinary Sciences, University of Pretoria, 0110 Onderstepoort, South Africa. e-mail krecek@opi.up.ac.ja

Paolo Lanfranchi, Istituto Patologia Generale Veterinaria, Università di Milano, Via Celoria, 10, 20100 Milano, Italy. e-mail p-lan@imiucca.csi.unimi.it

Karen Laurenson, Centre for Tropical Veterinary Medicine, University of Edinburgh, Easter Bush, Roslin, EH 25 9RG Midlothian, Scotland, UK. e-mail Karen.Laurneson@ed.ac.uk

Antonio Lavazza, Istituto Zooprofilattico Sperimentale della Lombardia e dell'Emilia-Romagna, Via Bianchi, 9–25124 Brescia Italy. e-mail alavazza@bs.izs.it

Maria Teresa Manfredi, Istituto Patologia Generale Veterinaria, Università di Milano, Via Celoria 10, 20100 Milano, Italy. e-mail mariateresa.manfredi@unimi.it

Hamish McCallum, Department of Zoology, University of Queensland, Brisbane 4072, Queensland, Australia. e-mail hmccallum@zoology.uq.edu.au

Angela McLean, IAH Compton, Newbury, Berks RG 20 7NN England, UK. e-mail angela.mclean@bbsrc.ac.uk

Graham Medley, Biological Sciences, University of Warwick, Coventry CV4 7AL, England, UK. e-mail medley@oikos.warwick.ac.uk

Stefano Merler, ITC irst, 38050 Povo, Trento, Italy. e-mail merler@itc.it

Dave Newborn, Game Conservancy Trust, Swale Farm, Satron, Gunnerside, Richmond DL11 6JW, N. Yorks, England, UK.

Rachel Norman, Department of Computing Science and Maths, University of Stirling FK9 4LA Scotland, UK. e-mail rachel.norman@cs.stir.ac.uk

Giovanni Poglayen, Istituto Malattie Infettive, Via S. Cecilia 30, 38123 Messina, Italy. e-mail malinvet@imeuniv.unime.it

Andrea Pugliese, Dipartimento di Matematica, Università di Trento, 38050 Povo Italy. e-mail pugliese@science.unitn.it

Sarah Randolph, Department of Zoology, University of Oxford, South Parks Road, Oxford OX1 3PS, England, UK. e-mail sarah.randolph@zoology.ox.ac.uk

Andrew Read, Institute of Animal Population Biology, University of Edinburgh, EH 93 JT Edinburgh, Scotland, UK. e-mail aread@holyroad.ed.ac.uk

Annapaola Rizzoli, Centro di Ecologia Alpina, 38040 Viote del Monte Bondone, Trento, Italy. e-mail rizzoli@cealp.it

Mick Roberts, Agresearch, Wallaceville Animal Research Centre, PO Box 40063, Upper Hutt, New Zealand. e-mail mick.roberts@agresearch.co.nz

Roberto Rosà, Centro di Ecologia Alpina, 38040 Viote del Monte Bondone, Trento, Italy. e-mail rosa@cealp.it

Fausta Rosso, Centro di Ecologia Alpina, 38040 Viote del Monte Bondone, Trento, Italy. e-mail rosso@cealp.it

Arne Skorping, Department of Zoology, University of Bergen, 5007 Bergen, Norway. e-mail arne.skorping@300.uib.no

Gary Smith, New Bolton Center, 382 W. Street Road, Kennett Square, PA 19348, USA. e-mail garys@vet.upenn.edu

Jonathan Swinton, Department of Zoology, Cambridge University, Downing Street, Cambridge CB2 3EJ, England, UK. e-mail js229@cam.ac.uk

Daniel Tompkins, Department of Biological Sciences, University of Stirling FK9 4LA Scotland, UK. e-mail dmt1@stir.ac.uk

Peter White, Department of Biological Sciences, University of Stirling FK9 4LA Scotland, UK. e-mail p.j.white@stir.ac.uk

Ken Wilson, Department of Biological Sciences, University of Stirling FK9 4LA Scotland, UK. e-mail kw2@stir.ac.uk

Rosie Woodroffe, Biological Sciences, University of Warwick, Coventry CV4 7AL. e-mail woodroffe@oikos.warwick.ac.uk

Mark Woolhouse, Centre for Tropical Veterinary Medicine, University of Edinburgh, Easter Bush, Roslin, EH 25 9RG Midlothian, Scotland. e-mail mark.woolhouse@ed.ac.uk

Enrico Zaffaroni, Istituto Patologia Generale Veterinaria, Università di Milano, Via Celoria 10, 20100 Milano, Italy. e-mail enrico@iconet.it

CHAPTER 1

Ecology of wildlife diseases

P. J. Hudson, A. P. Rizzoli, B. T. Grenfell, J. A. P. Heesterbeek, and A. P. Dobson

For real progress, the modeller as well as the epidemiologist must have mud on their boots. (David Bradley 1982)

1.1 Historical background to this book

The majority of living organisms are parasitic in one form or another, be they a virus invading a sea coral, a tapeworm within the guts of your dog, a cuckoo chick in a reed warbler's nest or a lion stealing a hyaena's kill on the African plain. The parasites are intimately linked with their host in a trophic interaction that has fascinating implications for both parasite and host. Most of the more traditional parasite species interact with their hosts at a range of organizational scales that stretches from the miniscule—the molecular battle between the host's immune response and the evading action of the parasite—through to the extensive—the structuring of communities and a driving force in the evolution of biodiversity and sex. These interactions also range over many orders of temporal scale, ranging from short time-scales of the history of an individual infection and an epidemic outbreak to time-scales of endemic co-exisitence and co-evolution. Few disciplines cover such breadth and few disciplines have developed so rapidly over the past 30 years.

Research areas evolve at different rates, at times needing periods of intense empirical work to set a foundation of understanding so that at other times, novel approaches and new theories can be formulated, and the discipline can expand rapidly. Modern epidemiology is such a discipline. For many years parasitologists focused on unravelling the complex and intriguing life cycles of parasites, describing in detail the stages of the life cycle, the parasitic mode of life, and banking the data in journals and texts, some of which were rather dry and repetitive descriptive lists. Then a series of workers started to realize that parasitology was not just a descriptive science but there were parallels with other disciplines, and the time had come to start synthesizing such disciplines. In our minds, an essential step was taken when Anderson and May (1978) synthesized our understanding of parasitology with the quantitative discipline of population biology. They pointed out that the parasite–host relationship was not simply the impact a parasite had on an individual, but an integral of these interactions at the population level, and at the same time a dynamic process where parasites were flowing from one host to the next; the rate at which this took place was determined by host behaviour and abundance. Where the hosts were suffering from the parasite insult with reduced survival and fecundity, they were also fighting back in terms of innate resistance and acquired immunity, leading to an evolutionary arms race where both were fighting for their lives and their fitness. They followed in the footsteps of major theoretical figures such as Ross, Macdonald, and Bartlett—but Anderson and May ignited the discipline at the right time. This was consolidated and extended in a meeting held in Dahlem in 1981 (Anderson and May 1982a). In many respects the Dahlem meeting was inspirational—the group of workers produced detailed group reports on what they felt needed to be done and which direction the discipline should take. While the original impetus was derived from population ecology and parasitology in natural host populations, much of the subsequent work over the next decade focused on applying population biology to human infections—notably in response to the AIDS epidemic (Anderson and May 1991).

At the same time, an increasing number of workers applied these ideas to explore the impact of diseases in naturally fluctuating wildlife populations, particularly in the applied context of conservation biology. This work was synthesized in a meeting organized in 1993 at the Isaac Newton Institute in Cambridge. The resulting volume (Grenfell and Dobson 1995) summarized the state of play of wildlife-disease modelling. There has been a large body of subsequent theoretical and empirical work on the ecology and evolution of infectious disease in natural populations. The volume and quality of this work now demands another synthesis—hence the meeting upon which this book is based.

We also now need a wider unification—wildlife disease ecology should bring evolutionary biology and parasite population dynamics closer together but also reach out to immunology, genetics, and molecular biology. In this respect there has been inspirational work by evolutionary biologists like the late Bill Hamilton, who has addressed questions of how hosts signal their disease status to prospective mates and how parasites have been important in the evolution of sex.

Following the excellent tradition laid down by the meetings in Dahlem and Cambridge we organized a meeting in a location that was both beautiful and inspiring. The Grenfell and Dobson (1995) volume was set in a mathematical institute and consequently started to look at the disease issues from a theoretical, mathematical viewpoint. In contrast we thought it would be timely to base the next volume in an empirical research station and take a more empirical and ecological view. We chose the *Centro di Ecologia Alpina*, a young and dynamic research centre, based close to the top of Monte Bondone, Trento in the Dolomitic Alps. The scenery was breath-taking, the facilities excellent and the hospitality of the president, director, and staff beyond belief. Ideal for quiet thought but a location that brought people close together and one that would result in synergy. We obtained funding and support, and invited 50 research workers, selected to cover the breadth of the discipline, and also for their ability to think, to write, and to communicate. We know we did not get everybody we should have but the mixture was good and the discussions buzzed. This volume is the result of that meeting. We organized the day into some formal talks, where we asked two people from each discipline to provide contrasting views on specific questions before we all retired to group discussions. During these discussions the chair and rapporteur called on others to give 5-minute presentations about models, data, or thoughts, and guided the discussion into stimulating areas of disease research. We then got a lead author for each chapter and asked them to produce a chapter for the volume, which we could edit into one coherent text. A great idea but not a simple task. We hope this text will provide another step in our understanding of wildlife disease ecology.

1.2 Subject area and content

This book summarizes recent developments and thoughts in wildlife disease ecology. We asked 50 of the leading workers to identify the interesting and important issues and to look at these from a range of perspectives, to try and mould the veterinarians view with that of the ecologists, and to stimulate empiricists and theoreticians to work closely together. We wanted to provide an understanding and an insight into disease dynamics that spans ecology and evolutionary questions and uses techniques from the molecular sciences through to mathematics. We do not believe that any one individual, or even a small group of workers, could achieve this. Moreover, we believe that this insight will not only provide intellectual satisfaction to ourselves but also provide guidance in the applied fields of disease control, conservation of animals threatened by disease, and help set future research agendas.

This volume addresses a series of what the editors thought were the most intriguing and stimulating areas, where we could bring together workers with disparate visions. First, we wanted to build on the volume by Grenfell and Dobson (1995) to see how the discipline was progressing but at the same time we did not want to fall into the trap of asking smaller and smaller questions that eventually run the risk of becoming insignificant. One key area for discussion was the role of spatial dynamics in disease transmission and the impact that the structuring of host populations has in influencing disease spread and persistence (Chapter 6). Second, we wanted to complement the earlier volumes and address areas of work that had not been fairly covered. One of these was tick ecology and research; here, improved molecular techniques and the use of satellite imagery have made remarkable progress and can make important predictions about how global climate change can affect disease prevalence (Chapters 7 and 6). Originally, workers thought that transmission rates were simply dependent on host density and that such diseases were unlikely to have an important effect on small, threatened populations. This is not the case—there are several species of canid that are currently under threat from diseases shared with reservoir hosts (Chapter 8). As we start to investigate multiple-host systems, parasite communities (Chapter 4), the role of reservoir hosts, and strain competition, we need to keep evaluating the role that parasites have in influencing host population dynamics and shaping community structure through apparent competition (Chapter 3). At the same time we need to develop new and tractable mathematical techniques to evaluate these questions and apply stochastic modelling techniques to problems of small populations and probabilities of disease persistence (Chapter 9). One area where there has been progress in recent years is in our understanding of the relative role of deterministic and stochastic forces in determining the onset of epidemics and their subsequent fate (Chapter 5).

Much of our understanding of disease is based on studies of individuals, particularly the very detailed studies that have been undertaken by veterinary workers. A major challenge is to link our understanding of individual level infections to how disease flows through host populations and so influences host dynamics. Such variation in parasitism between individual hosts are probably best understood for macroparasitic worms. We therefore start with a careful look at variations between individual hosts (Chapter 2), then examine how these influence the likelihood of regulating a host population (Chapter 3). Indeed, a central issue in the original models of Anderson and May (1978) was the impact of parasite aggregation, as captured by the negative binomial distribution. Understanding the origins and impact of aggregation are still a major focus for research. As ever, the relative role in determining aggregation of ecological and genetic heterogeneities, and the stochastic process of transmission, are key questions (Chapter 7).

We have tried in these first few pages to show you what excites us about our discipline and we hope encourage you to delve further into the book. Our publisher hopes you will get your money out and buy it.

1.3 How to use this book

This book is aimed specifically at those with biological interests—if you are interested in biology then this book will help to reveal the exciting new developments in wildlife disease ecology. We have a clear and simple objective: to unify studies of infectious disease biology and stimulate further growth in this field. So, if you are a medical or veterinary clinician, a molecular biologist, or a biology student just starting off on a career, this book is for you. The book is not a compendium of taxonomical life histories—there are already a number of excellent books on the market that cover these. One of our favourites, which focuses just on nematodes, is the outstanding volume by Anderson (2000).

The trick with any good non-fiction book is to learn how to find your way round it so that you can glean the details you need efficiently and effectively and enjoy doing it. We have tried to help you in this respect by structuring the chapters along the same lines and presenting the information in a way that we hope is easily accessible. Each chapter starts with a 'motivation statement' that identifies the key underlying questions that are addressed in the chapter and captures in a phrase the essence of the chapter. The background sets the scene and then we launch into the theory, the questions, what is known, and specific study cases to illustrate the issues. We have used boxes throughout the book for two main purposes. First, to provide background on a specific parasite–host system, for example phocine distemper virus in seals or helminths of Soay sheep. Second, to provide some of the more detailed technical details, of mathematical models or statistical procedures. We wanted to include both the biology and the theory without cluttering up the text and yet it is these very details that we know readers and students wish to find quickly, so they can use the book as a source of information. Finally, we have been brave and provided a vision about where the discipline is going and what the questions are that need to be tackled—Chapter 9. Actually, this is not quite true, we the editors were not brave—we asked our colleagues from parallel disciplines to look at our discipline and to identify what needed doing. Try it—we would like to think of this book sitting next to the computer of the next generation of bright young epidemiologists or sitting next to the armchair of our own retired professors as they cogitate on where their students took their discipline.

1.4 Who are the players in the parasite game?

Traditional parasitologists may well ask themselves if this book is about helminths or viruses? The answer is both. We follow the useful division of parasites into two groups: the macroparasites and the microparasites. We define these and then introduce a most important player in the world of epidemiology: R_0.

Macroparasites are the parasitic species where reproduction usually occurs via the transmission of free-living infective stages that passes from one host to the next. Direct reproduction rarely occurs within the definitive host, although asexual reproduction can occur in the intermediate hosts, e.g. Digenean trematodes often multiply within snails. It is true that compared to the microparasites they are relatively large, have long generation times, and are characterized by a great diversity of antigens so that immunity is transient and is a function of the history of infection. Infections tend to be chronic leading to morbidity rather than mortality. Our understanding of the system is based on the number of parasites per host, and models must capture the details of the intensity of these infections. They include the helminths (worms) and arthropods.

Microparasites in contrast tend to have rapid reproduction within a host and do not have a special infective stage. The generation time is short, such that populations rise rapidly within their host leading to a crisis that leads either to host death or the development of immunity. Antigens are usually simple and immunity is often life-long. The state of the host is the basis of our understanding, such that hosts can be classified as susceptible, latent (infected but not infectious), infectious, or recovered (immune). From this classification we can make compartmental models of disease dynamics—very different from the macroparasite models that explicitly consider the intensity of infection. The microparasites tend to include the bacteria, viruses, protozoa, and fungi.

$\boldsymbol{R_0}$ (read as 'R nought'). One of the fascinations of disease dynamics is that many aspects of the biology of the disease can be captured by this parameter and it would be rather unfair to progress much further into this book without introducing you to it. R_0 is fundamentally a measure of parasite fitness, since it is the number of new infections arising from an infected individual (microparasites) or number of female worms that are established from a female worm (macroparasite) in a population of fully susceptible hosts, where there are no density dependent constraints operating (macroparasites). R_0 influences many features of the epidemiology of an infection and wholly or partly determines whether a parasite can invade a susceptible host population, the subsequent pattern of disease dynamics, which strain or competing species dominates, the persistence pattern of an infection, and a range of other conditions. We use R_0 for all types of parasites. For an introduction see Anderson and May (1991) and Diekmann and Heeseterbeek (2000), which includes a detailed explanation of its calculation.

The introduction of R_0 encourages us to point out a further fascination of parasites—a duality—parasites act at a range of scales and actions that can be looked at in two ways. R_0 determines both the epidemiological patterns of the infection in the host population and the fitness of the parasite. Another duality: parasites are a threat to biodiversity, wiping out some host species through apparent competition (Hudson and Greenman 1998) but at the same time are considered a major selective force behind the evolution of biodiversity both within a host population and by selecting for hosts that are spatially distributed to avoid parasitism. Highly virulent parasites lead to rapid mortality of their host but by causing wounds and making a host cough and splutter they also increase transmission rate between hosts leading to the question: Where should individual parasite genotypes sit in the trade-off between virulence and transmission? Parasites inhabit hosts that can be considered suitable habitat patches, within a host population structured like a metapopulation, and yet the host population itself is structured as a metapopulation. How can we use these spatial structures to reduce spread of infections within and between host populations?

There is one final, if not obvious then somewhat ironic, duality that needs to be pointed out—the relative importance placed on parasitological studies in biology reflects the importance of disease issues to humankind. There is no question that disease has been a major force influencing the development of human society: countries like Italy were devastated by the plague and it took the human population more than 400 years to recover. We need only look back a generation to observe the devastating impacts that influenza and other infectious diseases had on human mortality and morbidity. However, during our lifetime there have been remarkable developments, the effects of penicillin application, pasteurization of milk, and the development of vaccines against the major childhood diseases have changed the face of the world. Indeed when we were young, smallpox was eradicated (or was it?) and a number of infectious diseases brought close to extinction. As such, when we went to study at university, we were told that parasites had little impact on human and wildlife host populations—parasites were dismissed by many ecologists as something the parasitologists had to teach so that we understood taxonomy. Now we have seen the emergence of infectious diseases such as AIDS that have made population growth rates negative in some parts of the world and diseases such as Ebola that wipe out whole communities of people in parts of Africa. We are concerned about the evolution of drug resistance against diseases such as tuberculosis and as the book goes to print, Europe has no idea about the true threat from nv-Creutzfeld Jacob Disease and Britain is recovering from a serious Foot and Mouth epidemic. These increasing threats have brought disease issues out of the closet and at the same time sparked off new and innovative studies of disease epidemiology and evolution in Biology.

That is where this book comes in.

CHAPTER 2

Heterogeneities in macroparasite infections: patterns and processes

K. Wilson, O. N. Bjørnstad, A. P. Dobson, S. Merler, G. Poglayen, S. E. Randolph, A. F. Read, and A. Skorping

Animals vary markedly in the number of parasites they harbour—most have just a few, but some have many. In this chapter, we ask why there is so much variation between individuals, how do we quantify this variation and what are the consequences of these heterogeneities for the dynamics of the host–parasite interaction?

2.1 Background

Exhaustive empirical surveys have shown that, almost without exception, macroparasites (parasitic helminths and arthropods) are aggregated across their host populations, with most individuals harbouring low numbers of parasites, but a few individuals playing host to many (Shaw and Dobson 1995). Heterogeneities such as these are generated by variation between individuals in their exposure to parasite infective stages and by differences in their susceptibility once an infectious agent has been encountered. Experimental studies have shown that the extent of spatial aggregation in the infective stage distribution is reflected in the level of parasite aggregation across hosts (Keymer and Anderson 1979). Moreover, in the absence of any heterogeneity in exposure, even small differences in susceptibility between hosts can rapidly produce non-random, aggregated distributions of parasites (Anderson and May 1978). What is unclear at present, is the relative significance of these different mechanisms, and the importance of interactions between mechanisms in accentuating individual differences in parasite loads. Mathematical models that examine these problems rapidly become intractable (Grenfell *et al.* 1995), while experimental studies and computer simulations also become rather complex.

Some of the variation in parasite loads we observe is predictable. For example, in mammals and some other taxa, males tend to be more heavily infected than females, perhaps due to differences in immune function (Poulin 1996; Schalk and Forbes 1997; McCurdy *et al.* 1998). Parasite loads tend to increase with age and may plateau in older animals, though if acquired immunity is important (or there is parasite-induced host mortality) then they may ultimately decline again, so reducing the degree of parasite aggregation. Genetic differences in susceptibility to infection may also be important, though their extent and direction are much more difficult to predict. Other factors that may contribute to the observed heterogeneities in worm burdens are the condition of the host (which may be a function of parasite load), host behaviour, parasite genetics, and seasonality. Comparative studies of aggregation suggest that the infection process and the habitat of the host may make significant contributions to the between-species pattern of aggregation (Shaw and Dobson 1995; Shaw *et al.* 1998).

Heterogeneities in parasite loads have many implications for epidemiological studies. One of the most important concerns the accurate determination of infection intensity—if there is a high degree of variability in the numbers of parasites per host, then a large number of hosts needs to be examined in order to obtain an accurate picture of parasite abundance in the host population. Parasite aggregation also presents some analytical problems as most standard methods of statistical analysis perform best when working with normal distributions—the skew in the parasite distribution means that either the parasite data have to be transformed prior to analysis, or special statistical methods, such as generalized linear modelling, must be employed (Wilson and Grenfell 1997).

Finally, parasite aggregation has important implications for the population and evolutionary dynamics of the parasite and its host (Anderson and May 1978; May and Anderson 1978; Poulin 1993). In macroparasites, host mortality and morbidity tends to be dose-dependent and so has most effect on individuals in the so-called 'tail' of the parasite distribution. The proportion of hosts in this susceptible tail will be relatively larger when parasites are randomly distributed across hosts (and the variance of the distribution is low), than when the distribution is highly skewed (and the variance is high). As a consequence, parasites are likely to be relatively more important as both a selection pressure (Poulin 1993) and a regulatory influence (Anderson and May 1978; May and Anderson 1978) in the former case than in the latter. Thus, a central theme of macroparasite studies over the years has been the development of a theoretical and empirical understanding of the stabilizing role of aggregation in the population dynamics of parasitic helminths and their hosts (Anderson and May 1978, 1982a).

In this chapter we provide a review of recent developments in studies of parasite aggregation, and highlight gaps in our current knowledge. We focus on empirical studies that provide new insights and theoretical developments that may provide new techniques for assessing the relative role played by different forms of heterogeneities in different host–parasite systems. We also draw on some classic empirical studies, especially where more recent studies are lacking. We begin by defining what is meant by an aggregated distribution, how best this can be quantified, and the pitfalls associated with measuring parasitism rates in wild animal populations. The majority of the chapter, however, focuses on the key heterogeneities in the host, parasite, and environment that promote heterogeneities in the distribution of parasites per host. We discuss the patterns that are observed, the mechanisms generating them and their implications for parasite epidemiology. Throughout we emphasize the gaps in our current knowledge and identify areas for future research.

2.2 An introduction to parasite aggregation

Parasites are invariably aggregated across the host population, with the majority of the parasite population concentrated into a minority of the host population (Fig. 2.1; Shaw and Dobson 1995; Shaw *et al.* 1998). In human communities, for example, generally less than 20% of individuals harbour 80% of the helminth parasite population. Thus, a relatively small number of individuals in the 'tail' of the parasite distribution are responsible for most parasite transmission and play an important role in the persistence of the parasite (Anderson and May 1985; Woolhouse *et al.* 1997).

In statistical terms, an aggregated distribution is one in which the variance/mean ratio of parasite numbers per host is significantly greater than 1. There has been much debate about how best to quantify the degree of aggregation and a number of related indices have been adopted (see Box 2.1).

However, regardless of which index of aggregation is used, comparing indices between sub-classes of hosts is generally problematic because they tend to covary with both the mean number of parasites per host and with the number of hosts sampled (see Box 2.2). Gregory and Woolhouse (1993) compared a number of these indices of aggregation and found that the corrected moment estimate of k (from the negative binomial distribution) varied least with mean parasite load and sample size, and this is now the index of aggregation most commonly used by epidemiologists. Not only are the vast majority of parasite datasets best described by the negative binomial distribution (Anderson and May 1978; May and Anderson 1978; Shaw and Dobson 1995), but its exponent k (an inverse measure of aggregation) is used to capture parasite overdispersion in the basic Anderson and May models (see Chapter 3). Hence this becomes the most appropriate parameter for empirical estimation. Box 2.3 discusses the statistical mechanisms that might generate negative binomial parasite distributions.

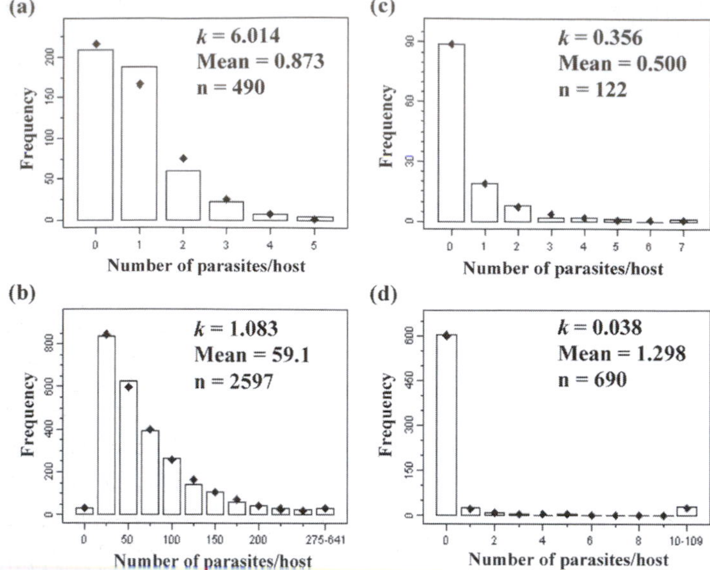

Figure 2.1 Observed parasite frequency distributions for four host–parasite interactions (after Shaw *et al*. 1998). In all cases, the bars represent the observed frequency distributions and the points are the fit of the negative binomial distribution. (a) host = perch *Perca fluviatilis*, parasite = tapeworm *Triaenophorus nodulosus*; (b) host = reindeer *Rangifer tarandus*, parasite = warble fly *Hypoderma tarandi*; (c) host = common starling *Sturnus vulgaris*, parasite = nematode *Porrocaecum ensicaudatum*; (d) host = pond frog *Rana nigromaculata*, parasite = nematode *Spiroxys japonica*. (For reference sources see Shaw *et al*. 1998.)

Box 2.1 Measures of aggregation

If the parasite population was distributed randomly amongst hosts, the variance (s^2) of the parasite distribution would be approximately equal to its mean (m), i.e. random distribution:

$$s^2 = m. \tag{2.1}$$

For an aggregated distribution, the variance is greater than the mean, i.e. aggregated distribution:

$$s^2 > m. \tag{2.2}$$

Thus, we can quantify the degree of aggregation simply as the ratio of the variance to the mean:

$$\text{variance-to-mean ratio} = s^2/m. \tag{2.3}$$

You will notice that this ratio varies from zero (when parasites are uniformly distributed amongst hosts), through unity (for a truly random distribution of parasites), to a number equal to the total number of parasites (for a maximally aggregated distribution).

Deviation from a random distribution can be tested by multiplying the variance-to-mean ratio by the number of hosts sampled (n) minus 1. This 'index of dispersion' (I_D) is then compared to the Chi-square distribution with $n - 1$ degrees of freedom (see Elliot 1977):

$$\text{index of dispersion, } I_D = s^2(n-1)/m. \tag{2.4}$$

A related index of aggregation is obtained by dividing the variance-to-mean ratio (or simply the standard deviation, s), by the sample mean. This is often referred to as the 'standardized variance' (SV):

$$SV = s^2/m^2 = s/m. \tag{2.5}$$

The standardized variance is often expressed as a percentage, simply by multiplying by 100, in which case it is then referred to as the 'coefficient of variation' (CV):

$$CV = s(100)/m. \tag{2.6}$$

A more general approach to the variance–mean relationship is given by an equation that has come to be known as 'Taylor's Power Law' (Taylor and Taylor 1977):

$$s^2 = a + m^b, \tag{2.7}$$

which can be re-arranged to:

$$\log(s^2) = \log(a) + b \cdot \log(m). \tag{2.8}$$

Here, aggregation is measured by the parameter b (parameter a depends mainly on the size of the sampling unit); b varies continuously from zero for a uniform distribution to infinity for a highly aggregated distribution ($b = 1$ for a random distribution).

continued

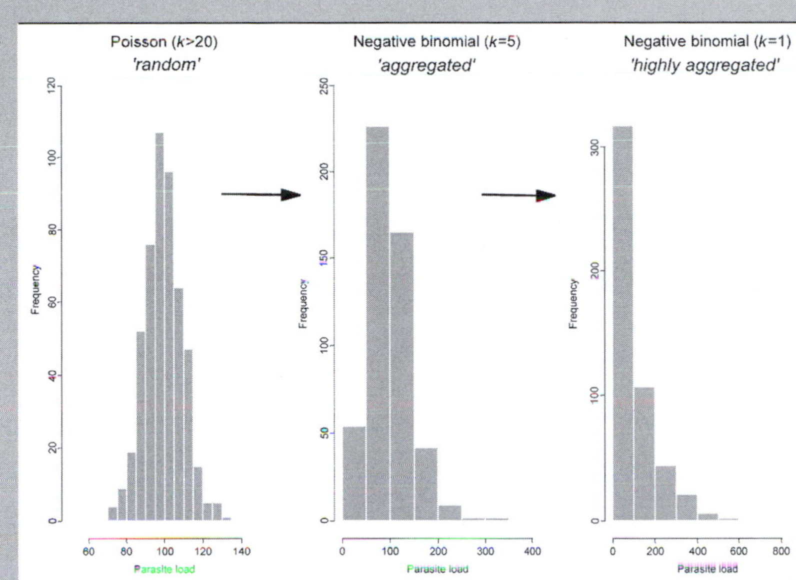

Figure 2.2 Effect of k on the shape of the negative binomial distribution. In all three graphs, the mean of the distribution is 100, but as k becomes smaller (*left to right*), so the distribution becomes increasingly skewed and the parasites become increasingly concentrated in fewer individuals. Note that the scales differ on the three graphs.

Taylor's Power Law cannot be used to quantify the degree of aggregation present in a single sample. However, it is useful when a collection of parasite samples is available from a number of different locations, populations or species (see Shaw and Dobson 1995). In this instance, log variance is plotted against log mean and the parameters a and b are estimated by the intercept (a) and the regression coefficient (b) of the regression line (see Fig. 2.2). A slope of unity ($b = 1$) implies a random or Poisson distribution of parasite counts. For most parasite datasets, the slope lies between 1 and 2 (average $b = 1.55$, Shaw and Dobson 1995), consistent with a negative binomial distribution (see Box 2.3).

The negative binomial distribution is defined as follows (Fisher 1941; Bliss and Fisher 1953):

$$s^2 = m + m^2/k. \tag{2.9}$$

Thus, the degree of parasite aggregation can be quantified by the parameter k. When k is large ($\gtrsim 20$), the distribution converges on the Poisson (i.e. $s^2 \to m$); as k gets smaller, so parasite aggregation increases until, as k approaches zero, the distribution converges on the logarithmic series (Fisher *et al*. 1943; Elliot 1977). For the vast majority of macroparasitic infections of wildlife hosts and humans, $k < 1$ (Shaw and Dobson 1995).

There are several methods for estimating k (Southwood 1966; Elliot 1977), the simplest of which is:

$$\text{moment estimate of } k = m^2/(s^2 - m). \tag{2.10}$$

This estimate is only approximate and can produce unreliable estimates when m is large, k is small or sample sizes (n) are low. A better estimate, which partially corrects for sample size, can be easily calculated (Elliot 1977):

$$\text{corrected moment estimate of } k = (m^2 - s^2/n)/(s^2 - m). \tag{2.11}$$

However, a more accurate estimate of k is obtained by applying maximum-likelihood techniques to the frequency distribution of parasites within a host population (Bliss and Fisher 1953; Anderson and May 1982a; Pacala and Dobson 1988). This can be achieved either by an iterative process (Bliss and Fisher 1953) or by maximizing the log-likelihood directly (e.g. Shaw and Dobson 1995).

Other estimates of aggregation have been used less frequently, but may be useful when studying aggregation from the parasite point of view. For example, Lloyd's (1967) 'Index of Mean Crowding' or 'Patchiness Index' (m^*) quantifies the degree of crowding experienced by an average parasite within a host:

$$m^* = m + (s^2/m - 1). \tag{2.12}$$

It can be seen that when the parasite distribution conforms to the negative binomial, the sample estimate of mean crowding is equal to $m(1 + 1/k)$ (Elliot 1977).

continued

A more recent index of parasite crowding is the 'Index of Discrepancy', D (Poulin 1993), which quantifies aggregation as the discrepancy between the observed parasite distribution and the hypothetical distribution in which all hosts are used equally and all parasites are in sub-populations (infra-populations) of the same size:

$$D = 1 - \frac{2\sum_{i=1}^{n}\left\{\sum_{j=1}^{i} x_j\right\}}{xn(n+1)}, \qquad (2.13)$$

where x is the number of parasites in host j (after hosts are ranked from least to most heavily infected) and n is the number of hosts in the sample. The Index of Discrepancy measures the relative departure of the observed distribution from a uniform distribution. Thus, D may range from zero (no aggregation) to unity (when aggregation is at its theoretical limit and all parasites are in one host), and these constrained limits potentially make it easier to compare aggregation across datasets that vary in their prevalence or mean parasite load.

The two indices of aggregation most commonly employed are s^2/m (variance-to-mean ratio) and k (of the negative binomial). Unfortunately, the relationship between these two indices is not simple (see Fig. 2.3). Scott (1987a) has argued that the variance-to-mean ratio is a better measure of the *degree of aggregation* (i.e. the length of the 'tail'), whereas k provides more information about the *spread of data around the mean*. Thus, she suggests that s^2/m should be used when the number of uninfected hosts (i.e. the zero class) is large, and the latter when the zero class is small. Since k is not independent of the mean, she also suggests that s^2/m be used in preference to k when comparing parasite distribution patterns across populations differing in their prevalence or abundance of infection, and when studying the dynamics of aggregation.

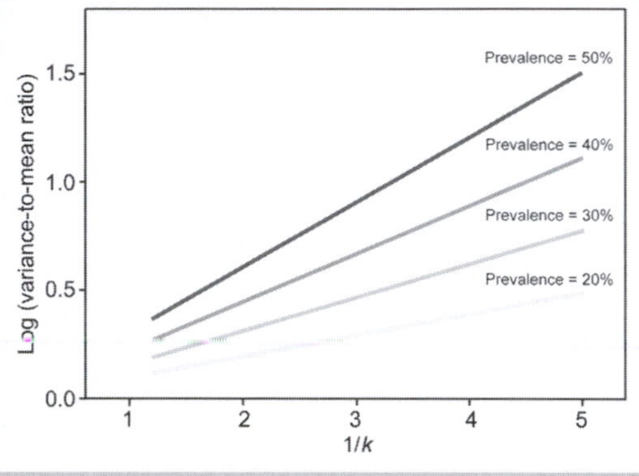

Figure 2.3 The relationship between variance-to-mean ratio and k of the negative binomial. Each line represents the function $\log_{10}(s^2/m) = (1/k) * \log_{10}(n/f_0)$, where n is the total sample size and f_0 is the number in the zero class of the distribution (after: Bliss and Fisher 1953; Pennycuick 1971).

Box 2.2 Sample-size biases: a simulation study

A simulation study by Gregory and Woolhouse (1993) illustrates the biases that may be introduced to sample estimates of average parasitism rates and indices of aggregation. When sample sizes are small, estimates of the arithmetic mean parasite load are consistently underestimated (Fig. 2.4(a)). The geometric mean burden (not shown) and the prevalence (Fig. 2.4b) are not biased in this way, but their accuracies are severely compromised (as measured by their 90% confidence intervals). All indices of aggregation, including the variance-to-mean ratio (Fig. 2.4(c)), moment estimate of k (not shown) and corrected moment estimate of k (Fig. 2.4(d)) tend to underestimate the degree of aggregation when too few hosts are sampled (see main text).

In field studies, sample sizes are frequently small, especially for individuals in the oldest age classes. As a consequence, sample estimates of aggregation may decrease with host age purely due to sample size biases (Fig. 2.5(a)). One might imagine that this problem could be resolved by combining parasite data from animals of similar age, so that each new age-class would have approximately equal numbers of hosts. However, this is not the case—combining age-classes in this way may result in artifactual increases in the estimate of parasite aggregation (Fig. 2.5(b)).

continued

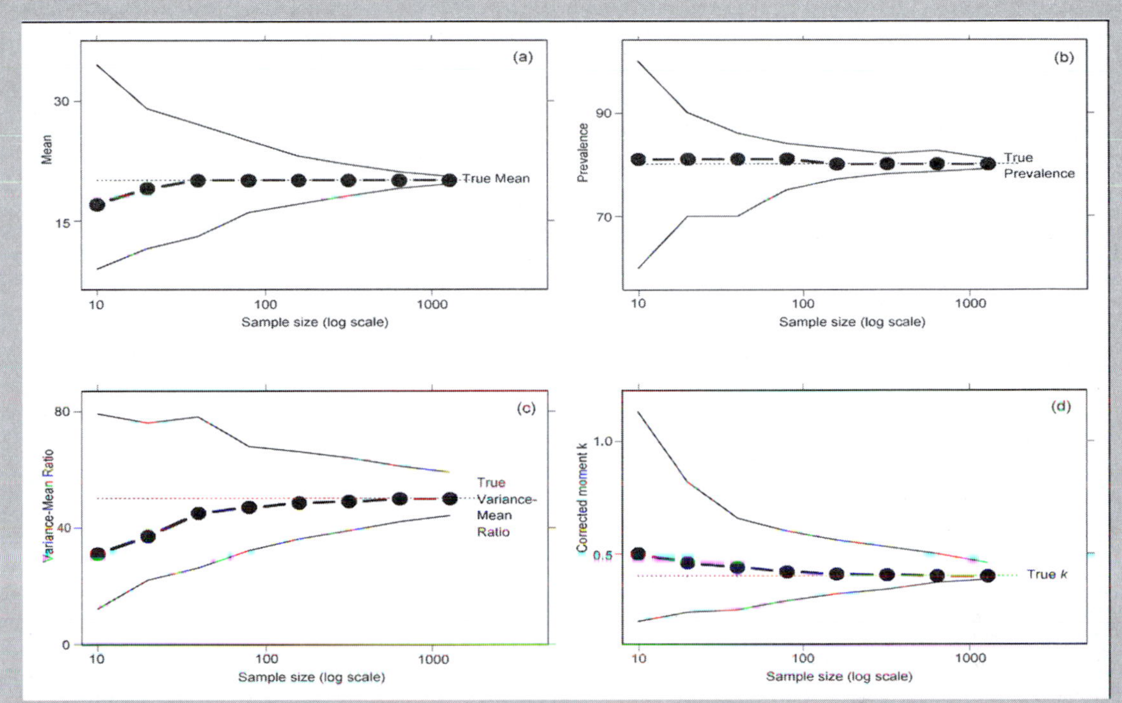

Figure 2.4 Simulation results showing the effect of sample size on estimates of (a) mean parasite load, (b) prevalence, (c) variance-to-mean ratio and (d) corrected moment estimate of k of the negative binomial (after Gregory and Woolhouse 1993). The *solid circles* represent the results of the simulations, the *solid lines* represent the 90% confidence intervals, and the *dashed line* represents the 'true' relationship between sample size and parameter values. For all of these Figures, the population mean and k for the simulations were 20 and 0.4, respectively.

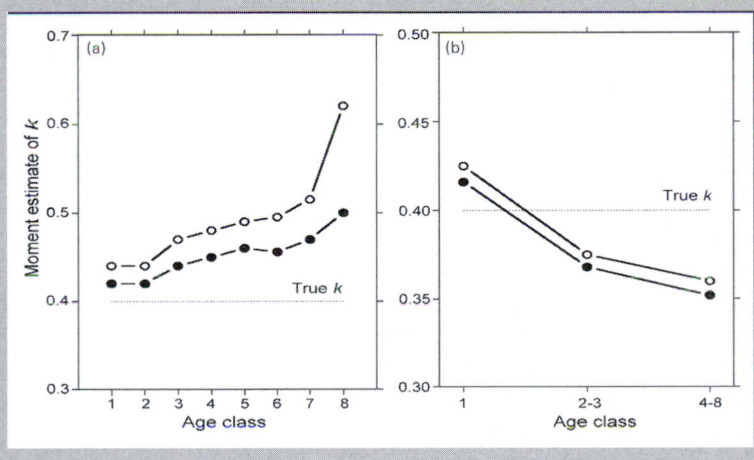

Figure 2.5 Simulation results showing the influence of sample size on age-dependent estimates of parasite aggregation, as measured by the moment estimate of the negative binomial k (*open symbols*) and corrected moment estimate of k (*closed symbols*) (after Gregory and Woolhouse 1993). In both simulations, the sample sizes for age classes 1–8 were: 100, 60, 40, 30, 25, 20, 15, and 10, respectively, and the means were 5, 20, 40, 50, 50, 40, 20, and 5, respectively. In (a) data are grouped into age classes of equal period, but unequal sample sizes; in (b) data are grouped into age classes of unequal period, but equal sample sizes. The *dashed line* represents the 'true' relationship between age class and k.

Box 2.3 The negative binomial distribution as a birth–death process

A universal trait of populations is that temporal variability increases with mean abundance, μ. The way the variance rises with the mean can therefore provide insights into the nature of population growth. The way the variance depends on the mean is called the variance function in statistics (e.g. McCullagh 1983). In its simplest form, the demographic stochasticity that arises from birth and death processes is related to Poisson variability (through the sum of binomial events: Bartlett 1956). In the simple density-independent death process, the number of individuals will be Poisson distributed, so that the variance, $V(\mu)$, is proportional to the mean: $V(\mu) = c\mu$. If we also include a birth process, the number of individuals will follow a negative binomial distribution (e.g. Kendall and Stuart 1963). A negative binomial process has a variance function that rises more rapidly with the mean than the Poisson, but more slowly than proportional to the squared mean (Box 2.1; Anderson and Gordon 1982):

$$V(\mu) = \mu + \mu^2/k. \tag{2.14}$$

In contrast, stochastic population growth, in a fluctuating environment, leads to a gamma (Dennis and Patil 1984) or log-normally (Engen and Lande 1996) distributed number of individuals. Here the variance is proportional to the squared mean. For more complicated patterns of population growth (e.g. with spatial or behavioural responses), the variance may rise even faster with the mean, an insight that led Taylor (1961; Taylor and Taylor 1977) to propose his famous 'Power Law' (see Box 2.1).

The log variance versus log mean plot is the most well known diagnostic to elucidate this relationship (Taylor *et al.* 1983; see Fig. 2.6). Poisson distributed numbers will have a slope of unity on such a plot, and log normal or gamma distributed numbers will have a slope of 2. In the case of negative binomial data, the slope is predicted to be non-constant: close to unity for small means and close to 2 for large means. Whether a change in slope is visible on a log–log plot will depend on the range in means observed and whether the clumping parameter (k) depends on the mean. Unfortunately, since the axes are so compressed, non-linearities in log–log plots can often be difficult to discern (see Tokeshi 1995).

A complementary tool to the log variance vs. log mean plot is to estimate the variance function directly (McCullagh 1983; Ruppert 1997). The error in the estimate of the variance is chi-square distributed, so that a generalized linear model (see Box 2.5) with a log-link and a 'quasi-gamma' error can be used to estimate the relationship. A non-parametric regression may be used to explore the variance-mean relationship without assuming *a priori* functional forms.

2.2.1 A comparative analysis of parasite aggregation

Shaw and Dobson (1995) examined previously published datasets from over 250 wildlife populations and attempted to determine which ecological and epidemiological processes generated variation across species in patterns of parasite aggregation and abundance. They found that there was a tight linear relationship between log variance and log mean (Fig. 2.6). Moreover the slope of the regression was significantly greater than 1 ($b = 1.55 \pm 0.037$ SD, $n = 269$), indicating that the parasite distribution was overdispersed (this compares with $b = 1.45 \pm 0.39$ for free-living animal populations (Taylor and Taylor 1977). The small degree of spread in the parasite data ($r^2 = 0.87$) is surprising and suggests that regardless of the infection process, mean parasite burden is the main determinant of the variance in parasite burden between hosts. Shaw and Dobson suggest that this is because of evolutionary constraints on the degree of aggregation. They argue that natural selection will lead to parasites with intermediate levels of pathogenicity and aggregation, because highly pathogenic parasites will generate high levels of parasite-induced mortality and hence lower parasite loads, a high degree of aggregation with reduced mating opportunities, and hence reduced fecundity (in microparasites, this is equivalent to the trade-off between transmission and virulence; Shaw and Dobson 1995).

Shaw and Dobson (1995) found that 13% of the variation in log variance was unexplained by log mean burden. However, much of this could be explained by the nature of the infection process. For example, trichostrongylids and other parasites that enter their hosts passively tended to exhibit relatively higher levels of aggregation, whereas dipteran (e.g. Hypoderma) infections of large mammals tended to exhibit relatively lower levels of aggregation. Whilst these trends are interesting

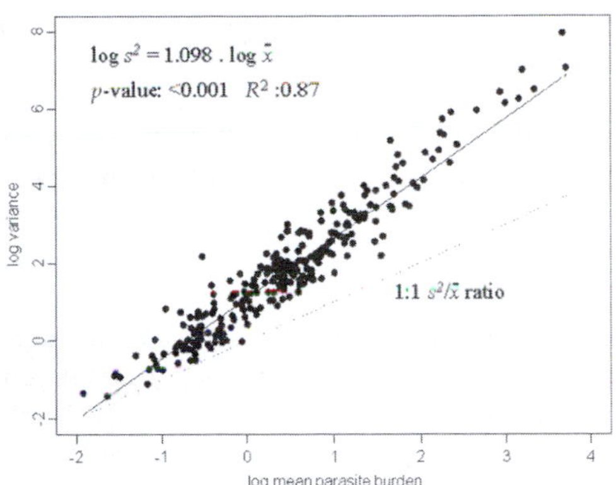

Figure 2.6 Log variance vs. log mean parasite burden for 269 datasets from a range of host–macroparasite interactions (after Shaw and Dobson 1995). The *solid line* is the fitted regression line (constrained to pass through zero), and the *dotted line* is the relationship predicted for a 1:1 variance:mean ratio (i.e. for a Poisson distribution).

and can identify potentially important processes generating variation in parasite loads between hosts, they are purely correlational and it is clear that an experimental approach is required if we are to unravel the relative importance of potential mechanisms.

2.2.2 Parasite aggregation and host dynamics

The role of parasites in host regulation is covered extensively in Chapter 3 (§3.3, §3.4). Here, we note only that the impact of parasites as a regulating force is critically dependent on the degree of parasite aggregation across the host population. As a rule, the stability of the host–parasite interaction will be enhanced by parasite aggregation (Anderson and May 1978; May and Anderson 1978). However, for highly overdispersed distributions (where k approaches zero and most of the parasites are living in just a few heavily-infected individuals), regulation of the host population will be difficult to achieve because too many parasites will be lost from the system by parasite-induced host mortality.

Clearly, parasite aggregation does not act in isolation and it is the interaction between aggregation, parasite virulence and transmission efficiency, and the host's population growth rate in the absence of parasites, that will determine whether the host–parasite interaction is stable or leads to cyclic or chaotic dynamics (see Box 3.2). For example, when parasite virulence is high, stability will be achieved only if parasite aggregation falls (i.e. k increases), otherwise too many parasites are lost from the system due to parasite-induced mortality. These factors also interact to determine the equilibrium host density. However, in general, as k increases and parasites become more evenly spread through the host population, so the net rate of parasite-induced mortality increases and the equilibrium host density declines.

2.3 Collection and analysis of parasite data

If we are to examine any of the patterns predicted by epidemiological models, it is important to accurately determine mean parasite loads and levels of variation within a host population. However, the collection and analysis of parasite data is fraught with difficulties, many of which are related to the shape of the parasite distribution. The long 'tail' of many parasite distributions means that unless a large number of hosts are sampled, inaccurate estimates of parasitism levels may be gained. Moreover, because the parasite distribution is not normal, classical methods of statistical analysis may produce biased estimates of parasite load and so alternative statistical methods have to be employed. The intimate association between parasites and their hosts only adds to the problems, because surrogate measures of parasitism, such as faecal egg counts, often have to be used, and these sometimes have unknown or variable levels of specificity and sensitivity. In this section, we review the sorts of difficulties encountered during data collection and analysis.

2.3.1 Sample-size biases

Field-based studies of host–parasite interactions often have to rely on limited and opportunistic sampling of the parasite population. It is therefore important that, wherever possible, samples are stratified and balanced, such that approximately equal numbers of hosts are sampled from all appropriate demographic groups (age classes, sexes, reproductive states, etc.) and sampling units (years, population densities, locations, etc.). The reason for this is that when only a small number of hosts are sampled, the probability of detecting the most heavily infected individuals in the population is low. Thus, when sample sizes are small there is a real danger of underestimating both the mean parasite burden and the degree of parasite aggregation (Box 2.2). However, this problem is minimized when aggregation is quantified using the corrected moment estimate of k (Box 2.1) or a maximum-likelihood estimate of k (Pacala and Dobson 1988; Gregory and Woolhouse 1993). The prevalence and geometric mean parasite loads are not biased by small sample sizes. However, the confidence intervals associated with both of these measures are inflated and so large sample sizes are always recommended. If the parasite population is highly overdispersed, large sample sizes are even more important.

A consequence of the reliance of population estimates on sample size is that if sampling is not stratified correctly in relation to host demography, artefactual patterns in mean parasite burden and aggregation may result (Pacala and Dobson 1988; Gregory and Woolhouse 1993). For example, sample sizes often decline with host age due to mortality and so if sampling effort is not directed at obtaining equal numbers of hosts in all age classes, then it might appear that average parasite loads decline in old animals and that parasite aggregation declines with age, purely due to sampling biases (Box 2.2).

This point is illustrated in Fig. 2.7, which shows data for cestode (*Diphyllobothrium ditremum*) infections of Arctic Char (*Salvelinus alpinus*) (for details see: Halvorsen and Andersen 1984). Examination of the age-intensity curve appears to show that the mean number of parasites rises to a peak in individuals aged 7–9 years, and then declines in older fish (as indicated by the dashed line in Fig. 2.7(a)). Halvorsen and Andersen (1984) fitted a two-parameter catalytic model to the running average parasite load (calculated over three consecutive age classes) and found that the best fitting model reached an asymptote at around 13 worms per fish. However, this model was biased by the small number of old animals sampled (>9 years old). When Pacala and Dobson (1988) fitted a three-parameter model to the same data using maximum-likelihood methods, the best-fit model indicated that average parasite loads continued to increase in old animals and reached an asymptote at nearly 50 parasites per host (solid line in Fig. 2.7(a)). The reason for the lower asymptote for the Halvorsen and Andersen model is that their

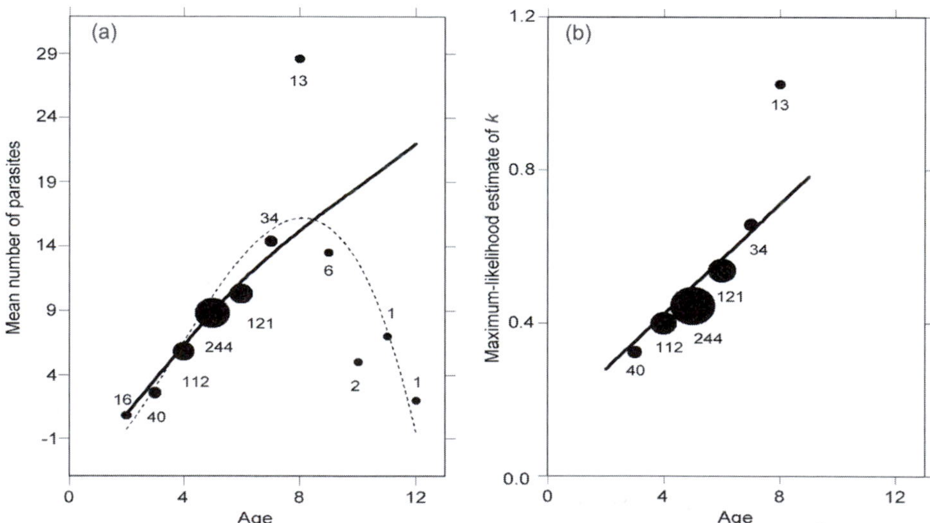

Figure 2.7 Age-intensity curve (a) and age-aggregation curve (b) for cestodes in Arctic Char, *Salvelinus alpinus* (original data from Halvorsen and Andersen 1984, after Pacala and Dobson 1988). In both Figures, the size of the symbol is proportional to the host sample size (indicated next to each point). In (a) the *solid line* is the maximum-likelihood best fit to the data; the *dashed line* is a least-squares polynomial regression to the means (not weighted for host sample size. In (b) the *solid line* is the maximum-likelihood best fit to the data (after Pacala and Dobson 1988).

procedure weighted equally the means of all the age classes (thus the low means for animals in age classes 10–13 are able to pull down the asymptote), whereas the Pacala and Dobson (1988) model weighted the means by their sample sizes (as illustrated by the size of the symbols in Fig. 2.7).

There is also a peak in the age profile of the variance-to-mean ratio, but this too appears to be due to sampling artefacts (Pacala and Dobson 1988). Robust evidence for a monotonic increase in aggregation with age was provided when the relationship between k (of the negative binomial) and age was modelled by a linear function fitted by maximum-likelihood (Fig. 2.7(b)).

2.3.2 Biased sampling

Much of the parasite data published in veterinary journals is collected opportunistically and comes from road kills, beach strandings, harvested animals, or animals found dead (e.g. Keith *et al.* 1985). All of these data are potentially biased samples of the true host population since parasitized animals are more or less susceptible to the sampling process than non-parasitized (or lightly parasitized) individuals. This may simply be due to the fact that the selection process avoids or selects animals with overt signs of parasitism. For example, vets primarily treat sick and diseased animals and hunters may target either the healthiest or the weakest looking animals in the population, which are likely to have unrepresentative parasite loads. Alternatively, there is evidence that parasites may manipulate the behaviour of their hosts in such a way as to maximize their rates of transmission (Moore 1984; Poulin 1994, 2000). For parasites in intermediate hosts, such behavioural alterations (such as reduced speed of locomotion, e.g. Zohar and Rau 1986) would result in the host becoming more susceptible to predation (by top predators, hunters or cars), whereas for parasites present in their definitive or only host, alterations (such as host castration) would tend to reduce predation risk at the expense of reproductive activities (Dobson 1988a).

Even apparently bias-free sampling methods may select an unrepresentative section of the host population. For example, rodents are known to vary in their degree of trap-shyness and trap-happiness (Courchamp *et al.* 2000) and it is likely that this will be influenced by their parasitological status. Similarly, parasitized birds or fish may be more or less likely to be caught in sampling nets. These sorts of biases are much less easy to quantify than those alluded to earlier, however it is obviously important to appreciate that such biases may exist when designing sampling programmes.

Other potential biases are associated with the counting of parasites, where the smaller life stages may be easily missed (Smith and Grenfell 1994). These effects are compounded in studies that use indirect measures of parasitism (such as faecal eggs counts), where there is the additional uncertainly of the exact nature of their relationship with parasite load (see below).

2.3.3 Specificity and sensitivity of indirect measures of parasitism

For most wildlife diseases, we are not in a position to make an absolute count of the number of parasites harboured by a particular host or population of hosts. Thus, we usually have to resort to inferences from coprological or haematological samples. Such methods include faecal egg counts (density of parasite eggs in the faeces), seroepidemiological measures (ELISA-based methods that quantify the responses of various antibodies, such as IgA, to immunogenic parasite antigens) and coproantigen detection (ELISA-based methods that quantify the amount of non-immunogenic parasite excretory/secretory antigens voided in the host's faeces, e.g. Hyde 1990; Johnson *et al.* 1996; Malgor *et al.* 1997; Nonaka *et al.* 1998). However, these estimates will often provide biased, or at least inaccurate, measures of parasitism. For example, faecal egg counts will be useless as a surrogate measure of worm burden if worm fecundity is subject to severe density dependent constraints. Similarly, the coproantigen detection methods will be useless if the assay fails to meet the required level of specificity. It is therefore important to determine the extent of any biases and misclassifications on a case by case basis before employing a particular technique on a large scale. Although most studies fail to do this, it is now clear that these may be substantial, particularly in coprological estimates of prevalence (Box 2.4).

Box 2.4 Sensitivity and specificity of indirect parasitological estimates

Two kinds of misclassification affect any indirect measure of parasitism: infected subjects appearing uninfected (false-negatives) and uninfected individuals appearing infected (false-positives). Lets define A = number of true-positives, B = number of false-positives, C = number of false-negatives, D = number of true-negatives, so the total recorded infected is $A + C$ and the total recorded uninfected is $B + D$ and $N = A + B + C + D$ is the whole population.

Sensitivity and specificity (Thrusfield 1995) are quantitative measures of such bias and may be calculated by matching the results gained by absolute parasite counts (obtained by dissection) with those from the laboratory technique applied:

$$\text{sensitivity} = \frac{100 \times \text{true-positive}}{\text{true-positive} + \text{false-negative}} = \frac{A}{(A + C)} \tag{2.15}$$

$$\text{specificity} = \frac{100 \times \text{true-negative}}{\text{true-negative} + \text{false-positive}} = \frac{D}{(B + D)} \tag{2.16}$$

$$\text{observed prevalence} = \frac{\text{true-positive} + \text{false-positive}}{\text{number examined}} = \frac{(A + B)}{(N)} \tag{2.17}$$

$$\text{true prevalence} = \frac{\text{observed prevalence} + \text{specificity} - 1}{\text{sensitivity} + \text{specificity} - 1} = \frac{-1 + A + B + D/(B + D)}{A/(A + C) - B/(B + D)}. \tag{2.18}$$

Knowledge of the sensitivity and specificity of a test provides better estimates of the true prevalence in comparison with the observed one. This point is illustrated in Table 2.1, in which the sensitivity and specificity of faecal egg counts are examined for two wild host species. These data clearly show how misleading coprological data can be and how large differences between observed and true prevalences can exist. It is important to determine the specificity and sensitivity of a diagnostic test on a case by case basis. These diagnostic errors may be amplified further by other systematic errors, particularly those arising from incorrect or biased sampling (Box 2.2).

Table 2.1 True and estimated prevalences along with sensitivity and specificity of the coprological test as determined by necropsy and coprological examination for gut helminths in (a) foxes and (b) roe deer

Parasite	True prevalence	Estimated prevalence	Sensitivity (±SE)	Specificity (±SE)
(a) Red Fox (*Vulpes vulpes*) **n = 208**				
Ascarids	51%	42%	79 ± 3%	97 ± 1%
Tapeworms	29%	2%	8 ± 2%	100%
Hookworms	25%	11%	29 ± 3%	96 ± 1%
Whipworms	3%	5%	43 ± 3%	96 ± 1%
(b) Roe deer (*Capreolus capreolus*) **n = 106**				
Strongyles	88%	31%	32 ± 5%	77 ± 4%
Whipworms	22%	9%	26 ± 4%	96 ± 1%
Nematodirus	9%	1%	13 ± 3%	100%
Capillaria	1%	1%	0%	99%

2.3.4 Repeatability of parasite counts

One way of assessing the variability associated with indirect measures of parasitism is to determine its repeatability (i.e. the ratio of the between-animal variance component to the sum of the between and within animal components; Falconer and MacKay 1996). For two samples with the same standard deviation, the repeatability is equal to Pearson's product moment correlation coefficient, r. Stear *et al.* (1995) determined the repeatability of faecal egg counts taken from domestic sheep infected with the nematode *Teladorsagia circumcincta* and found that the repeatability of duplicate egg counts was extremely high, 0.92—in other words, 92% of the variation in egg counts could be explained by differences between individuals. This result indicates that the actual counting process was very accurate. However, the repeatability of samples collected 2–3 days apart was not as high, averaging around 0.75. This indicates that individuals vary in their faecal egg production from one day to the next and that multiple samples over several days may be required to accurately determine heterogeneities in parasite loads. Over a longer time-scale, the repeatabilities continued to decline, which probably reflects variability in the worm burdens of the animals over time. Again, most studies of wildlife populations do not determine the repeatability of their parasitological measure (but see Gulland 1991a,b; Hudson and Dobson 1995). If this is low, then significant heterogeneities are likely to be obscured.

2.3.5 Statistical methods for quantifying parasite heterogeneities

As indicated earlier, parasite distributions are often empirically best described by the negative binomial. This can result in a problem for parasitologists wishing to describe their hard-earned data, because most classical statistical methods, such as linear regression and analysis of variance, are based on the assumption of a normal (or Gaussian) distribution. Traditionally, logarithmic transformation has been applied to such data, in an attempt to normalize the distribution. However, this transformation often fails, particularly when the distribution is highly aggregated or the mean parasite load is low (Fig. 2.8; Wilson and Grenfell 1997).

So which method should parasitologists and ecologists use to analyse their parasite data? As the study of wildlife diseases has become increasingly quantitative, and computers have become faster and more powerful, the number of statistical techniques available for analysing parasite data has grown, and new methods are being developed and refined all of the time. Two of the methods currently being used in the parasitology literature are generalized linear modelling and tree-based modelling (see Box 2.5).

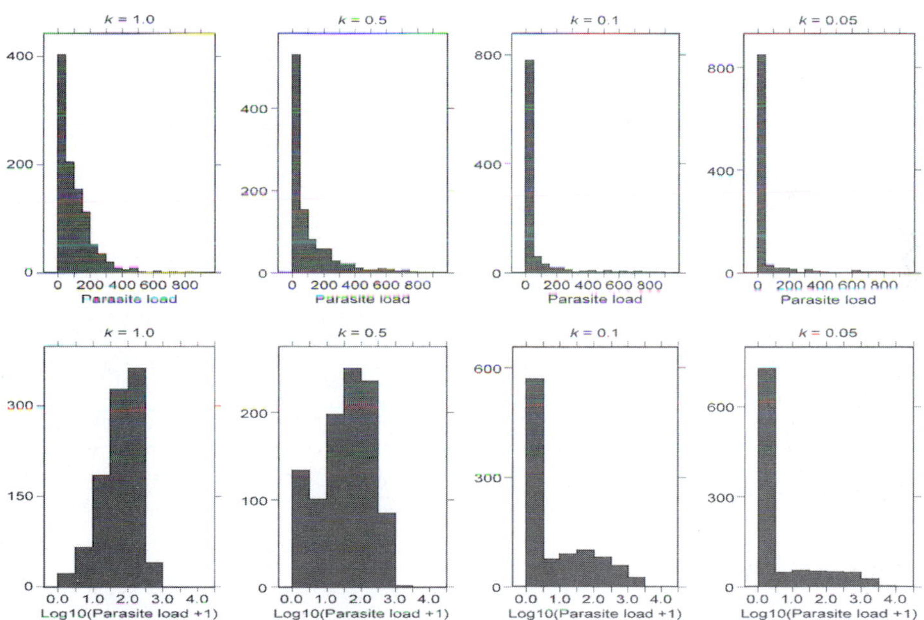

Figure 2.8 Effect of log-transformation on the distribution of parasite data (after Wilson and Grenfell 1997). The top row of Figures shows the frequency histograms for typical parasite data (1000 random samples taken from a negative binomial distribution with population mean equal to 100 and population k equal to 1.0, 0.5, 0.1, and 0.05, *left to right*). The bottom row shows how these same data look after $\log_{10}$-transformation (after first adding 1 to prevent zeros). Although the transformed distributions look approximately normal for low levels of aggregation ($k > 0.5$), Kolmogorov–Smirnov tests for goodness-of-fit to the normal distribution indicate that the transformation always fails to normalize the data ($P < 0.001$).

Box 2.5 Generalized linear modelling and tree-based modelling

Generalized linear modelling

Generalized linear models, or GLMs, offer a powerful alternative to logarithmic transformation and conventional parametric methods (Aitkin *et al*. 1989; Crawley 1993). GLMs are generalizations of classical linear models and allow the underlying statistical distribution of the data to be explicitly described. So, instead of assuming that the parasite data are normally distributed, they assume that they follow the Poisson or negative binomial distributions, as appropriate. As a result, the fit of a GLM is often better than the equivalent conventional linear models, even after the data have been log-transformed (Wilson *et al*. 1996; Wilson and Grenfell 1997). However, this will not always be the case, because negative binomial distributions do not necessarily 'add up' when combined (Grafen and Woolhouse 1993). In other words, if the distribution of parasites in each sub-class of host (age, sex, genotype, etc.) is correctly described by the negative binomial distribution, the overall (aggregated) distribution will be described by the negative binomial only if each sub-class has an identical mean (Dietz 1982; Pacala and Dobson 1988). Thus, when there is a high degree of sub-structuring in the data and component distributions differ markedly in there degree of skew (as measured by k), the estimated value of k for the aggregated dataset will fit none of the component distributions accurately and can sometimes lead to a badly fitting model.

In addition, if the data are sub-structured and the component distributions have different means but the same k, then the estimate of this 'common k' is always larger than the k estimated by lumping all of the data together (Shaw and Dobson 1995), i.e. combining sub-sets of data tends to exaggerate the degree of parasite aggregation (Hudson and Dobson 1995). Thus, wherever possible, maximum-likelihood estimates of the 'common k' should be used to describe the degree of aggregation within a host–parasite system (for details see: Shaw and Dobson 1995).

Tree-based modelling

A graphical alternative to GLMs, which is only just beginning to be employed by parasitologists and ecologists (for parasitological examples see: Shaw and Dobson 1995; Merler *et al*. 1996), is the use of tree-based models (Breiman *et al*. 1984). These are models that allow the structure of the data to be studied, by defining nodes, which divide the data into successive clusters with similar characteristics. The hierarchical nature of tree-based models allows the automatic selection of the most important predictor variables and, because the method is appropriate for both continuous and categorical data, it is very flexible and can be used for both classification and regression problems. Compared to linear models, tree-based models have the advantage that they are easier to interpret when the predictors are a mixture of numeric variables and factors, they handle missing data better, and they are better able to capture non-additive behaviour and multiplicative interactions. However, assessing the relative fit of different tree-based models is sometimes difficult.

Bagging (Breiman 1996) is a procedure that allows the tree-based models to be improved in terms of generalization error, i.e. prediction using novel data (Merler and Furlanello 1997). The bagging consists of generating a sequence of predictors, each of which is generated by producing multiple versions of the learning set by bootstrap re-sampling (Efron and Tibshirani 1986). The method then approximates the predictor by averaging over the bootstrapped datasets, so reducing the component of the generalization error due to the variability of the data relative to when a single predictor is used (Geman *et al*. 1992).

A case study of tree models with bagging

In this example, the bagging procedure is used to examine the factors important in generating heterogeneities in the number of *Ascaridia compar* worms found in rock partridges (*Alectoris graeca saxitalis*) in the Trentino region of northern Italy (see §3.6 for further details of the system). The dependent variable used in this analysis was the number of adult worms per bird (mean parasite burden $\pm$ SD $= 2.66 \pm 6.77$), which was found to conform to the negative binomial distribution ($k = 0.16$; $\chi^2_5 = 4.25$, $P > 0.5$). Predictor variables tested for inclusion in the model included host data (such as sex, age, and weight) and environmental data (such as mean rainfall) (for details see: Cattadori *et al*. 1999).

The bagging procedure indicated that the relative dryness or wetness of the habitat was the most important factor affecting the parasite distribution—in dry habitats, the mean parasite burden was higher (mean of logged data $= 1.0$) than in the wet (mean $= 0.44$). The model highlighted sex as the next most important factor affecting worm burden, with the mean parasite load being higher for females than males (mean $= 1.19$ for females and 0.76 for males, in dry regions; see Box 2.7). Finally, the young birds were more heavily infected than the adults, and in the wet regions adult males were completely parasite-free. Figure 2.9 shows a graphical tree representation of the bagging model.

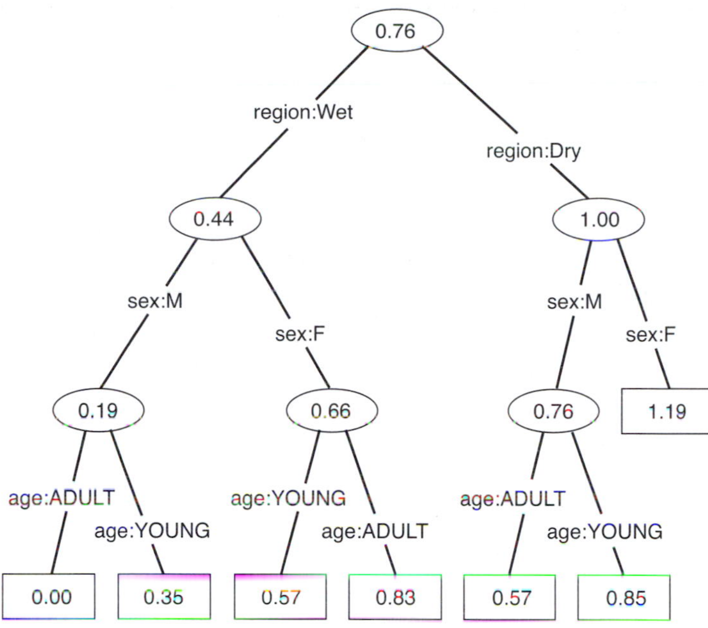

Figure 2.9 A graphical tree representation of a bagging model for the analysis of parasite loads from Rock partridge in Trentino, Italy. The mean of the log-transformed distribution is indicated for each partition of the predictor variable space. See Box 2.5 for more details.

2.4 Observed heterogeneities in parasite loads

In this section, we assess variation in parasitism rates associated with heterogeneities in the host population (including host genetics, age, sex, behaviour, and body condition), in the parasite population and in extrinsic factors. In each case, we aim to identify the general patterns observed, the mechanisms generating those patterns, and their epidemiological consequences.

2.4.1 Host age

What are the observed age-infection patterns?

Important epidemiological information, such as rates of parasite transmission and mortality, can often be obtained by analysing patterns of age-prevalence and age-intensity (Anderson and May 1991; Hudson and Dobson 1995). In the simplest case (referred to as Type I by Hudson and Dobson 1995), in which there is no vertical transmission and no reproduction within the host, parasites are acquired from the environment over time and mean intensities increase with host age (Fig. 2.10). If the rates of parasite acquisition and parasite mortality are constant, then the average number of parasites per host will increase towards an asymptote determined by the balance between these two rates (Type II). A number of empirical studies have reported age–intensity curves that either show a continual increase in parasite load or a gradual levelling-off of parasite burden with age (for examples see: Hudson and Dobson 1995). For other host–parasite interactions, the age-intensity curve is convex (Type III). In other words, rather than rising to an asymptote, parasite loads decline after an initial increase.

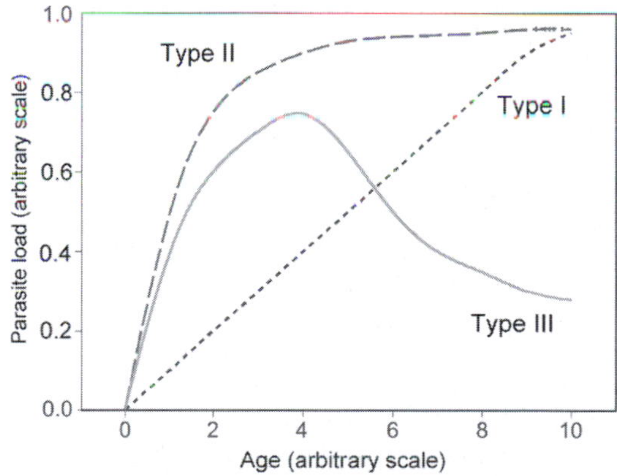

Figure 2.10 Hypothetical age–intensity curves, illustrating three general patterns: Type I (*dotted line*), Type II (*dashed line*), and Type III (*solid line*).

These different epidemiological patterns are highly specific to the host–parasite interaction under study, and may vary between populations. For example, Quinnell *et al.* (1992) studied infections of the nematode *Heligmosomum polygyrus* in the woodmouse *Apodemus sylvaticus* in an outdoor enclosure in Oxford, England. They observed that mean parasite intensity increased asymptotically with host age (Type II), whereas Gregory *et al.* (1992), studying a wild population of woodmice in Northern Ireland, found that mean intensity exhibited a convex (Type III) age-intensity profile. A similar difference was observed in red grouse infested with the tick *Ixodes ricinus* (Hudson 1992). On grouse moors where Louping ill virus (Box 7.5) was prevalent, the intensity of infection increased with chick age (up to 25 days), whereas on moors where Louping ill was absent, tick intensities peaked when the chicks were 8–14 days old and subsequently declined.

What are the mechanisms generating convexity in age–intensity curves?

There are a number of mechanisms that might account for convex age–intensity curves. These include parasite-induced host mortality, acquired immunity, age-related changes in predisposition to infection (e.g. due to the development of resistance mechanisms that are unrelated to previous exposure to parasites), age-dependent changes in exposure to parasites (e.g. due to behavioural shifts or seasonality), and age-related probabilities of accurately determining parasite loads (e.g. if older animals produce more faeces than younger ones, then a decline in faecal egg counts with age could be due to an increasing dilution of parasite eggs).

Distinguishing between the different mechanisms causing convexity in age-intensity or age-prevalence curves can be difficult, particularly when reliable data on age, mortality rates, and body weights are not available. This problem is exacerbated when parasite sampling is limited to a single time point (i.e. data come from a horizontal cross-sectional survey), because variation in exposure rate due to seasonal or yearly variation in the force of infection may distort the age-infection profile. Even when all of this information can be obtained and longitudinal surveys can be conducted, some of the potential mechanisms can be difficult to disentangle without experimental infections or the removal of parasites through treatment.

Is acquired immunity an important source of variation in wild host populations?

Acquired immunity develops in response to accumulated experience of infection and acts to decrease parasite establishment, survival, reproduction, and/or maturation. Although acquired immunity is believed to be an important factor causing convexity in the age–intensity curves for macroparasite infections of humans, domesticated ruminants, and laboratory animals (e.g. Anderson and Crombie 1984; Anderson and May 1985; Crombie and Anderson 1985; Lloyd and Soulsby 1987; Dobson *et. al.* 1990; Anderson and May 1991), there have been few clear demonstrations of acquired immunity in wildlife populations (but see Quinnell *et al.* 1992). At present, it is unclear whether this is because protective immunity does not generally develop or if sampling methods are just too crude to detect it. Even for interactions where effective immune responses may be observed under controlled laboratory conditions, acquired immunity may be difficult to detect in the field and may have little epidemiological significance if most of the hosts die before its impact is felt.

Is acquired immunity important in wildlife populations and can we learn anything from variation in age–infection profiles?

In theory, the answer to both of these questions is yes. Theory suggests that if acquired immunity is important then it should result in a negative correlation between peak levels of infection and the age at which the peak occurs—a phenomenon known as the 'peak shift' (see Box 2.6; Anderson and May 1985; Woolhouse 1998). Peak shift has now been demonstrated in a number of helminth infections of humans and in laboratory studies, but convincing evidence is still lacking for natural wildlife infections, due largely to logistical and statistical problems (see Box 2.6). However, these difficulties should not put us off trying to determine whether predictable patterns of age-prevalence occur across wild host populations.

Box 2.6 Peak shift

Acquired immunity develops in response to the accumulated experience of parasite antigens. Thus, in populations where transmission rates are high, the level of parasitic infection will rise rapidly and this will be followed by a rapid increase in the level of acquired immunity. As a result, parasite loads will peak at an early age and subsequently decline at a fast rate due to protective responses. In contrast, in populations where parasite transmission rates are low, parasite loads (and acquired immunity) will increase at a slower rate and the age at peak infection will be later (Anderson and May 1985). This will result in a negative correlation between peak levels of infection and the age at which the peak occurs—a phenomenon known as the 'peak shift' (see Fig. 2.11(a); Anderson and May 1985; Woolhouse 1998).

Peak shift has now been demonstrated in a number of helminth infections of humans (Fig. 2.11(b); Woolhouse 1991; Fulford *et al.* 1992; Mutapi *et al.* 1997), and in several experimental infections of laboratory mice (Crombie and Anderson 1985; Berding *et al.* 1986). However, evidence for peak shift in natural wildlife infections is lacking (for a possible example see: Müller-Graf *et al.* 1997). The main problem is often the logistical difficulties of collecting parasite data from a range of populations and accurately assessing the host age-structure. There may also be difficulties in statistically identifying the peak of such curves. Finally, even after patterns consistent with peak shift have been conclusively demonstrated, it may be difficult to exclude alternative interpretations, such as age-related differences in exposure to parasites or innate resistance (Gryseels 1994; Woolhouse 1998).

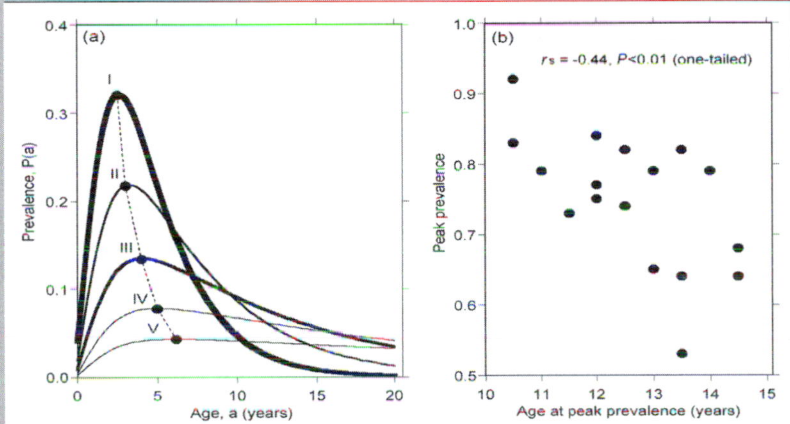

Figure 2.11 Predicted and observed relationship between the prevalence of parasite infection and age in populations subject to different transmission rates. (a) Predictions based on a two-stage catalytic model where the model is given by: $P(a) = \lambda/(\lambda - \nu)\cdot(\exp(-\nu a) - \exp(-\lambda a))$, where $P(a)$ is the prevalence of infection at age a, λ is the rate of infection (which, in this instance, takes the values 0.4, 0.2, 0.1, 0.05, and 0.025 for *solid curves* I–V, respectively), and ν is the rate at which infected individuals recover and become immune (set to $n = 0.5$ in this instance, meaning that protective immunity develops after a mean of two years). Peak prevalences for the different values of l are indicated by the *solid circles* joined together by the *broken curve*, which illustrates the predicted 'peak shift' with age. (b) Observed relationship between peak prevalence of infection and age at which the peak occurs for *Schistosoma haemotobium* infections of children from 17 schools in the Zimbabwe highveld (after Woolhouse 1991).

2.4.2 Host sex

Is there a sex-bias in infection levels in wild animal populations?

Epidemiologists have long recognized that males of vertebrate species, including humans, tend to exhibit higher rates of parasitism and disease than females (Alexander and Stimson 1988; Bundy 1988; Zuk 1990). Moreover, a number of meta-analyses have provided quantitative support for this assertion across a range of host and parasite taxa (see Box 2.7). Although these results appear reasonably robust, they are generated by data of highly variable quality, collected from a range of different sources.

Therefore, if we are to understand the significance of such results, there is a pressing need for well-designed, well-controlled experiments that address the following questions. First, are the observed biases genuine or do they reflect sampling or other artefacts (see Box 2.3)? Second, do the patterns of bias vary across host and parasite taxa, as comparative studies suggest (Box 2.7)? Third, if such patterns

Box 2.7 Sex-biases in parasitism rates: observed patterns

A number of recent comparative analyses have examined patterns of sex-bias in parasitism rates in wild host (and laboratory) populations (e.g. Poulin 1996; Schalk and Forbes 1997; McCurdy *et al.* 1998). In his analysis of 85 studies of free-ranging host populations published over the last 30 years, Poulin (1996) found that male mammals and birds had significantly higher parasite prevalences than females, and in mammals this relationship was true for parasite intensity also. However, there were no such relationships in other host taxa, including fish, amphibians, and reptiles. When these data were further divided by parasite taxon (see Fig. 2.12), male-biases were small, but highly significant for nematodes infections of birds (prevalence only) and mammals (prevalence and intensity), but there were no robust trends for the other parasite types. In fact, the intensity of cestode infections was higher in female birds than in males (though sample sizes were small).

Schalk and Forbes (1997) examined the sex differential in parasitism rates of mammals for a different dataset, in which both field and laboratory studies were included. They observed a similar bias towards males and, when they split their dataset by parasite taxon, found significant male-biases for arthropod and protozoan parasite loads but not for helminth burdens (Fig. 2.13). Interestingly, they also found that sex-biases observed in experimental studies (in which hosts were artificially infected) were much stronger than those detected in field studies (in which hosts were naturally infected), suggesting that the main differences may lie in the host immune responses rather than the infection processes (Box 2.8). Thus, quantitative support of sex biases in parasitism rates remains inconclusive and there is a pressing need for more experimental evidence.

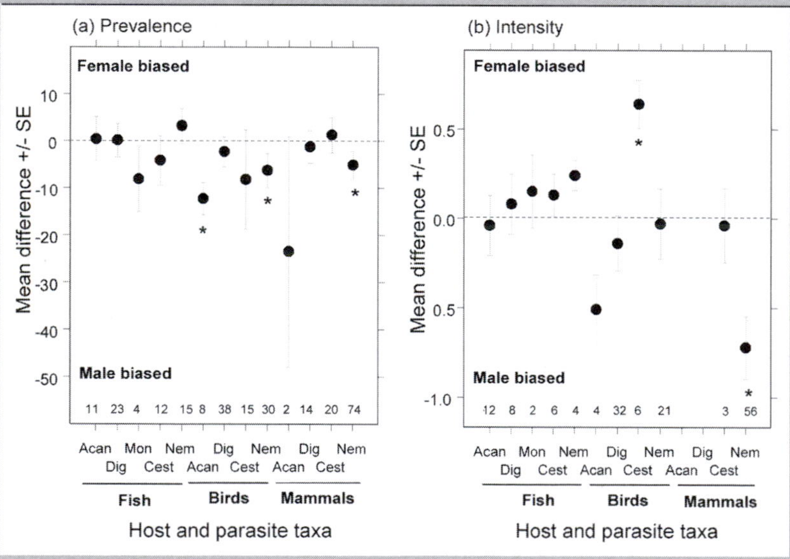

Figure 2.12 Mean difference in (a) prevalence and (b) intensity of parasitic infection between the sexes for three host taxa (after Poulin 1996). The *solid symbols* represent the mean difference (females minus males, weighted by a correction factor for sample size) and the *lines* are approximate standard errors. The *asterisks* indicate significant sex-biases. The numbers at the bottom of the Figure are the number of comparisons made. Abbreviations: Acan = acanthocephalans, Dig = digeneans, Mon = *monogeneans*, Cest = *cestodes*, Nem = *nematodes*.

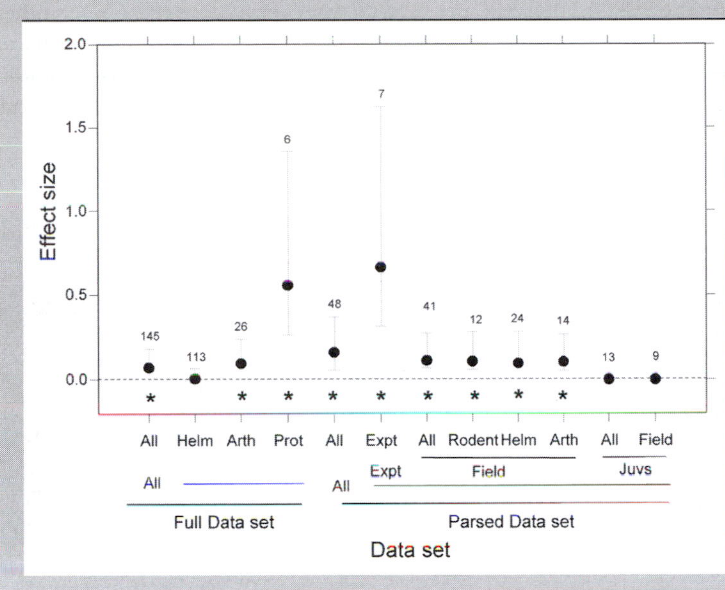

Figure 2.13 Mean effect sizes and 95% confidence intervals for sex-biases in parasitism rates in a selection of mammals (after Schalk and Forbes 1997). The *solid symbols* represent the mean effect size (a scale-free estimate of the average difference between the two sexes) and the *lines* represent the 95% confidence intervals. Significant male biases are present when the lower confidence interval fails to intercept zero and are indicated by the *asterisks*. The numbers above the bars are the sample sizes. Abbreviations: Helm = helminths, *Arth = arthropods*, *Prot = protozoa*, *Expt = experimental studies*, *Field = field studies*, *Juvs = juveniles* (all other data are for adults).

exist are they due to ecological differences between the sexes (e.g. in their behaviour, diet, etc.) or physiological differences (e.g. in the geometry of their immune system)? Finally, what are the epidemiological consequences of sex differences in parasitism rates?

What are the mechanisms generating sex-biases in parasitism rates?

There are a number of biological mechanisms potentially capable of generating sex biases in parasitism rates (for a recent review see: Zuk and McKean 1996). Often these causes are divided into ecological and physiological mechanisms. Ecological mechanisms include sex differences in behaviour, diet composition, and body size. For example, the male-bias in parasitism by the monogenean *Pseudodiplorchis americanus* observed in spadefoot toads (*Scaphiopus couchii*) is almost certainly due to differences in the reproductive behaviour of males and females. Whilst males spend long periods of time immersed in ephemeral pools exposed to the infective stages of the parasite, females visit the infected areas only briefly in order to lay their eggs (Tinsley 1989). Sex differences in diet are also likely to account for sex-biases in helminth infections of the marten, *Martes americana* (Poole *et al*. 1983), and in cestode infections of dace, *Leuciscus leuciscus* (Kennedy 1969).

Body size may also contribute to sex differences. In mammals, males are generally larger than females and there is good evidence that parasite load correlates with host size in a number of systems (Arneberg *et al.* 1998a), perhaps because large animals ingest more infective stages or offer them (or their vectors) larger targets. In birds of prey, females are often significantly larger than their mates and hence if this mechanism is important we might expect to find a reversal of the sex bias in parasitism. Significantly, a recent comparative analysis of blood parasitism rates in birds found no evidence for an effect of sexual size dimorphism on either the prevalence or intensity of infection (McCurdy *et al*. 1998).

Sex biases may also result from physiological differences between the sexes. For example, in vertebrates, there are often large sex differences in the levels of a number of steroid hormones, including testosterone, progesterone, and oestrogens. All of these hormones are known to have direct or indirect effects on components of the immune system and/or on parasite growth and development (Grossman 1985; Schuurs and Verheul 1990; Harder, A. *et al*. 1992; Hillgarth and Wingfield 1997). Testosterone depresses both cell-mediated and humoral immune responses, and has been invoked by some authors as a mediator of trade-offs between the development of sexually-selected traits and susceptibility to parasitism in males (see Box 2.8). Oestrogens, on the other hand, are believed to enhance humoral immunity while inhibiting cell-mediated responses. The production of stress hormones (e.g. corticosteroids), and the interaction between these hormones and the immune response, may also differ between the sexes (Klein *et al*. 1997; Klein 2000).

Box 2.8 Immunocompetence handicap hypothesis

Immunocompetence is a measure of the ability of an organism to minimize the fitness costs of an infection via any means, after controlling for previous exposure to appropriate antigens (Owens and Wilson 1999; see § 9.3.2).

In 1982, Bill Hamilton and Marlene Zuk proposed a role for parasitism in the evolution of sexually selected traits, such as colourful plumage and elaborate courtship displays (Hamilton and Zuk 1982). They suggested that these male traits had evolved to signal to females the bearer's good health and ability to resist the detrimental effects of parasitism. As a result, a female choosing a male with bright plumage and an elaborate courtship display would tend to acquire 'good genes' for parasite resistance for her future offspring. Since its formulation, numerous field workers have tested this controversial hypothesis with mixed success (reviewed by: Read 1990; Clayton 1991; John 1997; Hamilton and Poulin 1997) although some authors believe the hypothesis may not be testable (Read 1990).

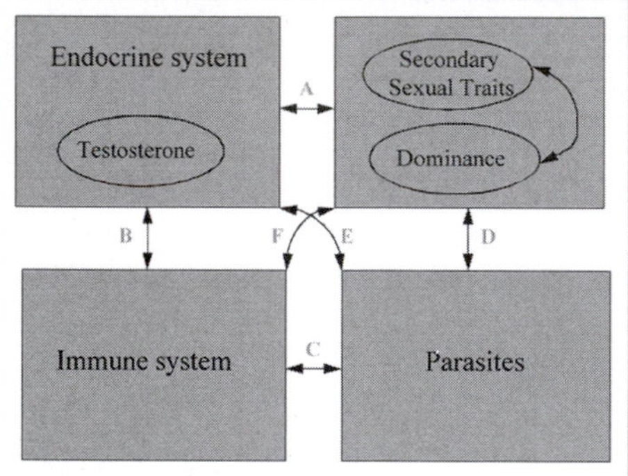

Figure 2.14 Model interactions included in the immunocompetence-handicap hypothesis (after Folstad and Karter 1992). Profiles of testosterone have a positive effect on the development of secondary sexual traits and dominance, while hampering the immune response (pathways A and B, respectively). Parasites interact with the immune system (pathway C), have a negative effect on secondary sexual development and dominance (pathway D), and cause reductions in testosterone profiles (pathway E). The development of testosterone-dependent secondary sexual characters also co-occurs with a reduction in immunocompetence (pathway F). A feedback system is postulated, operating through the direct and indirect relationships connecting model components, a feedback that links secondary sexual development to an individual's genetic resistance to parasites (Folstad and Karter 1992). Note that the IHH applies to 'any biochemical substance that is self-regulated and exerts a two-pronged effect of compromising the immune system and stimulating trait expression', and corticosterone has been suggested a possible candidate (Hillgarth and Wingfield 1997).

Evolutionary theory suggests that secondary sexual characters can act as honest signals of male quality only if they are costly to produce or maintain (Zahavi 1975; Kodricbrown and Brown 1984). Ten years after the Hamilton–Zuk hypothesis was published, Folstad and Karter (1992) proposed a proximate mechanism to explain the cost of male ornamentation. The original Immunocompetence Handicap Hypothesis (IHH) stated that because the primary androgenic hormone, testosterone, both stimulates the development of secondary sexual characters and reduces immune function (Grossman 1985), there is a physiological trade-off that both influences, and is influenced by, parasite burden (Fig. 2.14). Thus, only males with a high degree of genetic resistance to parasites will be able to produce high levels of testosterone and carry large ornaments.

The IHH was later modified to include the possibility that there was an adaptive role for immune suppression by testosterone in terms of resource re-allocation (Wedekind 1994). This version of the hypothesis proposed that because the production of secondary sex traits is costly, animals may have to shift energy and metabolites from other tissues in order to produce well-developed traits. Sex hormones, they argue, achieve this by shutting off energy from the immune system (and other systems), so that it can be re-directed for the production of the secondary sex traits. However, Hillgarth and Wingfield (1997) have argued that the energy savings accrued by such a strategy would be marginal and, given the potential costs of parasite infection,

continued

would be non-adaptive, especially if those resources could be drawn from other places, such as fat reserves.

A number of tests of the IHH have now been published, though most of these are purely correlative and so are potentially confounded (Hillgarth and Wingfield 1997). Some studies have manipulated testosterone levels and examined the consequences for parasitism and immune function (Saino *et al.* 1995; Hasselquist *et al.* 1999). However, there have been very few studies that have examined the possibility that there is a trade-off between sexual ornamentation and immunocompetence. One example is a recent study by Verhulst *et al.* (1999), who examined the evidence for a trade-off in selected lines of domestic fowl (*Gallus domesticus*). It is well established that comb size is important in both inter- and intra-sexual selection in chickens and its expression is testosterone-dependent. So, Verhulst and colleagues examined how comb size varied in males after 15 generations of divergent selection acting on primary antibody response to immunization with sheep red blood cells. There was a strong response to selection, with antibody titres varying significantly between all three selected lines (Fig. 2.15(a)). This was associated with a similar difference between the selected lines in their responsiveness to a variety of other antigens, including *Escherichia coli*, Newcastle Disease Virus, Bronchitis Virus, and Bursal Disease Virus, as well as with mortality following infection with Marek's Disease (suggesting that selection was operating on a significant component of the bird's humoural immune system). As predicted by the IHH, selection for enhanced immunocompetence, led to a reduction in both the degree of sexual ornamentation (comb size, Fig. 2.15(b)) and testosterone production (Fig. 2.15(c)). In other words, there appears to be genetic trade-offs between immune function an sexual ornamentation and between immune function and testosterone production.

Although consistent with the IHH, even this study has its problems. For example, recent experimental evidence has called into question the immunosuppressive properties of testosterone in birds (Hasselquist *et al.* 1999). Together with the fact that there is only one replicate of each selected line, this means that conclusions reached must remain tentative until further studies are conducted.

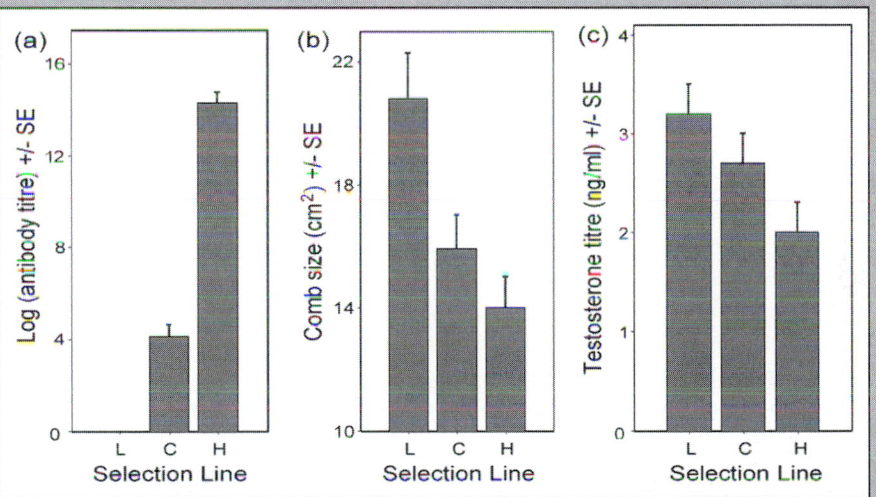

Figure 2.15 Response to selection in chicken lines divergently selected on response to sheep erythrocytes (after Verhulst *et al.* 1999): (a) $\log_2$ antibody titre to the challenge of sheep red blood cells; (b) comb size (cm^2); and (c) testosterone titre (ng/ml). In all Figures, L = low line, *C = control line and H = high line.*

In addition, there is evidence that the energetic costs of pregnancy and maternal care (Festa-Bianchet 1989), plus the immunosuppressive effects of some hormones produced during parturition and lactation, may increase the susceptibility of females to parasites at some times of year (Dobson and Meagher 1996). This may reverse any male bias observed outside the breeding season. Interestingly, a recent comparative study found that infections of *Haemoproteus* (a protozoan) were more common in breeding female birds than breeding males (McCurdy *et al.* 1998), even after controlling for phylogenetic effects.

Even if sex biases exist, and are relevant, determining the relative importance of the different mechanisms capable of generating them may prove extremely difficult, due to the fact that many of the ecological and physiological factors covary. Disentangling the relative importance of these factors will be possible only through careful experimentation in which each of the potential mechanisms is manipulated and/or controlled in turn. For example, useful insights may be obtained by examining sex differences in invertebrates, which do not produce sex steroids yet may exhibit male biases in susceptibility to parasitism (Wedekind and Jakobsen 1998). Valuable information may be gleaned from comparative analyses involving species or populations that buck the general trends (for a use of this approach see: Clutton-Brock *et al.* 1991). For example, it would be interesting to compare the patterns observed in related polyandrous and polygynous species that occupy similar habitats (e.g. a comparison of phalarope and sandpiper species in high latitude habitats). Although such comparisons have yet to be made, recent studies with *Peromyscus* mice and *Microtus* voles indicate that the situation may be far more complicated than had previously been envisaged (Box 2.9).

What are epidemiological consequences of sex-biased parasitism?

The sex biases detected in comparative analyses are usually rather small (<5%) and it has been suggested that, even if they do exist, they will generally have little impact on host evolution or parasite epidemiology. However, Poulin (1996) has argued that an increase of even a few parasites could be biologically meaningful. The epidemiological implications of sex-biased parasitism has rarely been discussed in the literature (but for discussion of the implications of female-biased parasitism rates in human populations see: Brabin 1992). If increased susceptibility to parasites during pregnancy results in parasite-induced reductions in fecundity relative to impact on survival, then this may destabilize the host–parasite interaction (see Chapter 3). The increased parasitism of males may, in part, explain the widespread observation of male-biased mortality, particularly in polygynous species of mammals and birds (Box 2.11; Promislow 1992; Promislow *et al.* 1992). Parasite-induced male-biased mortality would tend to stabilize the host–parasite interaction, except in species where males provide significant parental care or where males feed females during the breeding season. In these instances, parasite-induced male mortality or morbidity may be destabilizing if it results in reduced offspring production.

Klein and Nelson (1997) examined sex-differences in cell-mediated immunity in two species of *Peromyscus* mice. Contrary to expectation, they found no sex difference in cell-based immunity in the polygamous species (*P. maniculatus*), but they did in the monogamous species (*P. californicus*). This was despite the fact that the polygamous species had higher levels of testosterone, which is known to be immunosuppressive in mammals. These results are contrary to predictions based on the immunocompetence handicap hypothesis (Box 2.8). Subsequent studies on *Microtus* voles produced equally equivocal results and highlighted the importance of controlling for social factors when examining sex differences in immunity (Klein and Nelson 1997, 1998, 1999). We are clearly a long way from determining the mechanistic basis for sex biases in parasitism.

2.4.3 Host body condition

Host responses to parasites are of two general kinds: those that are directed against specific parasites, and more general ones affecting a number of different parasite species. We know very little about the relative importance of these two types of responses. What we do know is that hosts vary in a number of different characters (e.g. age, sex, reproductive status, and intensities of a number of various parasite species) and that this variation often affects host body condition. Non-specific host responses, such as fever or production of macrophages, are likely to be costly, but the cost will vary inversely with host condition. Body condition is also likely to affect the hosts' ability to compensate for damage inflicted by parasites,

Box 2.9 Sex differences in immune function

Any number of different mechanisms could produce sex biases in parasitism rates. Recent attention has focused on trying to determine whether these biases could be generated by sex differences in immune function. Measuring 'immunocompetence' (Box 2.8) is fraught with difficulties and a range of different measures have been used by workers in the field (reviewed for birds by Norris and Evans 2000). In a comparative analysis, utilizing information from more than 100 species of birds, Møller *et al.* (1998) assayed immune function by measuring the relative size of the Bursa Fabricius (which produces antibodies in juvenile birds) and the spleen (which produces lymphocytes in juveniles and adults). They predicted that if sex biases in parasitism rates were due to sex differences in immune defence, then one would observe that the more susceptible sex (i.e. males) would have smaller immune defence organs. As predicted, they found that whilst there was no sex difference in the size of the bursa, males had significantly smaller spleens than females (Fig. 2.16(a)). Moreover, this sex difference was more pronounced in adults than in juveniles (35% difference in adults, 7% difference in juveniles), which is consistent with predictions based on sexual selection theory (Fig. 2.16(b)). However, there are a number of difficulties with this study. For example, it is not clear whether there is a simple relationship between organ size and immune function, or even between immune function and disease susceptibility. Also, the health status of these birds was unknown, so it is possible that in this cross-section of birds, males had smaller spleens simply because they were healthier than females. As with many studies examining immune function in the field, the results are potentially confounded by the animals' previous exposure to pathogen and their current infection status (Owens and Wilson 1999).

This issue has been examined in captive reared birds where the immune function, as measured by the number of circulating leucocytes per unit volume of blood, were compared (Bennett *et al.* unpublished). As predicted, they found that in healthy animals, males had significantly lower numbers of heterophils, lymphocytes, monocytes, eosinophils, and basophils than females, and in some species females had more than twice as many circulating lymphocytes per unit volume as males (though the average difference was 5%). Thus, the higher parasitism rates generally found in males may be due to their lower concentrations of blood cells. However, Bennett *et al.* (unpublished) found that this trend was reversed in sick animals, with males producing significantly higher levels of all five types of leucocytes (by between 13% and 34%, depending on cell type). If robust, this result refutes the idea that males have lower blood cell counts simply because they are 'unable' to mount an immune response, and suggests that there may be a cost to healthy animals of maintaining high levels of circulating leucocytes. Moreover, that this cost is relatively higher for males than females. Again, these results are undermined by a lack of knowledge of the animals' previous infection history.

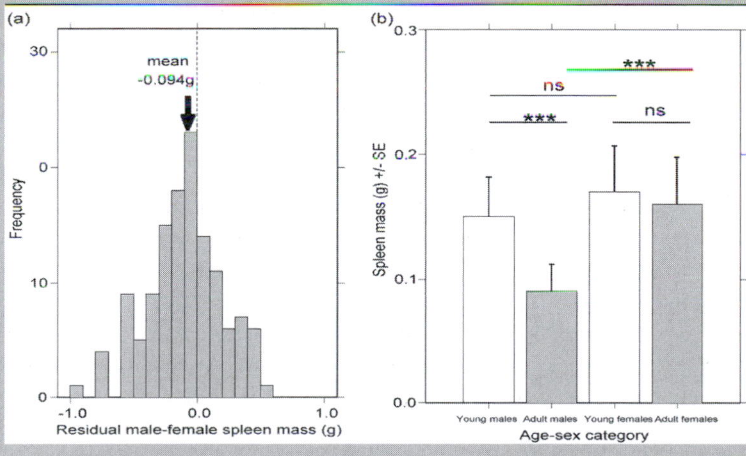

Figure 2.16 Sex differences in spleen size in 125 species of birds (after Møller *et al.* 1998). (a) Frequency distribution of sex differences in mass of spleen (after accounting for body mass differences), mean difference indicated by the *arrow*, is significantly different from zero ($P < 0.001$, $n = 125$). (b) Age-dependent spleen mass of male and female birds; values are means ± standard error (SE) for 25 species; *** $P < 0.001$; ns, not significant.

such as repairing tissues or replacing critical nutrients. Hosts in poor condition are therefore in a difficult bind: they have few resources available to spend on defence, but they cannot afford not to invest in them because parasites may induce more severe disease.

This situation should affect the distribution of the whole range of parasites within a host population. If differences in host body condition are of importance in generating observed infection patterns in wildlife, we may therefore expect intensities of different parasite species to covary. One way this can be studied is to sample a host population within a limited period of time and at a single location, count parasite numbers, and rank the host individuals with respect to intensities of different parasites. If general host responses matter, we would expect a significant agreement in the ranking of host individuals by different parasites. Indeed, this is what is observed in willow ptarmigan (Holmstad and Skorping 1998) and similar patterns have been observed in domestic (Stear *et al.* 1998) and Soay sheep (Wilson 1994). However, a lack of covariation has been observed in some other studies (e.g. Haukisalmi and Henttonen 1993; Nilssen *et al.* 1998). Caution needs to be exercised in interpreting these observations, however, because they could also be generated by extrinsic factors, such as covariation in exposure levels. For example, wet or humid areas might be good for both the free-living infective stages of nematode parasites and the juvenile stages of insect vectors of blood parasites, leading to some individuals having high levels of both gastro-intestinal nematodes and blood microparasites, independent of heterogeneities in body condition.

Hamilton and Zuk (1982) reasoned that if parasites and pathogens affect host body condition, then females could use a male's condition as a cue to his ability to resist disease-causing agents. They further postulated that because the parasites would be under selection to avoid host defences, a co-evolutionary arms race would develop between the parasite and its host, and hence host body condition might reflect a male's genetic quality. Females choosing males with elaborate plumage, complex song, or sustained courtship activity might therefore obtain 'good' resistance genes for her offspring (for reviews see: Read 1988; Møller 1990; Zuk 1992; Andersson 1994). Folstad and Karter (1992) developed this idea further by suggesting that the link between condition and parasitism might be mediated by the sex hormone testosterone or 'any biochemical substance that is self-regulated and exerts the two-pronged effect of compromising the immune system and stimulating trait expression' (see Box 2.8).

2.4.4 Host behaviour

By definition, parasites affect the fitness of their hosts. Therefore natural selection will favour individuals that evolve effective behavioural strategies to reduce the contact rate with the infective stages of parasites or their vectors. If individuals differ in their behaviour, then this can generate heterogeneities in parasitism rates—we have already discussed an instance where a sex difference in behaviour results in a sex bias in infection rates (spadefoot toads infected with the monogenean *P. americanus*; Tinsley 1989).

Behavioural strategies for avoiding parasitism or minimizing their impact are many and varied. For vertebrates these include grooming, grouping, selfish herding, avoidance of infested or infected conspecifics, and fly-repelling behaviour (reviewed by: Hart 1994, 1997). One strategy for reducing contact rates with directly transmitted macroparasites is to spatially segregate feeding and defaecating areas, thereby avoiding faeces contaminated with infective eggs or larvae (Hart 1994). Many animals have specialized latrines located away from their normal feeding areas, and there is anecdotal evidence from a number of species, which do not use latrines, that individuals will avoid feeding near faeces (Hart 1994, 1997). However, the avoidance of faeces will be an effective strategy only if it contains the parasites' infective stage (and, for most nematodes, it can take days or weeks for the free-living stages to become infective). There is therefore a need for behavioural experiments that not only establish that animals avoid areas where there are large numbers of parasites, but that any avoidance behaviour extends beyond the pre-patent period of the parasite.

Hutchings *et al.* (1998, 1999) have recently examined this problem in domestic sheep exposed to

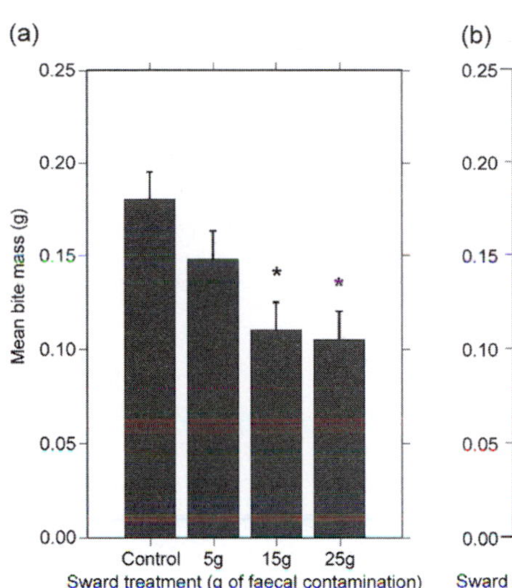

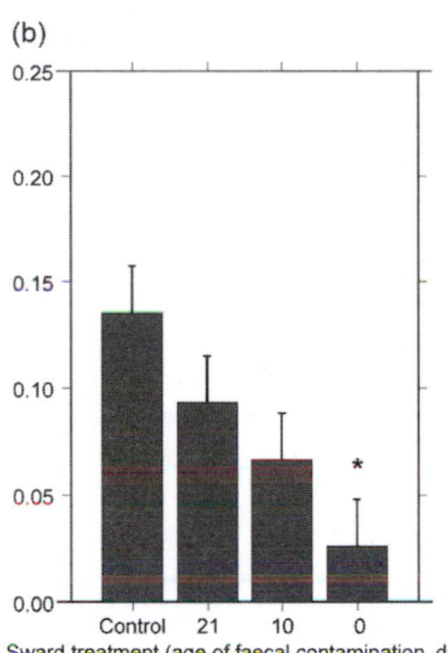

Figure 2.17 Avoidance of parasite-infested faeces in domestic sheep (*Ovis aries*). In both Figures, sheep were presented with two trays of grass varying in the amount or age of faeces (contaminated with the nematode *Teladorsagia circumcincta*) that contaminated it. The feeding propensity of the sheep was then scored by the mean mass ($\pm$ standard error) of herbage eaten per bite (after Hutchings *et al.* 1998). In (a) the amount of 21 day-old faeces contaminating the sward was varied, and in (b) the age of the 15 g of faeces was varied (in each case, the control group received no faeces). *Asterisks* denote significant difference in mean bite mass from the control group ($P < 0.05$). The number of infective larvae to which the sheep were exposed increased with amount of faecal contamination up to 15 g (in (a)), and increased with the age of the faecal contamination (in (b)).

grass swards contaminated with faeces infested with the nematode *Teladorsagia circumcincta*. They found that the sheep's avoidance of faeces increased with the amount of faeces present on the sward (Fig. 2.17(a)), but decreased with its age (Fig. 2.17(b)). Thus, although the sheep avoided contaminated swards, fresh faeces (which provide little risk of infection) presented the strongest stimulus for sward rejection. This may be because of a temporal decline in the strength of the cues used in faecal avoidance (e.g. olfactory stimuli), or may reflect a change in the relative costs and benefits of feeding on swards contaminated with faeces. The costs being associated with the risk of parasitism and the benefits being those associated with feeding on a sward fertilized by a rich nitrogen source (faeces).

Parasitism by *T. circumcincta* altered this behaviour further. Parasitized sheep fed at a reduced rate compared with non-infected animals and showed greater avoidance of contaminated swards (Hutchings 1998, 1999). Other factors shown to influence feeding behaviour were the sheep's immune status (animals with previous experience of *T. circumcincta* showed less aversion to contaminated swards) and their nutritional status (animals on a reduced diet were more likely to risk infection if this resulted in them feeding on a sward with a high nitrogen content; Hutchings 1999).

As indicated above, parasitized animals may alter their behaviour so as to reduce their risk of further infection or to minimize the impact of a current infection, via processes such as behavioural fever (Monagas and Gatten 1983; Florez-Duquet *et al.* 1998; Karban 1998), cold-temperature exploitation (e.g. Muller and Schmidhempel 1993), and anorexia (Kyriazakis *et al.* 1998). These responses to parasitism would tend to lead to a reduction in parasite heterogeneities over time. Alternatively, parasitized animals may change their behaviour in a way that leads to an *increase* in their risk of further infection. This may be because the host is forced by nutritional (or other) demands imposed by their parasites to forage in areas where there is an increased risk of further infection, or because of behavioural manipulation by the parasites themselves (Moore and Gotelli 1990, 1996; Poulin 1994; Thompson and Kavaliers 1994). Both of these processes would tend to accentuate heterogeneities by a process of positive feedback (§2.4.7) and could destabilize the host–parasite interaction (Dobson 1988a).

Behavioural strategies to reduce parasitism rates may extend to much greater spatial scales than those outlined above. For example, Folstad *et al.* (1991) showed that intensities of warble fly (*Hypoderma tarandi*) larvae in reindeer herds in northern Norway declined with increasing post-calving migration distance. Thus, by migrating out of calving areas, where there are high densities of the parasite's infective stages (adult flies), the reindeer reduce their level of infection; between-herd heterogeneities in infection levels may therefore be the direct result of behavioural differences between them.

Similar patterns have recently been reported by Altizer *et al.* (2000) working on infections of the protozoan *Ophryocystis elektroscirrha* in North American populations of the Monarch butterfly *Danaus plexippus*. They found that average spore loads were highest in a resident (non-migratory) population of Monarchs in Florida (70% heavily infected—spore classes 4 and 5), intermediate in moderately migratory populations in the western North America (30% heavily infected), and lowest in highly migratory eastern populations (less than 2% heavily infected) (see Fig. 2.18).

Consistent with the idea that migration allows the insects to escape parasitism in space and time, Altizer and colleagues also found that in western population of Monarchs, average spore loads declined as the distance between the summer breeding area and the nearest overwintering site increased. However, alternative mechanisms generating these patterns are also possible (Altizer *et al.* 2000). For example, they could be due to parasite-induced host mortality during migration or to differences between sites in environmental conditions. A further possibility is that they are due to genetic differences between populations in parasite infectivity and host resistance. Altizer (1998) found that individuals from highly migratory populations in the west were significantly less resistant to protozoan infections than the more sedentary eastern populations, and that western populations were associated with more virulent parasites.

2.4.5 Host genetics

How much genetic variation is there in susceptibility to parasitism?

There is a now wealth of information on the importance of host genetics in host–parasite interactions in crop plants (Burdon 1991; Kolmer 1996), in domestic and laboratory animals (Wakelin and Blackwell 1988; Kloosterman *et al.* 1992; Stear and Murray 1994; Stear and Wakelin 1998) and in human populations (Williams-Blangero and Vandeberg 1998; Williams-Blangero *et al.* 1999). There are many fewer good examples of genetic variation in disease resistance in natural host populations, and most of those come from plant–pathogen interactions (Jarosz and Burdon 1990; Burdon 1991; Thompson and Burdon 1992; Kolmer 1996; Yu *et al.* 1998; Glinski and Jarosz 2000). There is therefore a pressing need for more studies involving macroparasite infections of wild animals, particularly vertebrates (see Table 2.2).

One of the reasons for the lack of data from vertebrates is the difficulty of obtaining *unbiased* estimates of the genetic contribution to variation in parasite loads (Sorci *et al.* 1997). This is due partly

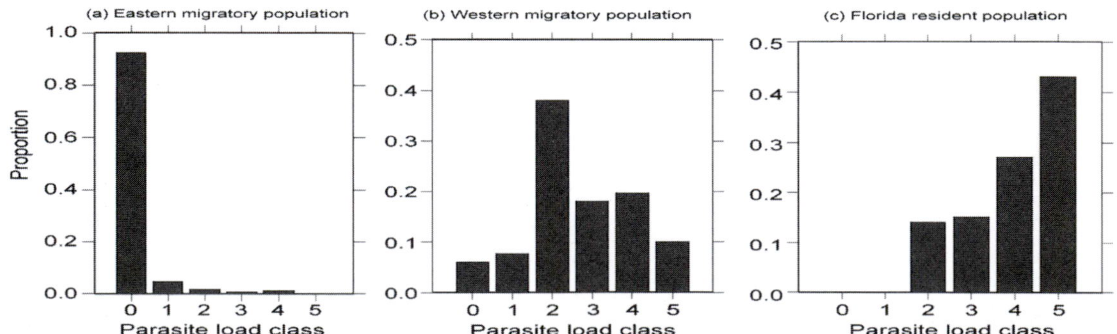

Figure 2.18 Frequency distributions of parasite loads in three North American populations of Monarch butterfly (*Danaus plexippus*) (after Altizer *et al.* 2000). The parasite is the protozoan *Ophryocystis elektroscirrha*. (a) The Eastern migratory population is from Sierra Chincua, Mexico. (b) The western migratory population from Pismo Beach, California. (c) The southern Florida population from Miami. Spore load classes are 0 = no spores, 1 = 1 spore, 2 = 2–20 spores, 3 = 21–100 spores, 5 = >1000 spores.

Table 2.2 Experimental and observational studies estimating the genetic contribution to macroparasite resistance in natural host populations: (a) heritability estimates, (b) variation between clones or strains

(a) Host species	Parasite species	Rearing method	Heritability analysis	Heritability (h^2)	Reference
Rhithropanopeus harrisii (xanthid crab)	*Loxothylacus panopaei* (sacculinid barnacle)	Common-garden	Full-sib analysis	$h^2 = 0.10$ (0–0.98 95%CI) ns[1]	Grosholz and Ruiz (1995a, b)
Transennells tantilla (bivalve mollusc)	*Parvatrema borealis* (trematode)	Common-garden	Full-sib analysis	$h^2 = 0.358 \pm 0.159$ (SE)***	Grosholz (1994)
Biomphalaria glabrate (snail)	*Schistosoma mansoni* (trematode)	Artificial selection[2] (R, C, S) for four generations	Response to selection – mean prevalence	Susceptible line mean $\simeq$ 80% Resistant line mean $\simeq$ 20%***	Webster and Woolhouse (1999)
Biomphalaria tenagophila (snail)	*Schistosoma mansoni* (trematode)	Artificial selection (R, C, S) for five generations	Response to selection – mean prevalence	Susceptible line mean $\simeq$ 100% Resistant line mean $\simeq$ 0%***	Mascara *et al.* (1999)
Drosophila melanogaster (fruit fly)	*Asobari tabida* (parasitoid wasp)	Artificial selection (R, C) for eight generations	Response to selection – proportion resistant	Resistant line mean $\simeq$ 60% Control line mean $\simeq$ 10%**	Kraaijeveld and Godfray (1997)
Drosophila melanogaster (fruit fly)	*Leptopilinia boulardi* (parasitoid wasp)	Artificial selection (R, C) for nine generations	Response to selection – proportion resistant	Resistant line mean $\simeq$ 50% Control line mean $\simeq$ 5%, $h^2 = 0.24$**	Fellowes *et al.* (1998, 1999)
Hirundo rustica (barn swallow)	*Ornithonyssus bursa* (mite)	Partial cross-fostering	Offspring–parent correlation	$r = 0.48$ (male parent)*** $r = 0.35$ (female parent)***	Møller (1990)
Rissa tridactyla (kittiwake)	*Ixodes uriae* (tick)	Observational	Offspring–parent regression	$h^2 = 0.720 \pm 0.232$(SE)**	Boulinier *et al.* (1997a, b)
Ovis aries (Soay sheep)	*Teladorsagia circumcincta* (strongyle nematode)	Observational	Half-sib analysis[3]	$h^2 \leq 0.688 \pm 0.287$ (SE)***	Smith *et al.* (1999)

(b) Host species	Parasite species	Assessment of variation	Observed variation between strains/clones	Reference
Tribolium confusum (flour beetle)	*Hymenolepis diminuta* (rat tapeworm)	Between 12 strains	Prevalence: 23%–100%*** Mean intensity: 1.43–6.88***	Yan and Stevens (1995)
Tribolium castaneum (flour beetle)	*Hymenolepis diminuta* (rat tapeworm)	Between 11 strains	Prevalence: 62%–100%*** Mean intensity: 4.22–25.69***	Yan and Norman (1995)
Acyrthosiphon pisum (pea aphid)	*Aphidius ervi* (parasitoid wasp)	Between 30 clones (from early 1989 only)	Prevalence: 0%–90%*** (broad sense heritability = 0.662, $CV_{clone} = 65.9\%$)	Henter and Via (1995)
Mus musculus (house mouse)	*Aspiculuris tetraptera* (pinworm)	Between 7 strains	Median intensity = 0–299 (males), 1–311 (females)***	Derothe *et al.* (1997)

[1] ns = non-significant, *$P < 0.05$, **$P < 0.01$, ***$P < 0.001$.

[2] R = selected for parasite resistance, S = selected for parasite susceptibility, C = control lines.

[3] Heritabilities based on offspring-parent regression were non-significant due to large standard errors, the quoted h^2 is for sire–female offspring correlation.

to the problem of distinguishing between genetic and non-genetic maternal effects in the field. One way round this problem may be to use cross-fostering experiments, in which hosts are taken out of their maternal (and paternal) environment and an assessment made of the genetic components of resistance in a neutral setting. Other powerful techniques (such as selection experiments) require the study animal to be brought into a laboratory setting, which may provide unreliable estimates of the genetic contributions to disease resistance by virtue of the fact that the non-genetic (environmental) component of variance has changed (though recent studies indicate a strong correlation between field and laboratory estimates of heritability; Weigensberg and Roff 1996).

There are several different levels at which the genetic contribution to parasite resistance can be examined. The heritability of a trait is the ratio of the additive genetic variation to the total phenotypic variation, and is a measure of the amount of genetic variation available to selection. Heritabilities provide an aggregate measure of the effects of all of the genes involved in parasite resistance, but tell us nothing about the specific genes involved. Due to the size of the host genome, identifying resistance genes is a difficult job and has rarely been attempted for diseases in wild animal populations, in marked contrast to the flurry of genome-mapping activity devoted to domestic and laboratory animals. By extrapolation from studies of macroparasitic infections of domestic animals (Schwaiger *et al.* 1995; Buitkamp *et al.* 1996; Stear and Wakelin 1998), the search for resistance genes is likely to focus on regions within and around the major histocompatability complex or MHC (Siva-Jothy and Skarstein 1998), part of the genome involved in antigen presentation to the vertebrate immune system (Apanius *et al.* 1997).

Another level at which genetics might impinge on parasite resistance is in terms of the overall level of genetic variation. Inbreeding depression is the decline in fitness attributable to the loss of genetic heterozygosity (Crnokrak and Roff 1999). By exposing recessive alleles for susceptibility to parasites in homozygous individuals, inbreeding might result in a reduction in parasite resistance.

To our knowledge, the only study of a wild host population that has examined the impact of genetic variation on macroparasite resistance at all of these levels is that of Soay sheep (*Ovis aries*) and their nematodes (mainly *Teledorsagia circumcincta*) on the Scottish island group of St. Kilda (see Box 2.10). This study indicates that there is substantial genetic variation in parasite resistance and that parasites are likely to be an important factor maintaining genetic variation within this insular host population. There is clearly a pressing need for similar studies in other wild animal populations. The logical starting point for such studies is likely to be where long-term and intensive monitoring of the host population is ongoing (a number of current bird and mammal studies fall into this category) and/or where closely related laboratory or domestic animal models exist. Among the mammals, candidate species are therefore likely to include feral populations of ungulates and wild populations of rodents. For many of these species there is a wealth of knowledge on particular host–parasite interactions, the immunological basis of parasite resistance is well-characterized and the appropriate genetic tools have been identified. However, it will probably only be through long-term, intensive collaborations between ecologists, geneticists, immunologists, and epidemiologists that inroads are likely to be made into determining the importance of host genetics in parasite epidemiological studies.

What maintains genetic variation in parasite resistance?
Given that parasites can be so detrimental to the fitness of their hosts, it is perhaps surprising to observe so much genetic variation in resistance to parasites in natural host populations. Why does natural selection not result in the fixation of genes conferring resistance? Evolutionary biologists have expended considerable effort musing over the potential mechanisms maintaining this genetic diversity, and a number of possible explanations have been advanced (Read *et al.* 1995).

One possibility is that genetic diversity is maintained by trade-offs between the fitness costs associated with resistance and those associated with parasitism (May and Anderson 1983). However, convincing evidence for such 'costs of resistance' in animal host population has been lacking until

Box 2.10 Soay sheep on St. Kilda

A feral population of Soay sheep has lived unmanaged on the St. Kildan archipelago, Outer Hebrides, Scotland for at least 1000 years. Historically, the population was restricted to the small island of Soay (99 ha), but in 1932, 107 sheep were moved to the much larger island of Hirta (638 ha). This population now comprises 700–2000 animals, and is characterized by periodic mass mortalities when up to 70% of the population may die overwinter (Fig. 2.20(a); Clutton-Brock *et al*. 1991; Grenfell *et al*. 1992a, 1998).

Although the proximate cause of death during these population crashes is protein-energy malnutrition (Gulland 1992), parasites have been implicated as a contributory factor (Gulland 1992; Gulland and Fox 1992; Gulland *et al*. 1993; Paterson *et al*. 1998; Coltman 1999). Since 1985, approximately 95% of individuals born in the Village Bay study area have been individually tagged and genetically sampled (Clutton-Brock *et al*. 1991; Bancroft *et al*. 1995a, b; Pemberton *et al*. 1996), and parasite loads have been estimated since 1988. Thus, it has been possible to both analyse genetic variation in resistance to parasites (see Box 2.11; Gulland *et al*. 1993; Paterson *et al*. 1998; Coltman *et al*. 1999) and to generate longitudinal profiles of parasitism by following individuals throughout their life (Boyd 1999).

Although the population plays host to 20 different species of macroparasite (Gulland 1992), the parasite that has the biggest impact on the sheep is the directly transmitted nematode *Teladorsagia circumcincta* (formally referred to as *Ostertagia*). *T. circumcincta* is also the most numerous of the trichostrongylids and each individual can harbour up to 20–30 000 of these small (7–12 mm) 'brown stomach-worms' (Gulland 1992).

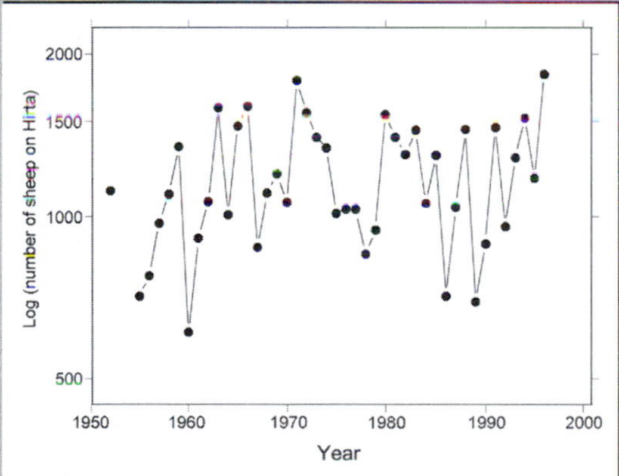

Figure 2.20 Population dynamics of Soay sheep on Hirta, St. Kilda (after Grenfell *et al*. 1998).

Symptoms of infection include poor weight gain or weight loss, loss of appetite and diarrhoea, and heavy infections can result in the sheep's death (Holmes 1985). On the mainland, farmers regularly dose their domestic sheep with anthelmintics, specifically to control this economically important parasite.

The dioecious adult worms live in the abomasum of the sheep where they mate and produce eggs that are voided in the faeces. It is the density of these eggs in the faeces that is regularly used on St. Kilda to measure parasitism rates ('faecal egg count' is strongly correlated with worm burden; Grenfell *et al*. 1995; Boyd 1999; Braisher 1999). On the pasture, the eggs may hatch within as little as 24 h, but they can survive on the pasture for several months prior to hatching, depending on environmental conditions. The emerging larvae moult twice before becoming the infective third stage, which the sheep ingest whilst feeding. Following ingestion, the L3 larvae migrate to the abomasum (fourth stomach) and enter the gastric glands, where they stay until they have completed two further moults and become mature adult worms. Eggs may appear in the faeces just 3 weeks following infection. However, larvae may become arrested at the early L4 stage for up to 3 months in a process known as 'hypobiosis'. The mechanisms determining whether, and for how long, a larva undergoes arrested development are not well understood, but probably include genetic, climatic, and density-dependent processes (Michel 1974; Gibbs 1986). It is probably the de-arrestment and maturation of these larvae in late winter and early spring that gives rise to the so-called 'periparturient-' or 'spring-rise' in faecal eggs counts.

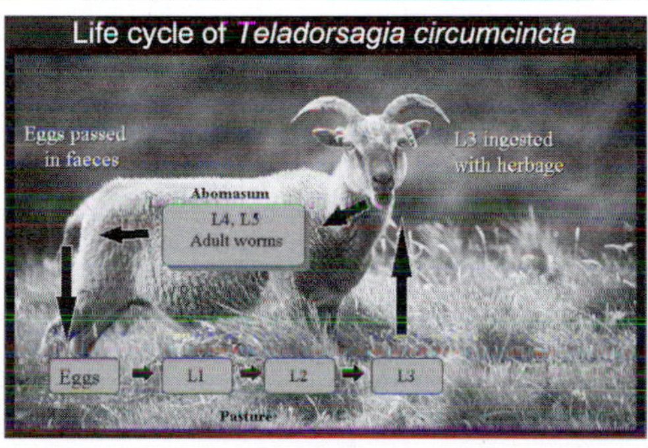

Figure 2.21 Life cycle of *Teladorsagia circumcincta* in Soay sheep. L1–L5 are the first–fifth larval stages.

relatively recently. One of the best examples illustrating a cost of resistance is a recent study by Kraaijeveld and Godfray (1997). They found that the fruit fly *Drosophila melanogaster* exhibited considerable genetic variation in resistance to attack by the braconid parasitoid wasp *Asobara tabida* (Fig. 2.19). Selected lines rapidly increased their cellular encapsulation response from 5% at the start of the experiment to greater than 50% after five generations of selection. However, after examining various life-history and other traits in the selected and control lines, the only apparent cost of resistance was a decline in the competitive ability of larvae in the selected lines when food was in limited supply.

Similar experiments with the eucoilid wasp *Leptopilina boulardi* demonstrated a similarly rapid response to selection (from <1% encapsulation at the

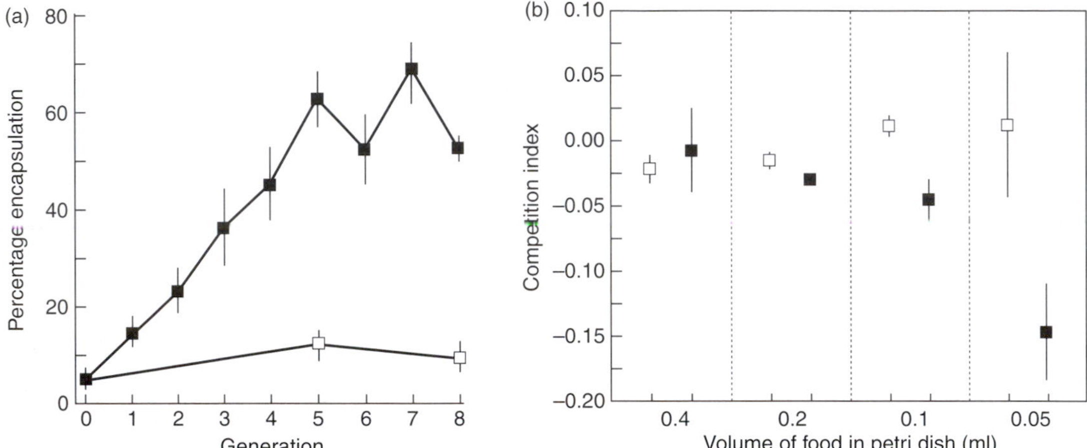

Figure 2.19 Encapsulation response and competitive ability of *Drosophila melanogaster* (after Kraaijeveld and Godfray 1997). (a) The frequency of encapsulation in flies belonging to the control (*open symbols*) and selected (*solid symbols*) lines. The Figure shows means and standard errors of the four selected and control lines. (b) The competitive ability of experimental flies in the two lines relative to a tester strain.

Box 2.11 Do parasites cause sex-biased mortality? A case study

In Soay sheep (see Box 2.10), males have consistently higher parasite loads than females (as measured by faecal egg count). Whereas the intensity and prevalence of infection show rapid declines with age in females (probably due to the development of acquired immunity), in males the declines are much less pronounced (Fig. 2.22).

Moreover, in common with many species of mammals, Soay sheep suffer male-biased mortality (see Fig. 2.23) and parasites have been implicated as a potential mechanism generating this bias.

Recent studies indicate that sex-differences in parasitism rates first appear as early as 12 weeks after birth and simple models suggest that this divergence is unlikely to be due purely to sexual size dimorphism (Boyd 1999). The two sexes differ markedly in their levels of testosterone, even at this young age, and it is possible that this plays a part in the difference between the sexes (see Box 2.9).

Observational data such as these can highlight potential relationships between sex and parasitism, and between parasitism and mortality. However, the only clear way to be sure of a causal relationship is to perform an experiment in which the parasites are chemically removed with anthelmintics. Short-term drenching experiments, which remove the current worm burden but allow immediate re-infection, indicate that males regain their parasites at a much faster rate than females, suggesting that males are predisposed to

continued

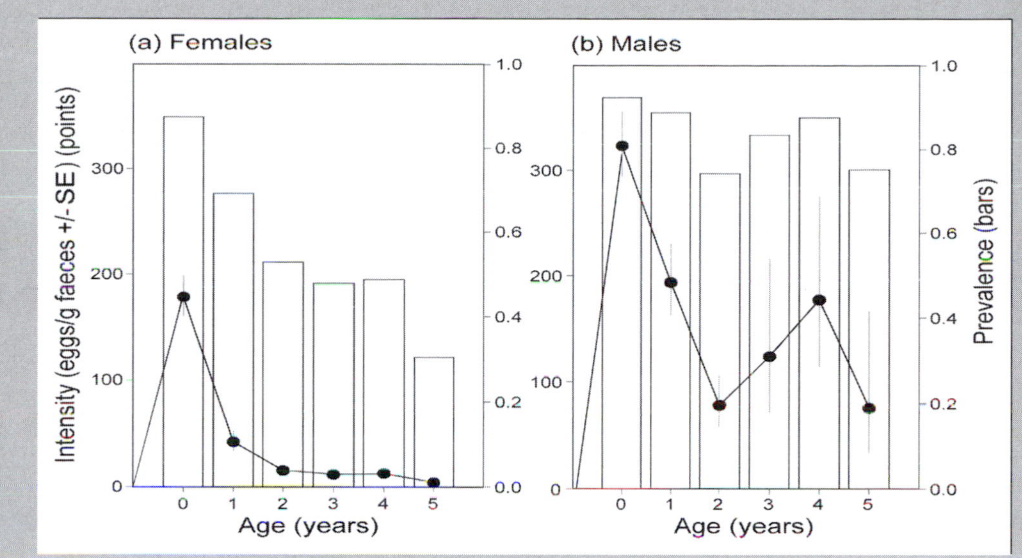

Figure 2.22 Sex difference in strongly prevalence and intensity in (a) female and (b) male Soay sheep (*Ovis aries*). Prevalence is indicated by the bars, and intensity (± standard error) by the points.

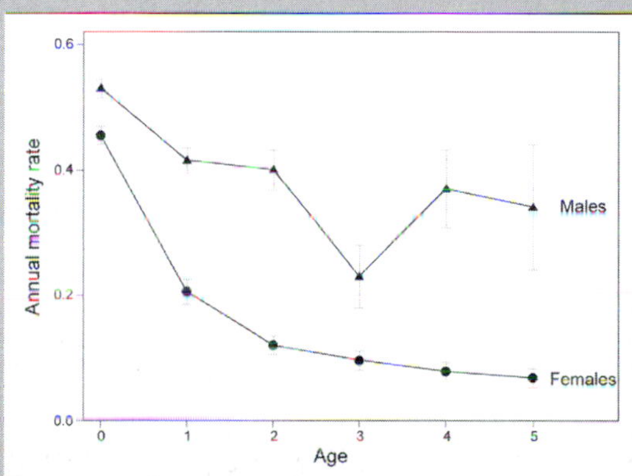

Figure 2.23 Sex difference in annual mortality rate in female and male Soay sheep (*Ovis aries*). Mortality rate is indicated by the *symbols* and the standard errors by the bars. Animals aged 0 are lambs aged 4 months old, animals aged 1 are yearlings aged 16 months, etc. (after Coulson *et al.* 2000).

high infection levels. More importantly, when parasites were removed for a longer period of time (i.e. several months) using anthelmintic boluses, the male-biased mortality observed in the control group was obviated in the treated group (Gulland *et al.* 1993). In fact, whereas parasite removal reduced the overwinter mortality of yearling females by less than 20%, in yearling males none of the 18 treated animals died, compared with nearly 50% of the 17 untreated males (see Chapter 3). Thus, not only do the two sexes appear to develop different levels of acquired immunity to their parasites but, in yearling males at least, parasites appear to be a much more important mortality factor.

Similar findings were reported during a recent epidemic of *Mycoplasma gallisepticum* in a population of house finches, *Carpodacus mexicanus*, in which disease-induced mortality resulted in a significant decline in the sex ratio from 1.08–1.44 males per female to 0.72 (Nolan *et al.* 1998). Interestingly, it was the males with the reddest plumage that survived the epidemic best, suggesting that plumage brightness might serve as an honest signal of disease resistance (see Box 2.8).

In terms of the epidemiological consequences of male-biases in parasitism rates, it appears that, although male Soays are producing substantially more parasite infective stages than females (arithmetic mean ± SD faecal egg count: adult males = 414 ± 382; adult females = 138 ± 235), the male-biased mortality means that these individuals constitute a relatively small proportion of the adult population (5–40%; Pemberton *et al.* 1996). Thus, the male contribution to parasite transmission may be much less than that of females, despite the higher egg production from males.

Box 2.12 The impact of host genetics on parasite distributions—a case study

A significant heritable component to faecal egg counts has recently been demonstrated in Soay sheep on St. Kilda (Box 2.10), using both offspring–parent regression and sib-analysis (Table 2.2; Smith *et al.* 1999). This result supports earlier studies, which showed that parasitism rates were associated with specific alleles at a number of protein loci, most notably the diallelic adenosine deaminase (Ada) locus (Gulland *et al.* 1993; Smith 1996). Gulland *et al.* (1993) found that individuals that were heterozygous at the Ada locus (FS) tended to have relatively lower parasitism rates in summer (females) or autumn (males), and that homozygous FF females had relatively higher faecal egg counts during the peri-parturient rise in spring. Consistent with the idea that Ada allele frequencies are maintained in the sheep population by parasite-mediated selection, overwinter mortality during population crashes was found to be highest for FF animals and lowest for FS animals.

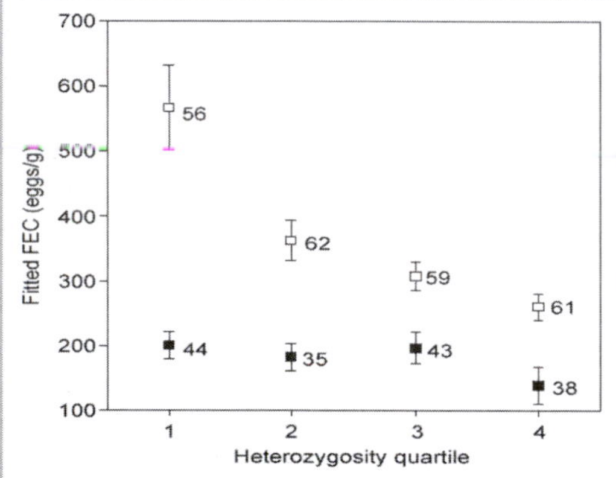

Figure 2.24 Effect of inbreeding depression on faecal egg counts in Soay sheep. *Open symbols* are the August faecal egg counts (± standard errors) from the best fitting model for high-density years; *closed symbols* are the fitted values for low-density years (after Coltman *et al.* 1999). The numbers next to the symbols indicates sample sizes. Heterozygosity is divided into quartiles—the 25% of individuals with the lowest heterozygosity scores (i.e. highest level of inbreeding) are in heterozygosity quartile 1.

Subsequent analyses have determined associations between parasitism and genotype at a number of microsatellite loci, the most interesting of which are located within the MHC (Paterson *et al.* 1998). Paterson and colleagues found that specific MHC alleles were associated with either low survivorship rates and high parasitism rates, or vice versa. For example, at the 'OLADRB' locus, lambs with the '257' allele had both low parasite resistance and low overwinter survival, whereas yearlings with the '263' allele had both high parasite resistance and high survival. This study is consistent with the proposition that parasites play an important role in the maintenance of MHC diversity in this population.

At a broader genetic level, recent work by Coltman *et al.* (1999) suggests that inbreeding depression may have a direct impact on parasite resistance. Across a range of microsatellite loci spread throughout the Soay sheep genome, individuals with low levels of heterozygosity not only had higher faecal egg counts (see Fig. 2.24), but also suffered higher overwinter mortality. Further, they observed that in sheep cleared of their parasites by anthelmintic treatment, overwinter mortality was independent of heterozygosity, providing experimental evidence for a role of parasites in selection against inbred sheep. Thus, parasite-mediated selection against inbreeding provides a mechanism retarding the loss of genetic variation in this population. There is clearly a need for similar studies to be undertaken on other naturally-regulated host populations.

start of selection to 40–50% after five generations) and the narrow-sense heritability of the trait was estimated to be 0.24 (Fellowes *et al.* 1998). Interestingly, in this study it also appears that increased resistance to parasitism is achieved by *D. melanogaster* only by sacrificing its competitive ability when food is in short supply, though the mechanism determining this trade-off remains unclear. Another result common to both studies is that the proportion of individuals encapsulating the two parasitoids never exceeded 60%, despite repeated episodes of selection. This suggests that there is an upper limit to the effectiveness of the fruit flies defence, though again the reason for this is unclear at present.

Despite the apparent similarity between these two selection experiments, it appears that they were selecting for different types of resistance. Fellowes *et al.* (1999) found that lines that had been selected for resistance to *L. boulardi*, showed a large correlated increase in resistance to *Leptopilina heterotama* and *A. tabida*, whereas lines selected for resistance to *A. tabida* showed little consistent cross-resistance to either of the *Leptopilinia* species.

This suggests that the attributes being selected for in the *L. boulardi* experiment were of general utility in resisting other parasitoids, whereas those selected for in the *A. tabida* experiment were more specific to that particular host–parasite interaction. Fellowes *et al.* (1999) discuss the potential importance of these different types of response for community structuring.

Other recent studies with invertebrates (snails, mosquitoes, and moths) have demonstrated that resistance may be traded-off against egg production, egg viability, larval competitive ability, adult body mass, and/or adult lifespan (Fuxa 1989; Boots and Begon 1993; Yan *et al.* 1997; Fuxa *et al.* 1998; Webster and Woolhouse 1999). There remains a lack of good evidence for trade-offs associated with parasite resistance in wild vertebrate hosts. However, a study with a captive population of domestic fowl (*Gallus domesticus*) suggests that there may be genetic trade-offs between the acquired and innate arms of the avian immune system (Norris and Evans 2000). Siegel and Gross (1980) used artificial selection to produce genetic lines that exhibited high (HA) or low (LA) antibody titres when immunized with sheep red blood cells. They subsequently found that HA birds were significantly less effective at controlling bacterial infections than LA birds (Gross *et al.* 1980). As bacterial infections are generally controlled by heterophils and other phagocytosing cells (Roitt *et al.* 1998), these results suggest that selection for enhancement of a component of the acquired immune system (antibody production) could lead to a reduction in the efficacy of a component of the innate immune system (phagocytes).

Another mechanism capable of maintaining genetic variation in parasite resistance is a co-evolutionary arms race between the host and its parasites, in which each party is continually responding and counter-responding to selection pressures imposed by the other (Stenseth and Maynard-Smith 1984). This process is sometimes referred to as the *Red Queen* after the character in Lewis Carroll's *Through the Looking Glass* who needed to run constantly in order to stay in the same place (Van Valen 1973; Lythgoe and Read 1998). Despite considerable theoretical activity in the past 25 years, there has been little direct empirical support for the hypothesis from wild populations. However, in a recent series of papers, Mark Dybdahl and Curtis Lively provide the best evidence yet supporting the Red Queen hypothesis (Dybdahl and Lively 1995, 1998; Lively and Dybdhal 2000). Over a 5-year period, they monitored the prevalence of a sterilizing trematode (*Microphallus* sp.) in 40 distinct clonal lineages of the snail *Potomopyrgus antipodarum* in a glacial lake in New Zealand. The frequency of the four most common lineages fluctuated markedly over time, and Dybdahl and Lively (1998) wanted to know whether these fluctuations were due to frequency-dependent selection imposed by co-evolving trematodes. A simple model predicted that if the Red Queen was responsible for the clonal dynamics, there should be positive correlation between the change in population clonal frequencies and the time-lagged change in clone-specific rates of trematode infection. As predicted, they found that rare clones had low levels of infection but, as they became more common, so they became over-infected and declined in frequency.

Although suggestive, an alternative explanation for these results is that there is a trade-off between competitive ability and resistance to infection, such that the best competitors are not only more common, but are also more susceptible to parasites. In order to test this hypothesis Lively and Dybdhal (2000) performed a series of experiments in which they exposed snail clones from two sources (Lake Poerua and Lake Ianthe) to two 'pure' sources of parasites from the same two lakes, as well as a 'mixed' source that comprised hybrid offspring from crosses from the two pure sources. As predicted by the Red Queen hypothesis (but not by the trade-off hypothesis), parasites were significantly more infective to sympatric sources of hosts than allopatric sources (Fig. 2.25). This 'local adaptation' is explained by the greater success of sympatric parasites on locally common host genotypes (Dybdahl and Lively 1998). This point is illustrated by the fact that Lake Poerua parasites infected common sympatric clones at a significantly higher rate than they did rare sympatric clones, whereas (allopatric) Lake Ianthe parasite clones infected rare and common snail clones at the same rates. Thus, the success of parasites on locally common host genotypes was due to commonness *per se*, rather than to

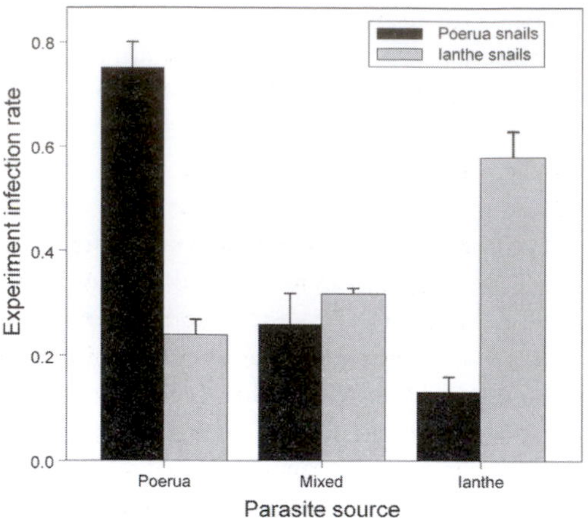

Figure 2.25 Local adaptation by a digenetic trematode (*Microphallus* sp.) to common clones of a Prosobranch snail (*Potamopyrgus antipodarum*) (after Lively and Dybdahl 2000). Local adaptation is shown by significantly higher infection rates for sympatric host–parasite combinations compared with non-sympatric combinations. Each bar represents the mean ($\pm$ standard error) of four replicates.

some correlated phenotypic character of the common genotype. This provides the best evidence yet that, as envisaged by the Red Queen hypothesis, local adaptation results from genetically based local co-evolutionary interactions. However, the ubiquity of this phenomenon remains to be determined.

What are the epidemiological consequences of genetic heterogeneities?

As discussed elsewhere, numerous theoretical studies have demonstrated that heterogeneities of any kind can have a significant impact on the dynamics of host–parasite interactions (e.g. Anderson and May 1982, 1991b). However, genetic heterogeneities differ from other types in the fact that co-evolutionary processes may result in the strength and direction of host–parasite interactions fluctuating over time. There are now a number of studies of plant–pathogen interactions (Alexander, H. *et al.* 1996; Thrall and Antonovics 1997), and interactions involving domestic animals and their pathogens (Woolhouse *et al.* 1998), in which host genetics and population dynamics have been synthesized in a single epidemiological model. These models generally show that unique epidemiological patterns can be observed when host genetics is explicitly defined within the model structure, and some experimental evidence is consistent with this view. However, we are aware of no equivalent studies of natural host-macroparasite interactions and there is certainly scope for such studies to be conducted.

2.4.6 Parasite genetics

There has been remarkably little work conducted on the importance of parasite heterogeneities in epidemiology, though in principal all of the heterogeneities previously considered for the host population apply equally well to the parasites themselves. In this section, we consider only genetic heterogeneities, the relevance of which is only just beginning to be realized. There is obviously a need for such studies to be extended to other forms of parasite heterogeneity.

How much genetic diversity is observed in parasite populations?

A number of studies on parasites of humans and domestic animals have quantified the degree of genetic variation within and between macroparasite populations (Blouin *et al.* 1992, 1995; Anderson *et al.* 1993; Anderson and Jaenike 1997; and reviewed by Anderson *et al.* 1998). These studies indicate that for many parasites, the within-population genetic diversity is extremely high (up to 10 times greater than for species in other taxa; Blouin 1992). Several recent studies of wild populations of host have observed similar levels of diversity. For example, Braisher (1999) studied populations of the nematode *Teladorsagia circumcincta* from two feral populations of Soay sheep living on the islands of St. Kilda and Lundy and found that there was just as much genetic diversity within populations of worms as there was between them. This sort of pattern is usually interpreted as indicating that the parasite has a very large effective population size (thousands of worms per host). However, there is also some evidence to suggest that DNA evolves faster in nematodes than in other taxa (Okimoto *et al.* 1994; Blouin *et al.* 1995), and this may contribute to the high genetic diversity.

Not all parasites have large effective population sizes and high levels of genetic diversity, however. For example, whereas trichostrongylids occur in their thousands, *Ascaris* worm populations of humans and pigs rarely exceed a few dozen per host. Anderson and colleagues (Anderson *et al.* 1993; Anderson and Jaenike 1997) studied *Ascaris* in Guatamala and elsewhere and found clear substructuring of the population at several different levels. For example, in one survey they found that 65% of nuclear genetic variation was found within host populations, 18% was accounted for by host species, while the remaining 17% was explained by geographical variation within host-associated populations. Despite the potential importance of parasite genetics to both evolutionary biology and epidemiology, we are only just beginning to examine geographical patterns of variation in natural populations. In particular, there are only a handful of studies that have attempted to map parasite genetic structure onto host genetic structure (but see: Mulvey *et al.* 1991; Dybdahl and Lively 1996). For example, Davies *et al.* (1999) studied the population genetic structures of the freshwater snail *Bulinus globosus* and its trematode parasite *Schistosoma haematobium* from eight river sites in Zimbabwe. They found that for the snail, genetic distance between populations was best correlated with proximity along rivers, whereas for the schistosome, genetic distance was better correlated with absolute geographical separation.

What can parasite heterogeneity tell us about the transmission process?

Genetic heterogeneities in worm populations can provide important insights into the transmission processes operating in wildlife host communities. For example, when a parasite is found in more than one sympatric host species, genetic studies can be used to determine whether a single transmission cycle is involved or if each host species is infected only by parasites derived from conspecifics (e.g. Anderson *et al.* 1993). This can have important consequences for chemical intervention strategies. Similarly, clustering of related parasites within hosts may indicate that there is a similar clustering of related infective stages in the environment or that parasite establishment is genotype-dependent (Anderson *et al.* 1995). If hosts are being infected by their own parasites (or those of their relatives), this could have important consequences for the evolution of parasite virulence. Parasites transmitted between close relatives can be expected to evolve lower virulence than parasites transmitted between non-related hosts (e.g. May and Anderson 1983; Ewald 1983, 1993; Clayton and Tompkins 1994; Herre 1995).

What are the epidemiological consequences of parasite genetic heterogeneity?

There is a substantial literature on the importance of host genetic heterogeneity in determining the outcome of helminth infections, but relatively little attention has been paid to the question of whether worm genetic diversity plays a role in immune evasion and parasite transmission. This is despite the very obvious importance of genetics in immune evasion by microparasites (e.g. influenza, malaria, dengue, etc.). One reason for the lack of interest is that the degree of antigenic polymorphism appears to be much lower in helminths than in most microparasites. However, since the antigens involved in protective immunity against helminths are largely unknown, levels of polymorphism in immunogenic loci may be uninformative.

Direct tests of the specificity of acquired protection involve experimentally immunizing hosts with particular strains or genotypes and then challenging with the same (homologous) or different (heterologous) parasites. Few such experiments have been performed within helminth species (tests of cross-species protection are much more common). Those tests that have been performed have typically involved poor experimental design, work with sublines of highly inbred lab strains, and non-natural hosts (Read and Viney 1996). As things stand at the moment, there is probably as much evidence that worm genetics matters as there is that it does not. Some of the best evidence that it does matter comes from the seminal work of Derek Wakelin and colleagues' on infections of the nematode *Trichinella spiralis* in laboratory mice, which undoubtedly shows that protective responses are stronger against homologous genotypes (e.g. Bolas-fernandez and Wakelin 1992; Goyal and Wakelin 1993a, b; Wakelin and Goyal 1996). However, the systematics of *T. spiralis* isolates remains in doubt and it is possible that the differences actually lie between incipient species, rather than within them. Work by the same group on infections of the nematode *Trichuris muris* in mice show similar patterns and are likely to be less confounded by phylogenetic problems (Bellaby *et al.* 1995, 1996). Evidence from other species remains mixed; in the fluke *Schistosoma japonicum*, for example, homologous immunity against different geographic isolates has been found in some cases, but not others. Perhaps the most extensive series of experiments has been conducted with the nematode *Strongyloides ratti* in rats, using isofemale lines (Viney 1999). Carter and Wilson (1989), working with two distinct isolates, found that homologous protection was stronger than that elicited by the heterologous line and she was able to replicate this effect with passive serum transfer. However, in an extensive series of experiments, A. F. Read *et al.* (unpublished) were unable to find any repeatable evidence that protective immunity was stronger against the immunizing genotype. Thus, at present we simply do not know whether genetic heterogeneity in the worm population is important in immune evasion (though it is likely that its will differ between host–parasite interactions).

This situation contrasts substantially with that in microparasites such as malaria, where there is little doubt that immune responses are at least in part genotype specific, with previously unseen genotypes having a growth advantage in semi-immune hosts (Brown 1999). This can have very interesting consequences for transmission. In experiments with the rodent malaria *Plasmodium chabaudi*, Taylor and Read (1997) found that mixed clone infections were more infectious to mosquitoes even though total parasite densities were no higher. Monoclonal antibody analyses of blood stage parasites, and PCR analyses of parasites in mosquitoes demonstrated that this was most likely a consequence of genotype-specific immunity against the numerically dominant clones within the infection (Taylor *et al.* 1998; Taylor and Read 1998). This should produce very interesting epidemiological dynamics: high rates of transmission will result in more mixed clone infections; in turn, this will generate more infectious hosts. It is unclear what the consequences of this will be when intervention strategies reduce transmission (a disproportionate drop in transmission?), or transmission rates are increased as a consequence of climatic change (a disproportionate rise in transmission?). Clearly, there is a need for some population level modelling of this phenomenon.

2.4.7 External heterogeneities

In this section, we consider heterogeneities that do not fall neatly into those that are attributes of the host or parasite. These include the spatial distribution of the parasite's infective stages in the environment, seasonal variation in infection levels, and heterogeneities generated by the parasites themselves.

Spatial distribution of external stages

The rate of acquisition of new infections often increases with the frequency of contact between the host and infective stages (Fig. 2.26). Thus, if there is spatial variation in the density of infective stages in the environment, and different hosts utilize different parts of their environment, then this will often lead to heterogeneities in parasite intensities across the population of hosts.

A particularly elegant set of experiments by Anne Keymer and Roy Anderson illustrates the role of the spatial distribution of infective stages in creating heterogeneity in the distribution of parasitic stages (Keymer and Anderson 1979). In these

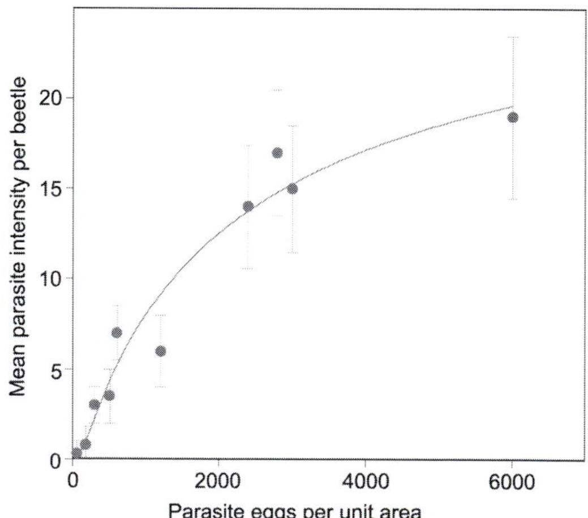

Figure 2.26 The influence of the mean density of tapeworm (*Hymenolepis diminuta*) eggs on the mean intensity of infections within the flour beetle (*Tribolium confusum*) (after Keymer and Anderson 1979). The bars represent the 95% confidence intervals. The line is a log-function fitted by eye. The effect of variation in the density of eggs on infection rate is illustrated in Fig. 2.27.

experiments, uninfected flour beetles (*Tribolium confusum*) were introduced into experimental arenas that contained the infective stages (eggs) of the tapeworm *Hymenolepis diminuta*. In each arena, the overall density of eggs was kept constant (at about 15 eggs/cm^2), but their spatial distribution was varied from approximately uniform ($s^2/m \sim 0$) to highly aggregated ($s^2/m = 2700$). After a fixed period of exposure (3 h), the beetle hosts were removed from the experimental arena and their subsequent infections examined. Although the spatial distribution of the infective stages did not appear to have a significant effect on the subsequent intensity of infection (Fig. 2.27(a)), it did have a marked impact and on the statistical distribution of parasites in the host population (Fig. 2.27(b)). In all cases, the parasite distribution within the host population was overdispersed, even when the spatial distribution of the infective stages was approximately uniform ($s^2/m \sim 0$). This illustrates the potential of behavioural or immunological differences to generate aggregated parasite distributions. Moreover, as the distribution of eggs in the environment became increasingly aggregated (s^2/m increased), so too did the distribution of parasites in the host population (i.e. s^2/m increased and k decreased; Fig. 2.27(b)). Interestingly, the degree of aggregation in the host population tended to an upper asymptote as the distribution of eggs became increasingly aggregated. This indicates that spatial heterogeneity in the infective stage distribution accentuates any behavioural or immunological differences between hosts.

This experiment is central to our understanding of the processes that determine heterogeneity in parasitic helminths, and it desperately needs to be replicated for other systems. In particular, it needs to be replicated in a system where the genetic (and perhaps the age) structure of the host population can also be controlled and manipulated.

Seasonality

Temporal variation in parasite loads appears to be common, particularly in highly seasonal aquatic systems (Shaw and Dobson 1995). For example, numbers of the copepod *Lepeophtheirus pectoralis* infesting plaice (*Pleuronectus platessa*) is highly seasonal (Boxshall 1974) and Shaw *et al.* (1998) found that all three measures of parasitism (the arithmetic mean number of parasites per host, the prevalence of infection, and the negative binomial k of the distribution) were at their lowest in late spring (mean < 1, prevalence $< 20\%$, $k < 1$) and peaked in early autumn (mean > 4, prevalence $> 90\%$, $k > 9$; Fig. 2.28).

Scott (1987a) examined temporal changes in the pattern of aggregation of the monogenean parasite *Gyrodactylus turnbulli* in free-running laboratory populations of the guppy *Poecilia reticulata*. Under conditions of regular immigration of uninfected guppies, the parasite undergoes recurrent epidemic cycles in the host population. She found that during the increasing phase of the epidemic cycle, there was an increase in the degree of parasite aggregation, presumably because of direct reproduction of the parasite on the surface of the host. As the peak prevalence and intensity were approached, the parasites became less aggregated, with the lowest degree of aggregation occurring during the declining phase of the cycle, presumably due to a density-

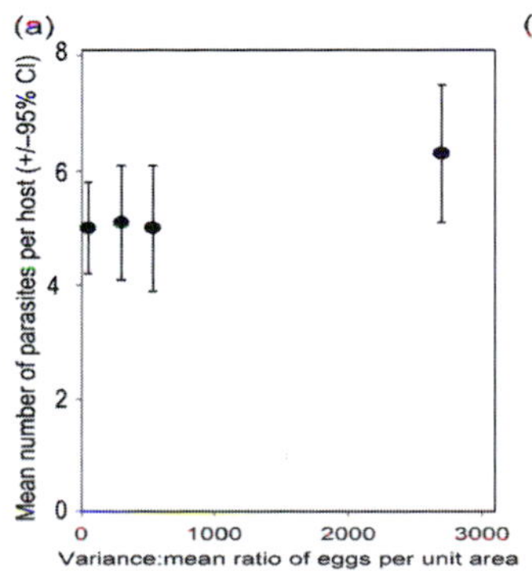

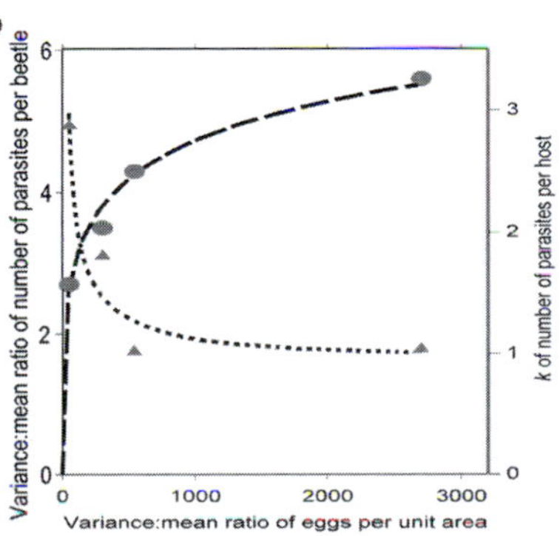

Figure 2.27 The influence of the spatial distribution of tapeworm (*Hymenolepis diminuta*) eggs on (a) the mean intensity and (b) distribution of infections within flour beetles (*Tribolium confusum*) (after Keymer and Anderson 1979). In (a) the bars represent the 95% confidence intervals. In (b), the distribution of infections is indicated by the s^2/m ratio (*circles*) and by k of the negative binomial (*triangles*). The *dashed line* is a log-function fitted by eye and the *dotted line* is a power function fitted by eye. The effect of mean density of eggs on infection rate is illustrated in Fig. 2.26.

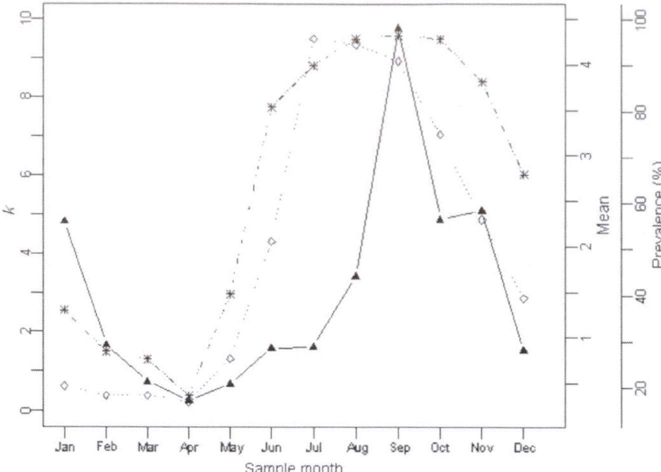

Figure 2.28 Temporal variation in the distribution of copepod infections in plaice (after Shaw *et al.* 1998). The data represent the arithmetic mean burden (· · ·▲· · ·), prevalence of infection (· · · * · · ·) and k of the negative binomial (· · · ◇ · · ·) for fish caught in 1972. (The original data were published by Boxshall 1974).

dependent death rate of infected hosts and/or a density-dependent reduction in parasite survival and reproduction on hosts that recover from infection. Thus, in all replicates of her experiment, there was a positive correlation between the mean intensity of infection and the degree of parasite aggregation, *as measured by the variance-to-mean ratio*. Interestingly, when she measured the degree of aggregation by the negative binomial k, aggregation appeared to *decline* with increasing parasite burden (i.e. k increased with parasite load). The reason for this discrepancy is that s^2/m and k measure different aspects of parasite aggregation—s^2/m is most sensitive to the length of the tail of the distribution, whereas k is most sensitive the overall shape of the distribution. Thus, Scott (1987a) has cautioned against using k to measure seasonal variation in parasite aggregation when the mean or prevalence of infection vary seasonally (Box 2.1). Note that in the plaice-copepod example described above, both s^2/m and k indicate similar patterns in parasite aggregation (D.J. Shaw personal communication).

Seasonal variation in parasite aggregation can be generated by variation in both host physiology (e.g. immune function) and host exposure to parasite infective stages. The latter is often due to the fact that the development and/or mortality rate of the free-living parasite stages (or their intermediate hosts) are temperature-dependent or sensitive to seasonal variation in humidity. Seasonal variation in exposure to parasites may also be driven by host-related factors. For example, on St. Kilda (Box 2.10), the density of infective strongyle larvae on the pasture exhibits two seasonal peaks: one in late spring (May), associated with the development of eggs voided by immuno-compromised peri-parturient Soay ewes, and a second in mid-summer (August), associated with the development of eggs produced by immunologically-naive lambs (Gulland and Fox 1992). The magnitude of this latter peak is dependent on the number of lambs produced in that year, but is also dependent on the prevailing climatic conditions. Whatever the mechanism generating seasonal variation in exposure to parasites, the end result is that any parasite sampling regime which lumps together data from individuals exposed to different seasonal regimes is likely to generate spurious estimates of parasite aggregation.

Heterogeneities generated by the parasites themselves

Not only are parasite loads heterogeneous, but the parasites themselves may cause these heterogeneities to become accentuated over time, by a process of positive feedback. Under such circumstances, deterministic or stochastic processes determine initial parasite loads, and these then become polarized as the parasites increase their host's susceptibility or exposure to further parasites (§2.4.3). This process has been investigated experimentally by Hoodless *et al.* (2001) working on tick infections of free-ranging pheasants (Box 2.13).

Under normal circumstances, natural variation within populations in some intrinsic factor, such as testosterone levels, may determine entry into one or other of two polarized positive feedback loops. In the pheasants, it seems likely that individuals with high testosterone levels establish their territories before questing ticks have reached their seasonal high abundance. Thus, although high levels of this hormone are usually associated with low immunocompetence (see Box 2.8; Folstad and Karter 1992), in this instance it is associated with lower parasite loads by reducing exposure rates. In this respect, pheasants contrast with rodents, in which high testosterone levels are associated with large home range size and high tick infestation levels on woodmice *Apodemus sylvaticus* (Randolph 1975, 1977), and with increased locomotory activity and reduced immunological resistance to tick feeding in bank voles *Clethrionomys glareolus* (Hughes 1998). The effect, however, is similar: one fraction of the host population, whether non-territorial cock pheasants or sexually active male rodents, feeds the majority of the tick population, thereby presumably supporting the majority of tick-borne parasite transmission. The importance of parasite-induced heterogeneities has probably been underestimated, but may be a powerful (non-genetic) factor reinforcing individual differences in parasitism rates.

Box 2.13 Heterogeneities generated by positive feedback—a case study

Hoodless *et al.* (2001) uncovered the possibility that parasites might themselves cause heterogeneities, which become accentuated over time, by a process of positive feedback. They did this using free-living pheasants (*Phasianus colchicus*) living in two woodlands in Dorset. They treated half the pheasants with long-lasting acaricide to reduce their tick infestations, and then monitored the pheasant's behaviour and parasite loads. Cock pheasants are territorial, setting up harems that they guard in small areas. The mating success of cocks depends on the brightness of their plumage and especially their wattles, the red fleshy swellings around the eyes. Reducing tick infestations on pheasants tended to improve wattle inflation and colour. The brighter the wattles, the better the cocks attracted hens (Fig. 2.29). Cocks without harems ranged more widely within woods than those with harems, whose small territories were usually on edges of fields. As a result, birds already carrying high tick burdens are likely to have a higher rate of contact with ticks questing in woodlands and therefore pick up even more ticks. Varying levels of tick infestation thus appear to introduce significant heterogeneity in the ranging behaviour of cock pheasants, which might in turn exacerbate the observed aggregated distributions of ticks amongst these hosts.

The additional significance of these observations is that these macroparasites, the ticks, are also vectors of microparasites, Lyme disease spirochaetes (see Box 7.2), so the impact of ticks on their hosts also determines the transmission dynamics of the Lyme spirochaetes. It is the non-territorial fraction of the pheasant population therefore that is likely to contribute most to maintaining natural cycles of Lyme disease in these habitats.

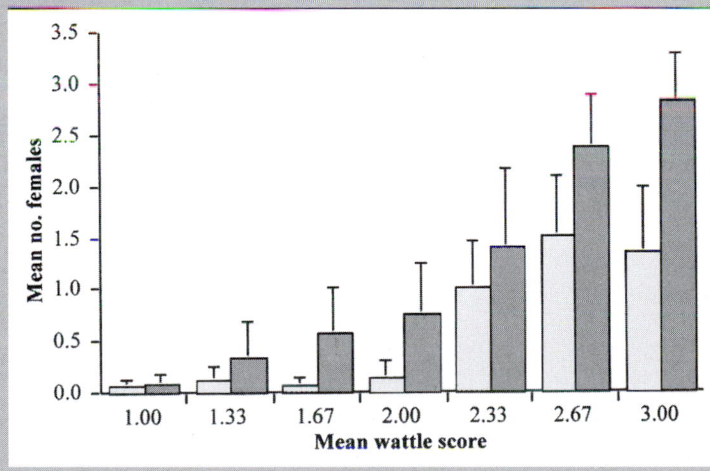

Figure 2.29 Mean harem size related to mean wattle score in male pheasants. Wattles were scored as follows: 1 = no wattle inflation; 2 = partial inflation; 3 = full inflation. Control birds = *light bars*; acaricide-treated birds = *dark bars*. Acaricide-treated females showed improved survival, higher nest survival rate and greater chick hatch rate. Male survival was not improved by acaricide treatment, perhaps because mate guarding by territorial males incurs an increased predation risk.

2.5 Synthesis

Developing a complete understanding of the processes that produce heterogeneity in the distribution of macroparasites in their host populations continues to be a central research area in parasite ecology. In this chapter, we have outlined our current understanding of this area. We conclude by suggesting some key research problems for the future. These can be divided into theoretical developments and empirical or experimental problems. Although we have a fairly complete understanding of the role that these heterogeneities play in determining the dynamics of host–parasite systems, there are still a number of unsolved problems in determining the relative importance of the different processes that determine observed degrees of aggregation. Central among these are genetic variability among hosts in their susceptibility to infection, spatial heterogeneity in the distribution of infective stages, and seasonal and diurnal variation in the risk of exposure. It is unlikely that the relative importance of these sources of heterogeneity will be the same in all systems. Yet, we still think it would be most instructive to examine their relative roles under controlled experimental conditions for two or three different well-studied systems. While comparative approaches have been useful in indicating the importance of other heterogeneities (see: Shaw and Dobson 1995; Shaw *et al.* 1998), there are simply no useful sources of comparative data for these major sources of heterogeneity. It may be possible to obtain some insight into their relative importance by some fairly detailed computer simulations, but in the absence of more experimental (or empirical) information, this may prove little more than a therapeutic exercise.

In a similar vein, purely analytical approaches to this problem have used moment-closure techniques to produce some new insights (Grenfell *et al.* 1995). These approaches have been particularly important in emphasizing the role of differences in host exposure in creating heterogeneities in immunological response to infection. Yet, the degree of parasite aggregation produced by these models is still significantly different from those observed in empirical systems. This is to be expected, as the degree of model complexity becomes totally unwieldy if it also has to consider genetic differences in immunological competence, spatial differences in the distribution and survival of infective stages and dynamic interactions between these. The upside of all of these shortcomings in our current understanding of the processes determining heterogeneity in host–parasite systems is that attempting to understand them in any one experimental system will produce new insights into the dynamics of host–parasite systems that will be useful for a range of other systems.

CHAPTER 3

Parasites and host population dynamics

D. M. Tompkins, A. P. Dobson, P. Arneberg, M. E. Begon, I. M. Cattadori, J. V. Greenman, J. A. P. Heesterbeek, P. J. Hudson, D. Newborn, A. Pugliese, A. P. Rizzoli, R. Rosà, F. Rosso, and K. Wilson

Parasites are ubiquitous and can clearly have an impact on the condition and survival of their hosts, but do they regulate the host population, what conditions lead to regulation, and how can this be demonstrated in the wild?

3.1 Background

The role that parasites have in influencing the dynamics of their host population has been a central question in the study of wildlife diseases (Kennedy 1975; Grenfell and Dobson 1995). Ecologists and wildlife managers have typically assumed that the answer to this question is 'No', believing that parasites are normally benign, specialized predators that live in a delicate balance with their hosts (Lack 1954). Whilst epidemic outbreaks of disease may cause massive host mortalities, they were explained as cases where environmental factors had disturbed the careful balance and thus rationalized as unusual exceptions to the general rule of no impact. It was even frequently argued that parasites were unlikely to impact on host population dynamics because if their actions led to the death of the host, then the parasite would also die. Theoretical and empirical evidence is now available that highlights the fallacy of this view. Since the lifetime reproductive success of parasites does not depend upon survival alone, but on the interactions between survival, reproduction, and transmission which determine the parasites ability to establish breeding offspring in new hosts, parasites can indeed have highly detrimental effects on their hosts (Box 3.1).

The manner by which parasites can influence host population dynamics was clearly demonstrated over 20 years ago in two theoretical papers published by Roy Anderson and Robert May (Anderson and May 1978; May and Anderson 1978). The simple mathematical model developed therein for macroparasites (those that do not reproduce directly within their host but which produce transmission stages that pass out of the host) clearly demonstrated that parasites could regulate host populations if they reduce host survival and/or fecundity in a density-dependent manner (Box 3.2). In many cases, however, the implications of this model have been misunderstood. Specifically, many empirical workers fail to realize that the identification of regulation is difficult when populations are at equilibrium. The hosts or parasites have to be experimentally perturbed to determine if regulation occurs—experiments that few workers have undertaken. Instead, the usual approach has been to examine dead animals to see if parasites are prevalent and if they are the principle cause of mortality. Some workers go on to look for correlations between measures of infection and host fitness and, in the absence of any clear relationship, take this as evidence that regulation does not occur. Such an approach will usually suffer from Type II statistical errors, with the null hypothesis of no regulation being mistakenly accepted.

To evaluate the importance of parasites to host regulation, we start by examining the importance of the basic reproduction number, R_0. We then consider the basic conditions under which either

Box 3.1 What selection pressures make parasites more virulent?

The conventional wisdom that well-adapted parasites evolve reduced virulence because their fitness depends on survival of the host is misleading since it fails to recognize the linkage of virulence to the dynamics of parasite transmission (Anderson and May 1982a; May and Anderson 1983; Ewald 1983). For example, in cases where parasites are transmitted to new hosts independently of host fitness (e.g. highly mobile parasites such as mesostigmatid mites) or parasites are transmitted by vectors (such as malarial protozoa), the parasite may not suffer a reduction in fitness by harming its host. These parasites can thus evolve a higher efficiency of host exploitation.

A good field study that tests these theories on the evolution of virulence is one conducted on the nematode parasites (*Parasitodiplogaster* spp.) of Panamanian fig wasps (*Pegoscapus* spp. and *Tetrapus* spp.). Here it is demonstrated (Herre 1993) that the wasp species characterized by population structures that result in increased opportunities for parasite transmission, harbour more virulent species of nematodes.

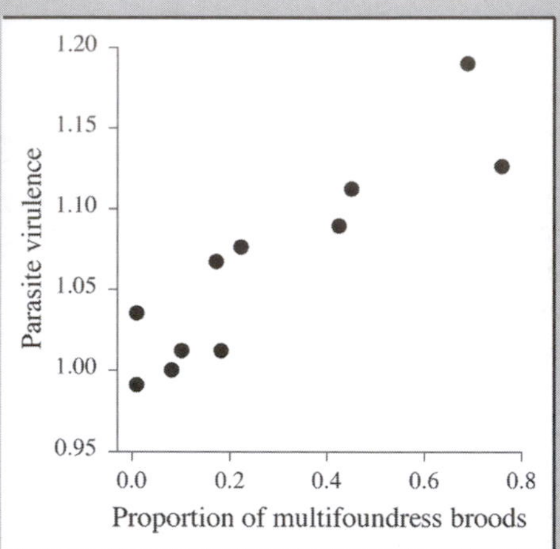

Figure 3.1 The relation between parasite virulence, measured as the lifetime reproductive success of uninfected relative to nematode-infected female fig wasps, and the proportion of multifoundress broods encountered in 11 species, where increased incidence of multifoundress broods provides increased opportunities for nematode transmission (after Herre 1993).

Box 3.2 What is regulation?

Regulation refers to the tendency of a population to decrease in size when it is above a particular level, but to increase in size when below that level (Begon *et al.* 1996a). Thus, any process that influences birth and death rates (and also immigration and emigration rates) in a density-dependent fashion has the potential to regulate a population.

Mathematical models can be used to predict when host regulation by parasites will occur. In the basic Anderson and May model, parasites can regulate a host population if the parasite birth rate exceeds the host birth rate plus parasite death rates, under the assumption that the death rate of infected hosts is altered by the number of parasites they harbour in a linear manner, that parasite transmission is virtually instantaneous, and that parasites are distributed randomly among hosts (Anderson and May 1978). The model takes the form of two differential equations, one of which describes the rate of change of the host population:

$$dH/dt = (a - b)H - \alpha P, \quad (3.1)$$

while the other describes the parasite population dynamics:

$$dP/dt = P(\lambda H/(H_o + H) - (b + \mu + \alpha) - \alpha P/H), \quad (3.2)$$

where H and P are the magnitudes of the host and parasite populations, a and b are the birth and death rates of unparasitized hosts, α is the host death rate due to parasitism, λ is the birth rate of parasite transmission stages, H_o is a transmission efficiency constant, and μ is the death rate of parasites within the host. Parasites are thus capable of regulating the growth of the host population in this model if $\lambda > \mu + \alpha + a$.

microparasites or macroparasites can regulate host populations, how such regulation can be demonstrated, and the use of mathematical models in conducting such demonstrations. We finish by investigating how parasitism can interact with other trophic interactions in the regulation of host populations, how such regulation may also result from indirect interactions among host species, and we consider what advances have been made in the modelling of host–parasite systems since the basic Anderson and May model was first formulated.

3.2 Infection dynamics and the basic reproduction number R_0

Parasitic infections of wildlife populations can occur as either epidemic outbreaks (sudden, rapid spreads, or increases in prevalence or intensity) or as endemics (parasites whose prevalence does not exhibit wide fluctuations through time in a defined location). Whilst macroparasites tend to occur as endemic infections (causing host morbidity rather than host mortality), microparasites (those parasites that undergo direct multiplication within their hosts) usually occur as epidemics where 'waves' of infection pass through populations after which the pathogen disappears as a result of hosts, previously susceptible to infection, either dying or becoming immune (Fig. 5.2). A particularly well-documented example of this phenomena is the 1988 epidemic of phocine distemper virus in the North Sea population of harbour seals (Boxes 3.3, 5.2, 6.5). In cases where microparasites tend to cause morbidity, rather than mortality, they can occur as endemic infections, e.g. lizard malaria (Schall 1983, 1990), rodent cowpox (Feore *et al.* 1997).

A critical factor in understanding epidemic outbreaks is being able to identify the initial source of the infection. For PDV, mathematical models predict that the 'critical community size' for PDV persistence, that is the minimum host population size (and density) required to allow the pathogen to persist in harbour seals, is several orders of magnitude greater than the known world population size (Swinton *et al.* 1998). Thus workers believe that the 1988 PDV outbreak in harbour seals was due to cross-species transmission from other host species (Hall 1995). Further modelling work suggests that a re-introduction of the virus into the harbour seal population would be unlikely to cause an outbreak on the scale of the 1988 epizootic until the seal population has recovered for roughly 20 years (allowing a build-up of susceptible individuals).

Box 3.3 The 1988 phocine distemper virus outbreak

During 1988 more than 18 000 harbour seals (*Phoca vitulina*) were found dead on beaches around the coasts of Northern Europe. The primary cause of death was identified as phocine distemper virus (PDV), a previously undescribed morbillivirus. See also Box 5.2 and Box 6.4.

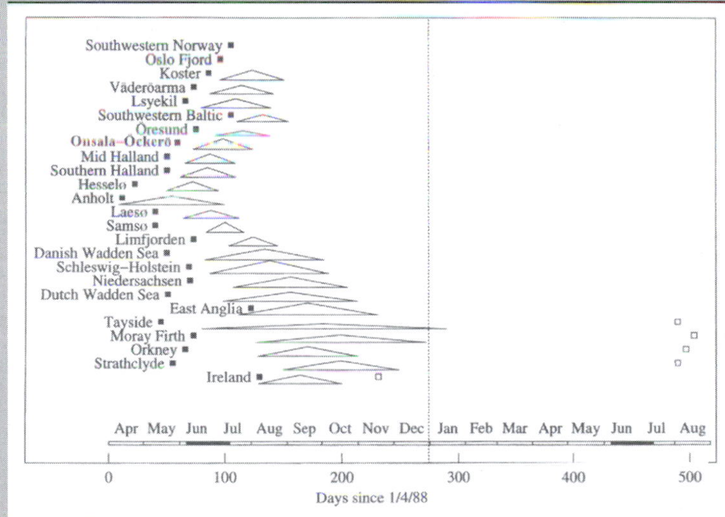

Figure 3.2 The first outbreak of phocine distemper virus (PDV) was observed on Anholt in Denmark. The disease spread round Scandinavia and to the Wadden Sea during the summer, arrived in Britain during June and by the end of the year, carcasses had been found at colonies all round the UK coastline. This spatial spread is illustrated here (after Swinton *et al.* 1998). The *closed boxes* denote the first recorded case in each location, the *triangles* are centred on the peak reporting time, with width equal to length of period in which mid-90% of cases recorded and height proportional to logarithm of total number of cases. The *open boxes* denote the last recorded case in those locations where known.

Whether epidemic outbreaks of infection are triggered, or whether endemic infections can persist, are both contingent upon the basic reproductive number (R_0) of a parasite in a host population. R_0 is a measure of the maximum reproductive potential of a parasite between one generation and the next for a given host population in a given environment. As such, it is arguably one of the most important and useful concepts in wildlife epidemiology, since it determines whether or not a parasite has the potential to spread in a host population. For microparasites, R_0 is defined as the average number of secondary cases which one case produces in a population consisting only of individuals susceptible to infection (Box 5.1). If $R_0 > 1$, i.e. if on average each infected individual is expected to produce more than one other case, a chain reaction of new cases will result leading to an epidemic outbreak (Fig. 5.2). If $R_0 < 1$, i.e. when an infected individual is expected not even to replace itself in the infected population, the pathogen is lost from the population.

For macroparasites the idea of R_0 is the same, but the definition and computation are subtly different. R_0 characterizes the ability of a parasite to invade a population where it was not present before, being concerned with the 'infectious output' of an individual—how much gets back into others. Since this depends on re-infection in the case of macroparasites, we conclude that R_0 for macroparasites cannot be characterized at the host level. Therefore, in contrast to microparasites, R_0 for macroparasites is defined as the number of new female parasites produced by an average female parasite when there are no density-dependent constraints acting anywhere in the life cycle of the parasite.

Box 3.4 Estimating the basic reproduction number R_0

R_0 can be estimated using relationships derived from host–parasite models. For example, the following relationship for macroparasites can be derived from the basic Anderson and May model detailed in Box 3.2:

$$R_0 = \lambda/(\mu + \alpha + b), \tag{3.3}$$

where λ is the birth rate of parasite transmission stages, μ is the death rate of parasites within the host, and α and b are the host death rates (due to parasitism and natural causes), whilst the following relationship for microparasites can be derived from the classic SI model detailed in Box 3.5:

$$R_0 = \beta C(N)/(\alpha + b) \tag{3.4}$$

where β is the rate of pathogen transmission, $C(N)$ is the contact rate between host individuals, and α and b are the host death rates (due to parasitism and natural causes).

Since obtaining good values for β is often impossible, workers have sought other methods of estimating R_0 for microparasites. Early methods for epidemic outbreaks involved using the initial exponential growth rate r of the infectious population to estimate R_0. However, if there is heterogeneity in the population that has a marked effect on transmission (as is likely), then there is no other relationship between R_0 and r than: $R_0 > 1$, if and only if, $r > 0$. Large values of R_0 could correspond to small values of r, and vice versa, so no definite conclusions about the value of R_0 can be based on this relation if only r is known. The reason is that R_0 measures growth on a generation basis, whereas r measures it in real time. For R_0 it therefore does not matter when during the infectious period the new cases are made, but for r it does since cases made earlier can generate new cases themselves at an earlier time and so make the infected population grow faster in real time.

If an epidemic outbreak has run its course, and has done so quickly compared to the time-scale of host demographic change, the final size of the epidemic can be more informative than the initial growth rate. If s_∞ is the proportion of the original susceptible individuals that remain susceptible at the end of the outbreak, then the final size of the outbreak is defined as $1 - s_\infty$. A key result from modelling is that there exists a relation between s_∞ and R_0 given by: $\mathrm{Ln} s_\infty = R_0(s_\infty - 1)$. Similar results for heterogenous populations exist, but special assumptions are needed to express them in terms of R_0. If only some fraction of individuals (s_0) are available to be infected at the start of an epidemic (e.g. due to some vaccination campaign or immunity from a previous outbreak), the formula becomes: $\mathrm{Ln}(s_\infty/s_0) = R_0(s_\infty - s_0)$.

If we look at endemic situations, one can estimate R_0 as the inverse of the steady state fraction of susceptible individuals. In a population where one knows the life expectancy L and the average age at infection A, one can calculate $R_0 = L/A'$ but only when the force of infection remains roughly constant.

Some ways in which R_0 can be estimated are presented in Box 3.4 with further extensions in Box 4.2. For further details see Anderson and May (1991), Dietz (1992), Heesterbeek and Roberts (1995a), and Diekmann and Heesterbeek (2000).

3.3 Host regulation by microparasites

For directly transmitted bacterial or viral microparasites, the parasite has the potential to regulate the host population when the per capita impact of the parasite exceeds the intrinsic growth rate of the host population, weighted by factors that determine the period that a host remains infectious or immune (Box 3.5). If the parasite is the only density-dependent mechanism in the system capable of regulating the host population then the proportion of the host population that suffer mortality as a consequence of infection will be directly proportional to the growth rate of the population. When the growth rate is low, then the parasite will be seen to cause hardly any mortality, even though it is still regulating the population. What is more, the prevalence of infection can also be extremely low. Anderson, R. M. (1995) provides a nice worked example of a fictitious respiratory viral infection in foxes and shows that a prevalence of infection of just 0.18% is needed to regulate the host population when the case mortality rate is 50%. If a wildlife ecologist undertook a field study at this point and found that less than 1 in 100 dead foxes had died due to infection, while the remainder had been run over on the road, then the ecologist may well conclude that parasites were of little significance. This would be wrong—the parasite would have been regulating the population. Either vaccinating the foxes to see if the fox population

Box 3.5 Regulation by microparasites

Population regulation by a microparasite requires the per capita impact of the pathogen to exceed the intrinsic growth rate of the host population (May 1983). The classic SI model captures these dynamics by classifying hosts as either Susceptible or Infected. This model again takes the form of two differential equations, one of which describes the rate of change of the susceptible host sub-population:

$$dS/dt = aN - bS - \beta C(N)SI/N, \quad (3.5)$$

while the other describes the infected host sub-population dynamics:

$$dI/dt = \beta C(N)SI/N - (\alpha + b)I, \quad (3.6)$$

where S and I are the magnitudes of the susceptible and infected host sub-populations, N is the magnitude of the total host population ($S + I$), a and b are the birth and death rates of unparasitized hosts, α is the host death rate due to parasitism, $C(N)$ is the contact rate between host individuals, and β is the rate of pathogen transmission. The expression $\beta C(N)SI/N$, being the fraction of the susceptible population that the infected hosts are able to contact and infect per unit of time, is known as the 'force of infection' (Anderson and May 1991). Parasites are thus capable of regulating the growth of the host population in this model if $\alpha > a - b$, under the assumption that contacts between susceptibles and infectives are described by the law of mass-action (for discussions regarding this law see: Boxes 5.3 and 5.4; De Jong *et al.* 1995; Begon *et al.* 1999).

In addition to the above requirement, the outcome of microparasite infection in this model will depend on the transmission rate β (Diekmann and Kretzschmar 1991). For increasing values of β:

1 the disease becomes extinct and the host population grows without bound;
2 both the S and I populations grow without bound, but the proportion that are infective tends to zero;
3 both the S and I populations grow without bound, but the proportion that are infective tends to a positive finite limit;
4 both the S and I populations tend to a bounded steady state (i.e. are regulated).

The SI model is a 'compartmental' model that can be easily modified to represent more complex disease systems, i.e. an SIR model includes a Recovered host sub-population (those that develop immunity), whilst SEIR models also include a compartment for host individuals in a post-infection latency period (Exposed). Acquired immunity, through decreasing the period of infectiousness, makes host regulation by microparasites less likely, whilst either delays in transmission, caused by long latency periods, or impacts on host fecundity can cause the host population and number of diseased animals to exhibit periodic cycles.

increased or manipulating the fox population to look at prevalence of infection would identify the important role of the pathogen.

3.4 Host regulation by macroparasites

While macroparasites are modelled in a different manner from microparasites, concentrating on the impact of individual parasites within the host population, a similar pattern holds. Once again it is quite possible for parasites to be regulating a host population when only a few individuals exhibit heavy infections. Macroparasites will regulate their host population when the relative per capita production of infective stages is greater than the weighted growth rate of the population (Box 3.6). Of particular importance is the degree of parasite aggregation within the host. When a large proportion of a macroparasite population is aggregated in a small proportion of a host population, then stable regulation is more likely, although there will be a critical threshold for the degree of aggregation below which the host population will escape the regulatory role of the parasite. Since the majority of macroparasite species exhibit aggregated distributions (Shaw and Dobson 1995; Shaw *et al.* 1998), it seems quite conceivable that these parasites may regulate their host populations. In this respect regulation may occur when infected host individuals are in the minority and only a few of which carry large parasite burdens. Field ecologists frequently expect regulating parasites to be abundant in any representative proportion of the population and in all dead individuals. The models tell us clearly that we should not expect this.

3.5 How can regulation be demonstrated?

Several experimental studies have been conducted which demonstrate the detrimental effects that parasites can have on fecundity and survival in wild animal populations (Table 3.1). However, simply showing that these effects occur in a density-dependent manner does not demonstrate

Box 3.6 Regulation by macroparasites

A major criticism of the basic Anderson and May model for macroparasites is that it is biologically unrealistic due to structural instability. The slightest alteration in the form of the various underlying biological assumptions will precipitate the system either to stability (disturbances damping back to the equilibrium point) or to instability (typically stable limit cycles). The model can be stabilized when the parasites are distributed in an aggregated manner among their hosts (see Chapter 2).

Aggregation was initially incorporated into the model by assuming that parasites have a negative binomial distribution among their hosts, with fixed clumping parameter $k > 0$ (where a smaller value for k means stronger aggregation):

$$dP/dt = P(\lambda H/(H_o + H) - (b + \mu + \alpha) - \alpha(k+1)P/(kH)), \quad (3.7)$$

although recent work shows that the variance-to-mean ratio of parasite distribution, and the way this depends on the mean of the parasite distribution, can be a more important determinant of model stability than k (Adler and Kretzschmar 1992).

With the incorporation of k, the condition for regulation becomes:

$$\lambda > \mu + \alpha + a + (a - b)(k + 1)/k, \quad (3.8)$$

i.e. host regulation is more likely when parasite fecundity is high, and when host fecundity, parasite mortality, the host death rate due to parasitism, and the degree of parasite aggregation are all low. However, if parasite reduction in host fecundity (δ) is also incorporated into the model:

$$dH/dt = (a - b)H - (\alpha + \delta)P, \quad (3.9)$$

the degree of parasite aggregation has to be relatively high for the model to be stable ($k < \alpha/\delta$), and thus for regulation to occur (May and Anderson 1978). Incorporating δ is essential when modelling macroparasite infections, since these parasites tend to cause morbidity rather than mortality in their hosts. Indeed, since the regulatory effects of δ tend to be greater than those of α, host regulation is more likely to occur with macroparasites than with microparasites. Macroparasites are predicted to cause host population cycles when δ is large relative to α (Dobson and Hudson 1992).

Table 3.1 The impact of specific parasite taxa on the fecundity and survival of wild animals, as demonstrated through the experimental manipulation of parasite loads (modified, with permission, from Tompkins and Begon 1999)

Host	Parasite	Impact
Anderson's gerbil (*Gerbillus andersoni*)	*Synoternus cleopatrae* (flea)	Reduced survival
Barn swallow (*Hirundo rustica*)	*Ornithonyssus bursa* (mite)	Reduced fecundity
Cliff swallow (*Hirundo pyrrhonota*)	*Oeciacus vicarius* (bug)	Reduced fecundity
European starling (*Sturnus vulgaris*)	*Dermanyssus gallinae* (mite)	Reduced fecundity
	Ornithonyssus sylvarium (mite)	Reduced fecundity
Great tit (*Parus major*)	*Ceratophyllus gallinae* (flea)	Reduced fecundity
House martin (*Delichon urbica*)	*Oeciacus hirundinis* (bug)	Reduced fecundity
Pearly-eyed thrasher (*Margarops fuscatus*)	*Philinus deceptivus* (fly)	Reduced fecundity
Purple martin (*Progne subis*)	*Dermanyssus prognephilus* (mite)	Reduced fecundity
Red grouse (*Lagopus lagopus*)	*Trichostrongylus tenuis* (nematode)	Reduced fecundity
		Reduced survival
Snowshoe hare (*Lepus americanus*)	*Obeliscoides cuniculi* (nematode)	Reduced survival
Soay sheep (*Ovis aries*)	*Teladorsagia circumcincta* (nematode)	Reduced survival

regulation, since reductions in host fitness due to the presence of parasites may only act in a compensatory fashion—that is, they may only affect individuals that would have lowered fitness anyway due to the effects of other regulatory forces such as competition or predation (Holmes 1982). Host regulation by parasites can only be demonstrated by perturbating the system away from an equilibrium, either by adding parasites to an uninfected host population and demonstrating a significant reduction in host density, or removing a parasite from a population and demonstrating a significant increase in density (Scott and Dobson 1989). One of the clearest successful examples of such an approach is that provided by Marilyn Scott working with free-breeding laboratory populations of mice (Scott 1987b). Here, introductions of the nematode parasite *Heligmosomoides polygyrus* reduced equilibrium host densities by almost 20-fold, after which removal of the parasite by anthelmintic treatment resulted in immediate increases (Fig. 3.3).

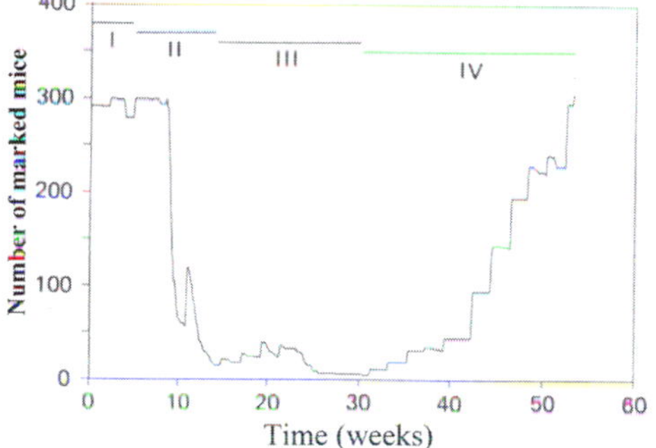

Figure 3.3 Regulation of a captive mouse population by the nematode parasite *Heligmosomoides poylgyrus*, as demonstrated over a 60-week period by a perturbation experiment. The number of marked mice (>2 weeks old) is plotted over four experimental phases: I = prior to parasite introduction; II = initial parasite transmission; III = high-level parasite transmission IV = after anthelmintic treatment (after Scott 1987b).

Although experimental perturbations have been carried out successfully with many captive host–parasite systems (e.g. Greenwood *et al.* 1936; Park 1948; Lanciani 1975; Keymer 1981; Anderson and Crombie 1984; Scott and Anderson 1984), such an approach has proved difficult to apply to field systems due to the practical complexities and ethical considerations involved. However, one field study that has accomplished it proves that parasites can indeed regulate wildlife populations (Hudson *et al.* 1998). Here, Hudson and colleagues demonstrate that parasitism by the gastro-intestinal nematode, *Trichostrongylus tenuis*, is a significant factor in regulating populations of red grouse (*Lagopus lagopus scoticus*) in Northern England.

The red grouse is a medium-sized gamebird that inhabits the heather moorland habitats of Britain. Extensive investigations of bag records from 175 individually managed grouse populations, coupled with detailed intensive demographic studies, have shown that 77% of populations exhibit significant cyclic fluctuations with a period of between 4 and 8 years (Potts *et al.* 1984; Hudson *et al.* 1985; Hudson 1992). Through the experimental reduction of the parasite burdens in grouse with an oral anthelmintic, replicated between populations, Hudson and colleagues were able to drastically reduce the extent of the cyclic population crashes (Fig. 3.4). This demonstrates that the interaction between the host and its parasite is the critical driving force behind the cycles in these populations (Tompkins and Begon 1999). The mechanism by which this occurs has been identified in previous modelling and individual-based experimental work as a density-dependent reduction in host fecundity that is large in relation to the impact of the parasite on host survival (Dobson and Hudson 1992).

In being the first and only experimental study to demonstrate that parasites can regulate free-living wildlife populations, the red grouse/*T. tenuis* story is insufficient evidence for a 'general' or 'typical' role of parasites in the regulation of their wildlife hosts. Other examples are needed. One promising system in which a second demonstration could potentially be made is that of Soay sheep (*Ovis aries*) and the gastro-intestinal nematode *Teladorsagia circumcincta*. Like red grouse, the Soay sheep population on Hirta (St. Kilda, UK) undergoes population crashes, in which the parasitic nematode is thought to play a role due to the occurrence of high parasite burdens in crash years (Box 2.10) and experimental demonstrations of parasite impact on host survival (Box 3.7). However, since the Hirta population is unique, any experimental manipulation of population dynamics would require control periods where no manipulation occurred both prior to, and following, the period of parasite reduction. With many years often passing between population crashes, and their unpredictable nature (Grenfell *et al.* 1998), such an experiment may take at least a decade to conduct.

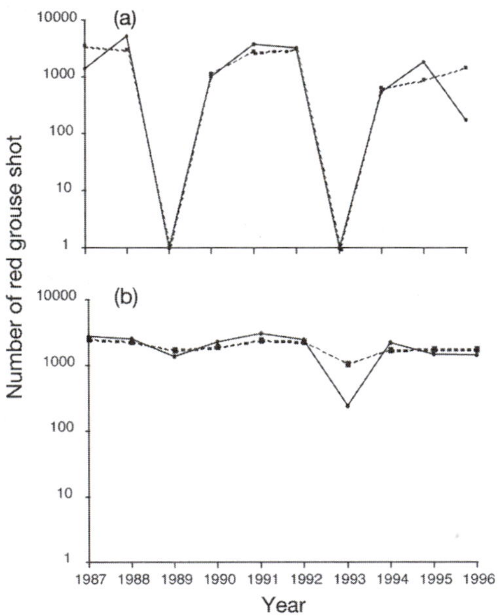

Figure 3.4 Population changes of red grouse, as represented through hunting records, as (a) two control sites and (b) two sites where grouse were treated with an oral anthelmintic in the years when population crashes were predicted to occur (after Hudson *et al.* 1998).

A system that may be more conducive to a demonstration of regulation is that of cowpox in natural island populations of wild rodents, where populations could be manipulated with both controls and replication. However, although a clear impact of the virus on host fecundity has already been documented (Feore *et al.* 1997), it would be premature to conduct manipulations without first constructing models showing that such an impact may be sufficient to cause regulation.

Whilst the regulation of wildlife hosts has been experimentally demonstrated for macroparasites, no such unequivocal demonstration has yet been made for microparasites. The best evidence that is currently available for such regulation has been obtained by examining natural and human-introduced pathogens in areas where host populations can be carefully monitored. A good example is seen in the spread of rinderpest in East African wildlife where, at the end of the last century, the pathogen was introduced into wildlife populations that had no prior exposure to the infection (Plowright 1982). Following the introduction into the Horn of Africa, the subsequent epidemic caused massive mortalities in wild and domestic ruminants that took 10 years to spread from Somalia to Cape Town. Rinderpest was eventually eradicated from much of East Africa following the development of an effective vaccine. The initiation of a vaccination scheme in cattle surrounding the Serengeti in the

Box 3.7 Do parasites regulate Soay sheep?

One of the well-studied host–parasite interactions is that of Soay sheep and the nematode *Teladorsagia circumcincta* on St. Kilda (Gulland 1992; Gulland and Fox 1992; Grenfell *et al.* 1995). The Soay sheep population on St. Kilda is characterized by unstable dynamics and periodic mass mortalities. Mortality is strongly density-dependent, but the depth of any crash is critically dependent on the weather in late winter (Grenfell *et al.* 1998). The immediate cause of death is protein-energy malnutrition. Parasites, however, have been implicated as a factor in the mortality of the sheep by the observation that dead animals have large numbers of macroparasites (mainly *T. circumcincta*) and exhibit severe parasite-induced pathology. In addition, there is a strong negative correlation between parasite burden and fecundity in all age classes (Wilson, unpublished data). However, correlational data like these cannot be used to infer that the parasite is regulating the host population.

Since population manipulations with Soay sheep are logistically difficult, the only experimental evidence to date that supports a role of parasites in the Soay sheep population crashes is individual-based work. Gulland *et al.* (1993) removed the parasite loads from a group of lamb and yearling Soay sheep, using long-lasting intrarumenal anthelmintic boluses, and demonstrated that their overwinter survival was greater than a comparable group of control animals (Fig. 3.5). The treatment also affected the reproduction of their host. See Box 2.10 for further details.

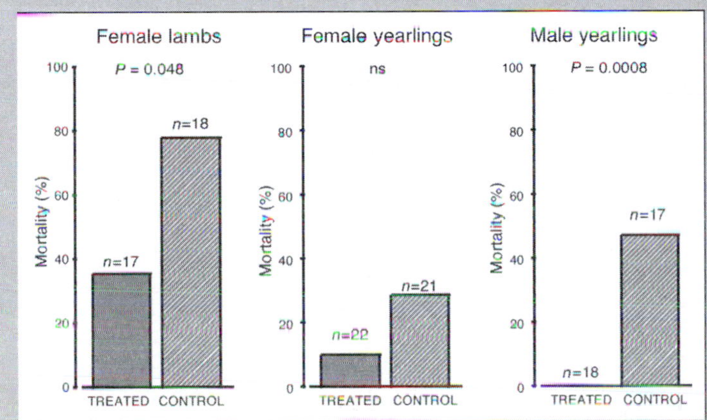

Figure 3.5 Mortality of Soay sheep on St. Kilda treated with anthelmintic boluses compared with an untreated control group. Treatment reduced survival in each age and sex cohort demonstrating that parasites were reducing survival (after Gulland *et al.* 1993).

late 1950s was followed by an increase in wildebeest numbers in the Serengeti from around 300 000 to around 1.5 million (Fig. 3.6). Increases also occurred in numbers of buffalo and the two main predators of these two species, the lions and hyenas. These findings implied that rinderpest was holding the ungulate populations at a much lower density than the habitat could support and that predator numbers are dependent upon prey density, rather than acting as a regulatory constraint.

3.6 Applying models to empirical data

Although mathematical modelling cannot provide the definitive proof of host regulation by parasites, since this requires careful experimental studies, such models can be important in improving understanding and making predictions when planning the manipulation of wildlife populations. Since large-scale manipulations in the field can be difficult to accomplish, due to logistical and ethical constraints, models can be invaluable in two ways. First, they can identify wildlife systems in which regulation by parasites may be occurring and, second, they can give insights into how specific experiments should be designed to successfully demonstrate regulation. These benefits of mathematical models are apparent in the study conducted by Hudson and colleagues on red grouse and *T. tenuis* (see above). Here, a model constructed and parameterized prior to population manipulations being conducted showed that it was indeed

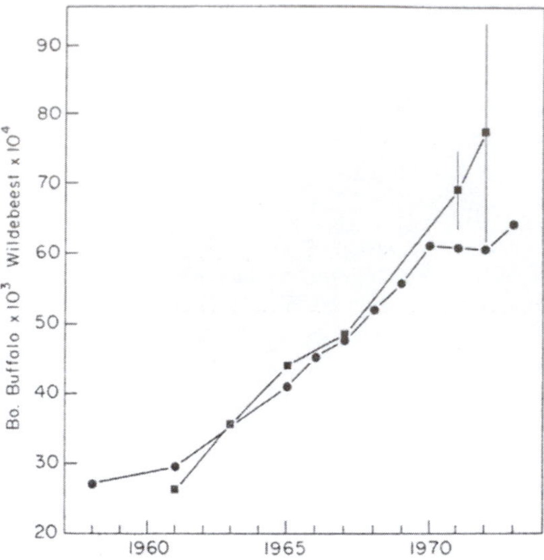

Figure 3.6 The observed increase in buffalo and wildebeest populations (*circles* and *squares*) in the Serengeti National Park, following the initiation of a rinderpest vaccination scheme in cattle in the late 1950s. Vertical lines are 95% confidence limits (after Sinclair 1977).

possible for parasites to be the cause of the observed cycles in the red grouse populations and the model was used to predict the 'crash' years when the effects of experimentally removing parasites would be observed. Furthermore, the model was also used to predict the proportion of grouse populations that would need to have their parasites removed for the manipulation to be effective (May 1999).

Another gamebird in which parasites may be causing population cycles is the rock partridge (*Alectoris graeca saxatilis*), an upland bird inhabiting Alpine prairie and meadows. Hunting records from the Italian province of Trentino show cyclic dynamics in about 40% of populations, with cycle periods of 4–7 years (Cattadori *et al.* 1999). The populations also fall into two distinct groups, those in wet habitats and those in dry habitats, with the fluctuations in abundance being more synchronous within than between groups, and the populations in dry habitats tending to be more cyclic (Fig. 3.7). Modelling of the population dynamics suggests that the cyclic behaviour would not be maintained in these populations without a driving density-dependent force, most likely a trophic interaction. That parasites may be this driving force is suggested by a higher prevalence and intensity of nematode infection (principally *Ascaridia compar*), and a lower degree of parasite aggregation in the cyclic than in the non-cyclic populations (Rizzoli *et al.* 1997, 1999). Moreover, laboratory experiments have demonstrated that parasites reduce the breeding production of partridges and modelling has demonstrated that these features are destabilizing and will tend to lead to population cycles. The link between cyclic partridge population dynamics and habitat type may be because the survival and availability of *A. compar* infective eggs is greater in dry habitats, leading to higher parasite burdens (which impact on fecundity) and lower parasite aggregation. This system thus has the potential for an experimental manipulation to demonstrate regulation, although logistical constraints may not permit a large scale experiment.

In many instances, obtaining parameter estimates for models of wildlife hosts and their parasites can be almost as difficult as conducting population manipulations (McCallum and Scott 1994; McCallum 2000). One method of tackling this problem is to obtain first estimates of some of the key parasitological and host demographic parameters by applying a Bayesian updating method. As a case study, consider the chamois population (*Rupicapra rupicapra*) in the central Italian Alps, for which parasitological investigations were carried out after a population crash (Rizzoli 1995) and modelled using a 4-equation deterministic model (Box 3.8). This is an extension of the basic Anderson and May model for macroparasites, incorporating several elements of biological realism—a free-living parasite stage is included, parasite aggregation is modelled as a dynamic variable, a host carrying-capacity is included, and it is assumed that infections will generally occur with several larvae at the same time ('clumped infections'). The model was fitted to the parasitological data collected, assuming that parasite effects on host fecundity and mortality are equivalent to those seen in Soay sheep, by minimizing the sum of squared deviations between model predictions and observed data in a Bayesian framework. Starting from a prior distribution of parameters, reflecting initial uncertainties, and a likelihood function of the observed data, the

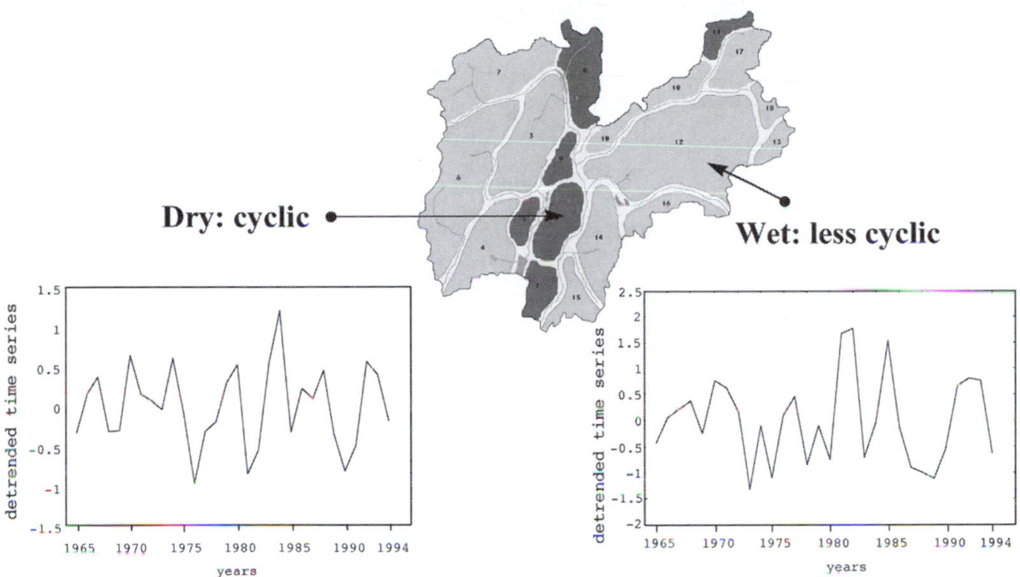

Figure 3.7 The province of Trentino, showing the 18 major mountain groups and the two groups of rock partridge populations: wet habitats (*light*) and dry habitats (*dark*) (after Cattadori *et al.* 1999).

application of the Bayes' formula obtains the posterior distributions of parameter uncertainty. In this specific case, the initial parameter distributions were chosen on the basis of literature data and likelihood functions computed for a large number of parameter combinations. A Latin hypercube sampling scheme (McKay *et al.* 1979; Blower and Dowlatabadi 1994) was used to improve the efficiency of the computation of posterior distributions. This illustrated that the model constructed could not account for the amount of parasite aggregation measured in the population studied, and that some other way of modelling aggregation (as opposed to the variance-to-mean ratio used) must be considered (Quinnel *et al.* 1995).

3.7 Parasites and other trophic interactions

Models of host–parasite systems are often analysed to determine their equilibrium states, that is the numbers or densities of hosts and parasites at which they equilibrate. Such analyses are usually static, in that steady states are determined with the assumption that a locally stable steady state is in fact globally stable. This is not always the case, and long-term simulation studies should be performed to check for cyclic behaviour (as in red grouse). Since computer simulations appear to be more readily accepted than analytical results, it should be routine to present analytical and computational results that support each other. Also, when dealing with parasite communities, there is a further complication: the analytical results could indicate that two steady states are simultaneously locally stable, indicating the presence of bistability. For systems with this property the equilibrium will depend on initial conditions, such as the order in which the species in the community come together.

With only two species, there is essentially only one way to construct the system: the two species are brought together. A study of the dynamics of the Indian meal moth (*Plodia interpunctella*), its granulosis virus and the parasitoid *Venturia canescens* (Begon *et al.* 1996b) illustrate the possible complications of including a third species. The host alone, and both of the two species systems (host–pathogen and host–parasitoid), exhibited effectively indefinite cycle dynamics, with cycles approximately one host generation in length. However, whilst the dynamics of the various three-species systems were consistent in their lack of persistence and loss of

Box 3.8 The chamois/Trichostrongilidae model

The model by Anderson and May (1978) for macroparasites with direct life cycles has been adapted to examine the population biology of Trichostrongilidae in chamois (Pugliese *et al.* 1998). The main refinements over the Anderson and May model include modelling aggregation as a dynamic variable (Adler and Kretzschmar 1992), introducing a carrying capacity for the host (Pugliese and Rosà 1995), and assuming that infections will generally occur with several larvae at the same time. The resulting model consists of four coupled differential equations describing changes in the host population size, N, the mean adult parasite burden, x, the aggregation of parasite distribution, A (defined as the ratio of the variance to the mean), and the number of free living larvae, L:

$$\frac{dN}{dt} = N(\beta[1 + (A - 1)(1 - \xi)]^{-\frac{x}{A-1}} - \mu - \nu N - \alpha x) \tag{3.10}$$

$$\frac{dx}{dt} = x(-\sigma - \alpha A - \beta[1 + (A - 1)(1 - \xi)]^{-\frac{x}{A-1}}) + \theta\psi L \tag{3.11}$$

$$\frac{dA}{dt} = -(A - 1)\left(\sigma + \alpha A + \frac{\theta\psi L}{x}\right) + \beta x[1 + (A - 1)(1 - \xi)]^{-\frac{x}{A-1}} + \frac{\theta\psi L}{x}\lambda \tag{3.12}$$

$$\frac{dL}{dt} = hNx - \delta L - \theta LN \tag{3.13}$$

where β and μ are the birth and death rates of hosts, ν is the density-dependent increase in host mortality, σ is the death rate of adult parasites within the host, λ is the mean number of free-living stages forming a single infecting 'parcel', h is the rate of production of infective larvae, δ is the death rate of infective larvae, ψ is the proportion of ingested larvae that develop into adult parasites, α is the rate of parasite-induced host death, ξ is the parasite-induced reduction in host fecundity, and θ is the average rate of host infection by parasites. The changes to the model came up with two interesting findings relative to the Anderson and May model. First, the region of instability was increased leading us to suppose that macroparasites, particularly when they have an impact on morbidity rather than mortality, will destabilize host numbers. Second, the infection process provides a reasonable, but not yet satisfactory, description of the aggregated distribution.

generation cycles, their detailed patterns differed according to how they were constructed. The host–parasitoid populations to which the pathogen was added displayed no consistent pattern, but the host–pathogen system to which the parasitoid was added showed clear evidence of multi-generation cycles (Sait *et al.* 2000). This suggests that sequence cannot be easily ignored.

The interaction between parasitism and other trophic processes, such as predation and competition, can magnify apparently small effects of parasites on host morbidity (Price *et al.* 1986; Minchella and Scott 1991; Holmes 1995). An example of such an interaction in a wildlife population is that provided by Murray *et al.* (1997), working on the predators and nematode parasites of snowshoe hares (*Lepus americanus*). Here, free-ranging individuals that had their worm burdens reduced by anthelmintics were significantly less likely to be caught by predators than a similar group of control animals (Fig. 3.8). Predation rates were reduced still further in animals that also received food supplements, suggesting an important interaction between parasite-induced susceptibility to predation and the host's plane of nutrition. The importance of parasitism/nutrition interactions is highlighted in many other wildlife populations, although a conclusive experimental demonstration is still required (Murray *et al.* 1998).

Macroparasite models indicate that, in general, parasite-induced reductions in host survival will tend to stabilize host dynamics, whereas parasite-induced reductions in host fecundity will tend to destabilize numbers generating cycles. However, the dynamic consequences of some of the more subtle effects of parasites on the host are often less straightforward. For example, models indicate that parasite-induced susceptibility to predation is destabilizing in the snowshoe hare system (Ives and Murray 1997), but is stabilizing in the red grouse-*T. tenuis* system (Hudson *et al.* 1992). The key difference between these two systems appears to be the coupling of the predator's dynamics to that of its prey in the former case, but not in the latter. Whatever the reason, these models make it clear that apparently small effects of parasites on

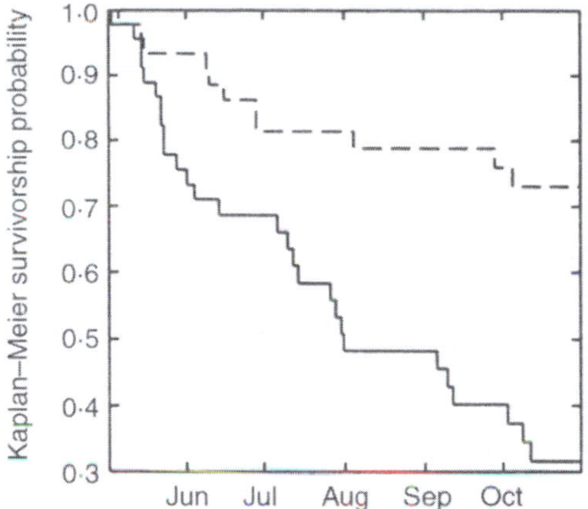

Figure 3.8 Kaplan–Meier survivorship curves for parasite-normal (*solid line*) and parasite-reduced (*dashed line*) snowshoe hares during May–October 1992. The principle cause of hare mortality during the study was predation (95% of mortalities) (after Murray *et al*. 1997).

the morbidity of their hosts can have important dynamic consequences.

3.8 Parasite-mediated 'apparent' competition

Parasite regulation of wildlife populations may also result from indirect interactions among host species. For example, when two host species, which do not compete for resources, share the same parasites an increase in the density of the shared parasites, as a consequence of an increase in density of just one of the two species, has the potential to reduce the population growth rate of both species present, thus regulating them (Holt and Lawton 1994; Hudson and Greenman 1998). Since each of the two species will suffer as the other increases in abundance, exactly as is found in competitive situations, this type of interaction has been termed 'apparent competition' (Holt 1977). Apparent competition between two host species can lead to the rapid extinction of one of the two hosts, with the species that persists being the one that can support the higher densities of shared parasites (Holt and Lawton 1993). Indeed, shared parasites have been implicated in the population extinctions of several wildlife species (Tompkins and Wilson 1998).

Cases where host exclusion can be attributed to the presence of shared parasites are often cited as examples of apparent competition (e.g. Settle and Wilson 1990; Grosholz 1992). However, conclusively demonstrating that host exclusion is due to apparent competition, as opposed to either direct competition or other parasite effects, can be difficult to accomplish and has led to much confusion (Hudson and Greenman 1998). For example, the classic study undertaken by Park, on the competitive interactions between two *Tribolium* beetles and the influence of a shared sporozoan parasite (*Adelina tribolii*), has been cited as one of the clearest demonstrations of apparent competition. Here, whilst one beetle species dominated mixed cultures run in the absence of the parasite, the outcome was invariably reversed when the sporozoan was present (Park 1948). However, although these results may indeed be due to apparent competition, they may also be due to the parasite reducing the competitive ability of the previously dominant host. From the experiments conducted there is no way to distinguish between the two mechanisms.

A recent study using laboratory populations of two moth species (*Plodia interpunctella* and *Ephestia kuehiella*) and their shared parasitoid (*Venturia canescens*) has explicitly demonstrated the action of apparent competition mediated via a shared parasite (Bonsall and Hassell 1997, 1998). Here, the exclusion of *E. kuehiella* that occurred when the parasitoid was present could be attributed solely to apparent competition with *V. canescens*. This was accomplished, first, by using an experimental design that prevented direct competition between the two host species and, second, by comparing with controls of each host species with just the parasitoid. Had the exclusion not been due to apparent competition between the two host species, patterns observed in the controls would have been similar to the patterns observed when both hosts were present (Fig. 3.9).

As is always the case with proving parasite regulation of host populations, explicitly demonstrating apparent competition mediated via shared parasites for wildlife populations is far harder to accomplish due to the logistical constraints involved—in many circumstances the controlled experiments required are unworkable. The majority of field studies to date have thus been descriptive, often failing to disentangle parasite effects from resource competition among hosts, let alone distinguish between apparent competition and other parasite effects (e.g. Schall 1992; Schmitz and Nudds 1994; Hanley *et al.* 1995). Moreover, a number of experimental manipulations conducted in the field have failed to demonstrate parasite-mediated apparent competition (Hanley *et al.* 1998; Rott *et al.* 1998). Mathematical modelling is required to fully understand the role of apparent competition in natural systems, and to identify systems where experimental demonstrations could possibly be conducted. Until recently, the theory of apparent competition mediated via shared parasites has centred on the deterministic Susceptible–Infected models (see Box 3.5) that have been applied so successfully to the dynamics of single host–pathogen systems (Holt and Pickering 1985; Begon and Bowers 1995). Transferring this theory from one to multiple host species, however, has been hampered by problems in identifying model equilibria due to the intractability of the algebra. Further theoretical developments have only been possible through the application of bifurcation maps to the analysis of deterministic models (see Box 3.10).

A second difficulty in extending epidemiological studies to systems with even two hosts is in the collecting of dynamical data from multi-species systems. The aim of a current study is to collect such data for a system of two gamebird species and their shared parasites. Numbers of wild grey partridge (*Perdix perdix*) have declined dramatically in the UK within the past 40 years and this has been attributed to the indirect effects of agricultural intensification (Potts 1986; Tapper 1992). However, populations in non-agricultural areas have also declined and some have not recovered following

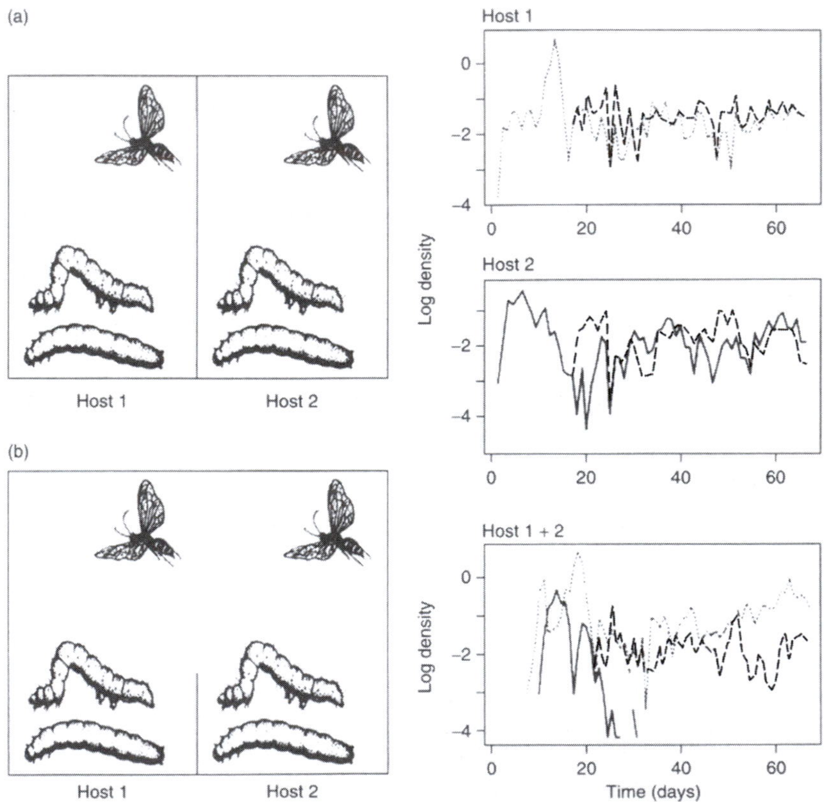

Figure 3.9 Apparent competition between two moth species, *Plodia interpunctella* (host 1, *dotted line*) and *Ephestia kuehniella* (host 2, *unbroken line*), mediated via the shared parasitoid wasp *Venturia canescens* (*dashed line*). (a) When parasitoids were kept within single-host systems, the parasitoid and each host coexisted with stable dynamics. (b) When the parasitoid had access to both host species, host 2 showed diverging oscillations in abundance and invariably went extinct (after Bonsall and Hassell 1998).

extensification. Recent studies indicate that this may be partly due to apparent competition with released pheasants (*Phasianus colchicus*), mediated by the shared caecal nematode *Heterakis gallinarum* (Wright *et al.* 1980; Tompkins *et al.* 1999, 2000). Specifically, evidence suggests that while the pheasant is primarily responsible for the spread of infection, the negative impact of this shared parasite is

Box 3.9 Modelling the pheasant–partridge system

The macroparasite model used to describe the two-host-shared-parasite system is defined by the following equations (for $i, j = 1, 2; i < j$):

$$dH_i/dt = r_iH_i(1 - H_i/K_i) - (\alpha_i + \delta_i)P_i \quad (3.14)$$

$$dP_i/dt = \phi_i\beta_iWH_i - (\mu_i + b_i + \alpha_i)P_i - \alpha_ik'_iP_i(P_i/H_i) \quad (3.15)$$

$$dW/dt = \lambda_1P_1 + \lambda_2P_2 - \gamma_0W - \beta_1WH_1 - \beta_2WH_2 \quad (3.16)$$

where W denotes the number of free-living stages in the common infective pool while, for the ith host, P_i denotes the adult parasite and H_i the host populations. Parameter estimates were obtained from a combination of infection and transmission experiments, together with sources in the literature and unpublished data (Table 3.2).

The natural host mortality rate (b_i) takes account of density dependence, whereby:

$$b_i = b_{i0} + r_iH_i/K_i \quad (3.17)$$

and the net natural birth rate at low population levels is given by:

$$r_i = a_i - b_{i0}. \quad (3.18)$$

Parameter k'_i equals $1 + 1/k_i$. Figure 3.10 illustrates the result of multiple runs of the model at varying levels of $\alpha_{\text{partridge}}$ and $\delta_{\text{partridge}}$, where '1' represents pheasant and partridge coexistence, and '0' represents partridge exclusion.

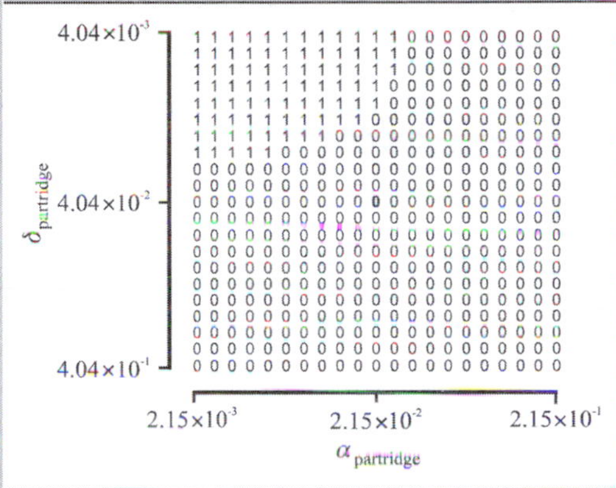

Figure 3.10 Grey partridge exclusion by parasite-mediated competition as predicted by simulation runs of the model. The figure plots impact of *Heterakis gallinarum* on partridge fecundity against survival such that low impacts are observed in the top left hand corner. The number 1 denotes coexistence and the 0 elimination of partridges. The number in *bold* indicates that the predicted model outcome at the empirically determined parameter values is one of partridge exclusion (after Tompkins *et al.* 2000b).

Table 3.2 Parameters for the pheasant–grey partridge *Heterakis gallinarum* model

Parameter	Symbol	Pheasant value	Partridge value	Units
Natural host fecundity	a	1.55	1.50	year^{-1}
Natural host mortality	b	0.65	0.80	year^{-1}
Host carrying capacity	K	6	3	home range^{-1}
Mortality of parasite eggs	γ	0.90	0.90	egg^{-1}year^{-1}
Ingestion of parasite eggs	β	6.70×10^{-5}	5.58×10^{-5}	egg^{-1}host^{-1}year^{-1}
Parasite establishment	ϕ	0.590	0.065	egg^{-1}
Parasite fecundity	λ	26666	2761	eggs year^{-1}
Parasite mortality	μ	4.15	4.17	eggs year^{-1}
Aggregation of parasites in hosts	k	0.30	0.30	
Parasite increase in host mortality	α	0.00	2.15×10^{-2}	worm^{-1}year^{-1}
Parasite reduction in host fecundity	δ	8.28×10^{-4}	4.04×10^{-2}	worm^{-1}year^{-1}

greater on the partridge. An impact on body condition, observed for the partridge but not for the pheasant (Tompkins *et al.* 1999, 2000a), has recently been confirmed by controlled experimental infections (Tompkins *et al.* 2001). The parameterization of a deterministic two-host, shared-parasite model of the system predicts that this impact will result in partridge exclusion when the two host species occur in the same area with the shared parasite present (Box 3.9). The potential thus exists for an experimental demonstration of apparent competition in the pheasant-partridge system via the manipulation of host and parasite populations.

3.9 Advances in the modelling of host–parasite systems

Since the basic Anderson and May model was formulated, predicting how parasites may regulate host populations, much work has been conducted to test the assumptions that the models of host–parasite systems make (for microparasites see Boxes 5.3, 5.4). For example, one of the more crucial assumptions in epidemiological theory is that the probability that a parasite egg or larva (or infected individual for microparasite infections) contacts a host is assumed to increase with increasing host population density, thus providing a positive link between host density and parasite transmission rate (Anderson and May 1978; May and Anderson 1978). This assumption also plays a critical role in models of intervention strategies against parasites in humans and in agricultural systems (Anderson and May 1978, 1991; Grenfell 1988, 1992; Dobson and Hudson 1992; Coyne and Smith 1994). However, is the assumption of density-dependent transmission generally true?

A comparative study of the abundance of gastrointestinal strongylid nematodes (a large and important group of directly transmitted macroparasites) in mammals suggests that the answer is yes. Looking across communities of nematodes harboured by different mammalian species living at widely different population densities, parasite abundance increases with increasing host density, once confounding factors are controlled for (Arneberg *et al.* 1998b). For example, host species living at high population density for their body weight generally harbour nematode populations with higher abundance than comparable host species living at lower densities (Fig. 3.11). This is exactly the pattern we expect from epidemiological theory if parasite transmission rates are positively affected by host population density (Dobson 1990; Arneberg *et al.* 1998b). Thus, empirical evidence strongly supports the assumption that macroparasite transmission rate is indeed generally a positive function of host population density, although further comparative studies of other groups of parasites are needed to support this finding.

Aside from the work conducted to justify the use of certain assumptions, most of the developments made in the modelling of host–parasite systems have added relatively little to the basic determinis-

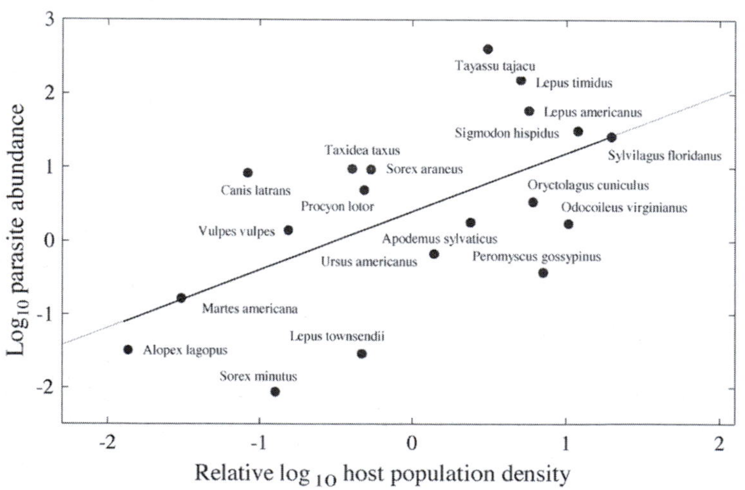

Figure 3.11 Relationship between relative host population density and abundance of strongylid nematodes across 19 mammalian species, representing examination of 6670 individual hosts. Host density is plotted as residuals from a correlation with $\log_{10}$ host body weight. Parasite abundance is the within host averages of all strongylid nematodes. Names are given for host species. Similar relationships are seen if parasite density is measured as a prevalence (fractions of hosts infected) or intensity (average number of parasites per infected host) (after Arneberg *et al.* 1998a).

tic models constructed 20 years ago. Although much of the minor details have been modified, such as using different techniques to model parasite aggregation, or different laws of parasite transmission and impact (see, for example: Begon *et al.* 1999; Bouloux *et al.* 1998), the fundamental structure and behaviour of the models has remained unaltered (Grenfell and Dobson 1995). One advance that has been made is the application of bifurcation theory to these systems, allowing for an additional route in the analysis of shared-parasite multi-host systems (see Box 3.10 for an example). This theory provides tools to study the effects on the dynamic behaviour of systems of equations when key parameters are 'slowly' varied (e.g what happens to equilibrium values for key variables?—see Fig. 4.1 for an example). This theory has contributed a lot in the last 10 years to the understanding of 'complex' models in population dynamics and microparasitic infection and has grown itself in the process. By now standard software is available for bifurcation analysis of complicated systems (see, for example, Kuznetsov 1998 and the software program CONTENT).

3.10 Synthesis

A general consensus among empirical workers on wildlife disease is that the gap between theory and sound experimental evidence has remained or even widened over the past five years:

Box 3.10 Applying bifurcation theory to multiple host–parasite models

Bifurcation maps can be drawn in parameter space that detail how model states will alter depending on the parameter values. Each point in parameter space is related to a set of points in state space, in which the stationary states of the system are defined by equilibrium points. If we follow a path through parameter space, then in general these equilibria will move, tracing out trajectories in state space. As a consequence of the hierarchical structure, collisions between these moving equilibria take place. In such a collision two events typically occur—either there is stability switching between the equilibria involved in the collision or one of these equilibria becomes biologically relevant.

Such a collision of two equilibria is called a transcritical bifurcation (Wiggins 1990) and the point in parameter space for which the model generates the collision is called a bifurcation point. The set of bifurcation points can be collected together to define bifurcation curves and these curves in turn define bifurcation maps. Each equilibrium has its own bifurcation map. Consider, for example, the map shown in Fig. 3.12 for the infected coexistence equilibrium of a two-host macroparasite model.

Fig. 3.12 is drawn in the 2D cross-section of parameter space defined by the inverse basic reproductive numbers R_{0i} for the two hosts. The shaded area is the area in which the equilibrium is relevant. As path P → Q is followed it is at point R that equilibrium first appears, emerging through the uninfected coexistence gate and picking up the stability of that equilibrium by stability switching. The infected equilibrium remains stable along path P → Q until point S is reached, at which point it exits through the infected host 1 gate passing on its stability to this equilibrium. The stable state of the system is then the infected state where host 2 has been excluded by the virulence of the infection.

When investigating shared-parasite multi-host systems, lower order equilibrium properties can be analysed analytically and translated into properties of higher order equilibria for which they act as gateways. The stability boundaries for the lower order equilibria are typically the relevance boundaries for the higher order equilibria and the stability of the latter is inferred from the stability of the former just prior to collision. These bifurcation curves provide the framework in which additional structural features (saddle nodes and Hopf bifurcations) can be identified and located numerically, allowing a comprehensive catalogue of system behaviour to be constructed (Greenman and Hudson 1997).

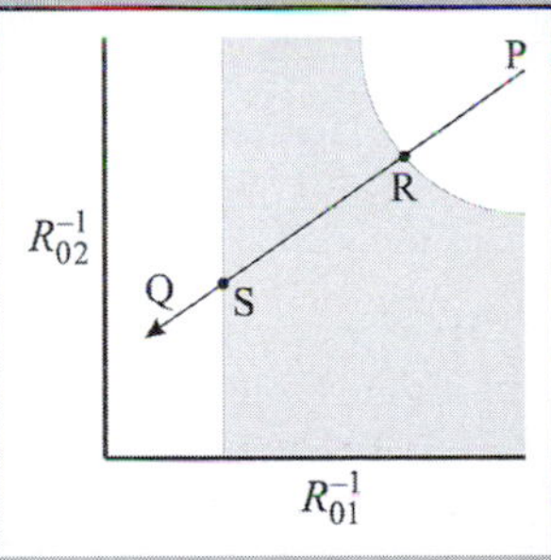

Figure 3.12 Bifurcation map for the infected coexistence equilibrium of a two-host macroparasite model. See text for further details.

1 Twenty years after the potential for host regulation by microparasites was clearly illustrated (May and Anderson 1978), an experimental field demonstration is still required.

2 Although we now have experimental proof that macroparasites can regulate wildlife populations, more work is needed before a 'general' or 'typical' role can be formulated.

3 It is increasingly clear that trophic interactions can greatly influence the outcome of interactions between wildlife populations and their parasites. However some major interactions, such as that between parasitism and nutrition, still require experimental demonstration in the field.

4 The development of techniques for the analysis of shared-parasite–multi-host models has provided the means by which experimental work with these more complex systems can be planned and efficiently executed.

In addition to requiring more experimental demonstrations of parasite regulation of host populations, more basic experiments detailing parasite impacts on fecundity and survival in wild animal populations are also required. The studies conducted to date are biased towards the effects of avian ectoparasites and impacts on host fecundity (Table 3.1). This is most probably due to avian–ectoparasite systems being easier to manipulate and monitor than other systems, and impacts on fecundity being easier to document in short-term studies than impacts on survival. More examples have been published recently of how parasites can reduce the survival of wild host individuals, but they failed to distinguish the effects of particular parasite taxa, and again were conducted with avian–ectoparasite systems (Brown *et al.* 1995; Clayton *et al.* 1999). If a 'general' rule of parasite involvement in wildlife population dynamics is ever to be obtained, these biases need to be addressed—more diverse systems need to be investigated and more longer term studies need to be carried out.

CHAPTER 4

Parasite community ecology and biodiversity

M. G. Roberts, A. P. Dobson, P. Arneberg, G. A. de Leo, R. C. Krecek, M. T. Manfredi, P. Lanfranchi, and E. Zaffaroni

Parasites coexist in communities, but what determines the structure of the community, which species are able to invade and which species get squeezed out? What can community ecology tell us about parasite communities, and what lessons can be drawn from parasite communities and applied to community ecology as a whole?

4.1 Background

A wild-animal population is typically host to a whole community of parasites of different species. These parasite species may be sufficiently independent that their population dynamics can be investigated individually. In reality it would be more usual for multiple species to interact such that the presence of any one species may influence the presence and abundance of another, either directly through competition or indirectly through the host immune system. The interactions between parasite species in a community are fundamental in determining the community structure, the viability of each species within the community, and the potential for new species to invade the community. Models of the interaction of parasites in a community have revealed fundamental determinants of community structure, most notably the important role of R_0, the basic reproduction number, for each species (Dobson 1985; Dobson and Roberts 1994; Roberts and Dobson 1995). But this is not the whole story, if we take a step back we can see that each individual host provides a habitat patch for its community of parasites and that parasite-community dynamics are a miniature representation of community ecology at the larger scale. As such, the study of parasite communities has the potential to reveal ecological principles with wider implications for the preservation of biodiversity, the response of whole ecosystems to habitat destruction, or the biological control of pest species.

4.2 The community ecology of parasites

Communities of parasitic helminths have a number of features that should intrigue ecologists. In some ways the study of parasite communities can be considered as one of meta-community dynamics: the communities consist of meta-populations of n parasite species, each member of which colonizes a patch of habitat (the host) containing resources to be exploited.

The study of parasite communities presents a number of opportunities that are not usually available for ecologists studying the structure and diversity of free-living animal and plant communities. In contrast to these systems, it is easy to define a host as a 'patch' of habitat for a parasitic helminth (§6.4). The system allows us to address general questions that are important to our understanding of biodiversity. For example, in free-living communities of plants, examination of the trade-offs between colonization and competitive ability has helped explain why some species are abundant and others relatively rare. In parasite communities, the birth, death, transmission, and parasite-induced host-mortality rates of each species are crucial in determining the potential membership of a parasite community, the interactions between parasite species within any individual host then modify the outcome of these primary processes.

In this chapter we initially examine parasite communities from South African zebras who are host to a bewildering range of parasite species. Does this represent a system that has escaped human intervention, and what is the function of the microbes that infect the parasites? We then use helminth parasites of *Anolis* lizards from different islands in the Caribbean to briefly review the main features of models of parasite community dynamics. This system allows us to concentrate on the importance of the basic reproduction number, R_0, for determining a parasite species' relative abundance in a community. We then describe some simple mathematical models that provide a connection between community stability and community structure before presenting some case histories where interesting, and sometimes counter-intuitive, dynamics have been observed. In an example from a farmed host, a tapeworm-control programme succeeded in eradicating one species but increased the abundance of another because of the community interactions between parasite species. In a more complex example, we describe data for six species of ruminants living in the European Alps that share generalist parasite species, but also have specialist parasite species that are limited in their host range. We then discuss possums, which were introduced into New Zealand earlier this century and lack some of the parasite species found in Australian possums: is this because of founder effects? Would the missing parasite species be viable under New Zealand conditions, and if they were introduced would they act as agents for possum control?

Our approach illustrates important similarities between these different parasite communities, yet some types of data are consistently difficult to obtain, and the parameters needed to model parasite transmission are always difficult to estimate. We therefore discuss some alternative methods of estimating the key life-history rates that determine the composition and reltive abundance of the parasite community.

4.2.1 The parasites of zebras in Southern Africa

The diversity of habitats occupied by zebras in South Africa varies from the Namibian desert, with an annual rainfall of 100 mm, where Burchell's and Hartmann's mountain zebras live, to the South African savannah with an annual rainfall of 1200 mm where only the Burchell's zebra is found. The zebra and its helminth fauna are an example of a

Box 4.1 The zebra's amazing parasite fauna (= the wandering zoo)

Why are cyathostomes important? Knowledge of the effect of these nematodes in the zebra and donkey is anecdotal, but in horses they are associated with clinical disease. For example, when the encysted stages in the gut wall emerge all at once (i.e. spring in yearlings in Europe), haemorrhage, colic, and death can occur. Zebras have large proportions of these larvae in their small and large intestinal gut walls. It has not been established whether clinical disease occurs in the zebra or if they have developed adaptation mechanisms (i.e. bone marrow has been observed to be darker red in zebras than in horses suggesting more red blood cells are produced).

The large intestine of the equid is unique with its four compartments differing in function, pH, and nutrient content, as well as cyathostomid species. Associated with these parasites are microbial communities attached to, and predominantly colonizing, the vulva of the females (Krecek *et al.* 1987b). The most prevalent cyathostome species are often the most abundant and show the microbe association. Efforts to cultivate these organisms have not been successful (Mackie *et al.* 1989; Krecek *et al.* 1992).

Are these microbes a type of biological control, or is their presence an indicator that interventions, such as the application of anti-parasitic remedies, have not been used? How are these various populations of nematode and helminth and microbe species interacting? Cyathostomes cause clinical disease in horses and current interventions using chemotherapy are faced with mounting anthelmintic resistance. Are there some keys in the zebra host: parasite relationships?

Table 4.1 The parasitic fauna of zebras. The hosts include Burchell's zebra (*Equus burchelli antiquorum*), the Cape mountain zebra (*Equus zebra zebra*), and Hartmann's mountain zebra (*Equus zebra hartmannae*)

Parasite	Number species
Nematodes	
Strongyloidea	
Cyathostominae	18
Strongylinae	10
Habronematoidea	7
Spiruroidea	2
Oxyuroidea	1
Rhabditoidea	1
Trichostrongyloidea	1
Filarioidea	1
Cestodes	
Anoplocephalidae	1
Trematodes	
Paramphistomidae	1
Arthropods	
Gasterophilidae	6
Oestridae	2
Bovicola	2
*Haematopinus**	3
Rhipicephalus	29 of 75

* In *E. burchelli*

pristine single host containing multiple parasite infections. At a glance the diversity of parasitic fauna in zebras includes 41 nematode species, two cestode and trematode species, and more than 42 species of arthropods (see Table 4.1). Large numbers of pinworms (Atractidae) with burdens of 100 000 000 *Crossocephalus viviparus* and 3 800 000 *Probstmayria vivipara* are not uncommon in individual animals (Krecek *et al*. 1987a; Malan *et al*. 1997). In addition, the cyathostomid species are associated with microbial communities, presenting another layer of inter-species interaction. At first sight, data such as these seem overwhelming. Yet any attempt to explain them requires us to ask the classic questions of community ecology: 'Are there any underlying patterns of diversity and abundance?' and 'Which processes are responsible for generating and constraining diversity in this system?'. The only rational way to attempt to understand such complexity is to examine much simpler parasite communities and to see if the rules that govern invasion, persistence, and abundance in these systems, can be scaled-up to explain communities of greater complexity.

4.2.2 The parasites of *Anolis* lizards

Islands have consistently provided opportunities for ecologists to examine how communities have evolved. Closely situated islands tend to contain communities with slightly different species composition and this occurs due to differential colonization, extinction, and the evolution of new species. The *Anolis* lizard species complex that inhabits the Caribbean islands has radiated through a complex mix of slow lizard dispersal between islands and even slower dispersal of islands through the movement of the underlying Caribbean tectonic plate (Roughgarden 1995). *Anolis* lizards contain a relatively simple parasite community, but the lizards on each island have been isolated for sufficiently long that a different mixture and diversity of parasite species are found on each island (Dobson *et al.* 1992; Dobson and Pacala 1992; Staats and Schall 1996). Examination of lizard parasites from different islands provides important insights into the factors underlying the structure of more complex parasite communities.

The clearest pattern to emerge from the lizard data is a relationship between habitat and parasite diversity. Xeric islands (and habitats) tend to have depauperate parasite communities, while lizards that live in more mesic habitats tend to support a more diverse parasite community (Dobson *et al.* 1992; Dobson and Pacala 1992). There is no simple relationship between host abundance and parasite community diversity that correlates with this dominant pattern of parasite abundance. *Anolis* lizards are abundant in some xeric habitats, and may be uncommon in mesic areas. This may reflect the confounding influence of introduced predators on *Anolis* abundance. Mongoose were introduced into several Caribbean islands to control rats; while failing to do this, they have reduced the density of many endemic reptile species. However, the parasite data do illustrate an interesting and important relationship between parasite species diversity and abundance: an increase in the numbers of parasite species observed in each host species is matched by increases in the abundance of individual parasite species. Moreover, parasite species do not enter the community in random order. Instead their relative abundance appears to be 'nested', with those species that are common in the diverse communities being the only species present in the depauperate communities.

This nested structure implies that parasite transmission success is crucial in determining whether a parasite can establish in the community of species utilizing a host population. Transmission success will be determined by three main factors: the life history of the parasite, host density, and the external environment (particularly with respect to survival of free-living infective stages). The differences in parasite diversity and abundance in mesic and xeric habitats reflect the influence of the external environment. There are insufficient data on *Anolis* density to examine the role played by host density in this system (but see §4.8.1). The life histories of the different parasite species determine not only the conditions for when any species can establish in a host population, but also the relative abundance of each species in the community. Those species that can still invade hosts, even when densities are low, are even more ubiquitous at high host-population density or under conditions that favour increased survival of free-living infective stages.

A simple graphical explanation of this phenomenon was originally suggested by Dobson (1990) and formalized by Roberts and Dobson (1995). We suggest it provides an important 'null model' for the underlying structure of a wide range of parasite communities. We make this argument partly to offset the huge emphasis on inter-specific competition in shaping parasite communities. While we certainly agree that competition plays an important role in shaping the structure of parasite communities, parasites have first to establish before they can compete. Moreover, the stability and persistence of any individual parasite species is strongly dependent upon its statistical distribution in the host population (Anderson and May 1978; May and Anderson 1978 chapter 2). As the covariance in the distributions of two potentially competing parasite species is crucial in determining the potential for any competition to occur, then establishment and persistence may supersede competitive interactions in determining the structure of parasite communities (Dobson 1985). We will therefore derive the structure of this parasite community model before using it to examine other examples of more complex parasite communities.

4.3 A model for parasite communities

What conditions determine when a parasite species can invade a community and, conversely, when it will be squeezed out and driven to extinction? Simple models can be used to provide an answer to these questions. The ability of any parasite species to colonize a host is dependent upon its basic reproduction number, R_0. When this is less than unity, the parasite species will be unable to invade and its numbers will decline to extinction. Lurking behind this simple insight is a deep suspicion that the structure of ecological communities is determined and constrained by the life history and demographic attributes of the species that can potentially coexist in the community.

Parasite populations are regulated by interactions with their host population, and by interactions with other parasite species in the community mediated through their host population. The basic reproduction number (R_0) determines a species' ability to invade a community, or its danger of being squeezed out. One must be careful to calculate R_0 under the correct conditions: for example, in the presence of other species, when determining the ability to invade a community. The invasion and persistence criteria are related to the equilibria of the equation describing the dynamics of the parasite community. In this chapter we do not discuss the details of the models, but we explore the connections between their properties and the dynamics of parasite communities. Our arguments concentrate on the gastro-intestinal parasites of mammals, as these are the most widely studied. For further reviews of the population dynamics of nematode parasites of farmed or wild animals, see Smith and Grenfell (1994) and Roberts *et al.* (1995), respectively.

Box 4.2 Threshold quantities for parasite populations

The basic reproduction number (R_0) of a macroparasite is the average number of female offspring (or offspring for hermaphrodite species) produced throughout the lifetime of a mature female parasite, which would themselves achieve reproductive maturity in the absence of density-dependent constraints on parasite population dynamics (Anderson and May 1982b; Roberts 1995; Roberts and Heesterbeek 1995).

A single parasite species in a farmed animal host where there is no parasite induced mortality:

$$R_0 = \frac{N}{H+N}\frac{p(0)q\lambda(0)}{\mu(0)} \qquad (4.1)$$

Multiple parasite species/strains in a farmed animal host:

$$R_0 = \frac{N}{H+N}\frac{p(r)q\lambda(r)}{\mu(r)} \qquad (4.2)$$

A single parasite species in a wild animal host with parasite induced mortality:

$$R_0 = \frac{K}{H+K}\frac{pq\lambda}{\mu+\alpha+d} \qquad (4.3)$$

Multiple parasite species in a wild animal host:

$$R_0 = \frac{N}{H+N}\frac{pq\lambda}{\mu+\alpha+d} \qquad (4.4)$$

Definition of symbols:

- α increase in host mortality due to the presence of a single parasite (yr^{-1});
- β rate at which, larvae are eaten by host animals ($ha\ yr^{-1}$);
- λ rate at which a single adult parasite produces eggs (yr^{-1});
- μ rate at which adult parasites die (yr^{-1});
- ρ rate at which larvae are lost from pasture due to other causes (yr^{-1});
- d mortality rate of host population in the absence of parasites (yr^{-1});
- p probability that an ingested larva become an adult parasite;
- q probability that an egg hatches into a larva;
- r level of herd immunity;
- H coefficient governing saturation of parasite transmission ($= \rho/\beta$);
- K host population carrying capacity (ha^{-1});
- N host population density (ha^{-1}).

4.3.1 Regulation and the basic reproduction number

Gastro-intestinal nematodes are ubiquitous parasites of all animal species, and a major cause of production loss on sheep and cattle farms. Parasites of wild animals may increase the mortality of the host, and therefore have the ability to regulate the host population (Chapter 3; Anderson and May 1978). In contrast, it would be unusual for a farmer to allow his animals to die from parasitism, and the regulation of their parasite populations is via protective acquired immunity (Roberts 1995).

A simple differential equation model (Roberts and Grenfell 1991) may describe the dynamics of nematode parasites. Analysis of the model shows that the parasite population can establish itself if $R_0 > 1$, where R_0 is the basic reproduction number (see Box 4.2, eqn 4.1). The 'parameters' λ, p, and μ in that equation are in fact functions of a variable r, which is a measure of the host (herd) immunity that has been acquired due to larval challenge and/or adult parasites. Hence the zeros in parentheses signify that R_0 is calculated when the herd-acquired immunity is zero or, in practical terms, when regulation of the parasite population by the host is minimal. Acquired immunity may reduce the establishment of adult parasites and/or their production of eggs, and/or increase parasite mortality (Roberts and Grenfell 1991; Woolhouse 1992a).

Host animals are usually infected with communities of different parasite species, some of which will generate reciprocal immunity. Consider a community of n parasite species coexisting within a host population in a stable equilibrium, and attempt to introduce a species (number $n+1$) that is not part of the community. The new species can invade the community if $R_0 > 1$ for that species, but now R_0 must be calculated at the level of herd immunity that has been generated by the existing n-species community, as perceived by species $n+1$. Hence, the potential invader can only succeed if it can establish in the host population at the levels of acquired immunity produced by the existing community. The appropriate formula for R_0 is therefore that given by Box 4.1, eqn 4.2. Results of this nature are important when considering selection for resistance to chemotherapy, where the different parasite strains play parts analogous to those played by competing species (Roberts and Heesterbeek 1995).

Interactions between nematode species that parasitize wild hosts are less likely to be due to acquired immunity, and more likely to be due to the effects of parasites on host mortality and fecundity, and competition for niches within the host. Dobson (1985) classified these as 'exploitation' and 'interference' competition, respectively. In the next section we discuss a framework for the existence and stability of viable parasite communities of wild animals, and the potential for other species to invade the community. This is based on models similar to those in Dobson and Roberts (1994) and Roberts and Dobson (1995), which are multispecies extensions of the classical models of Anderson and May (1978).

A central feature of the Anderson and May model is the assumption that each individual parasite increases the mortality of the host, and therefore heavily-parasitized animals die at a faster rate than those with lighter parasite burdens. In order to make the model mathematically tractable, the second assumption is that the distribution of parasites among hosts is negative binomial and the exponent remains constant as the host population density changes. Although there is extensive evidence to support both these claims (for example: Chapter 2; Shaw and Dobson 1995; Shaw *et al.* 1998), they are contradictory as the differential death rate among hosts changes the distribution of parasites (Chapter 3). Nevertheless, this approximation has led to many useful insights to parasite dynamics and control.

The characteristic dynamics of the Anderson and May model are of a parasite population either becoming extinct ($R_0 < 1$) or existing in equilibrium with a regulated population ($R_0 > 1$) (Anderson and May 1978; Roberts 1995; Roberts *et al.* 1995). The expression for R_0 now includes host mortality due to parasitism and is calculated when the host population is at its parasite-free carrying capacity (see Box 4.2, eqn 4.3). When considering the potential of another parasite species to invade a parasite community in a wild host, we find that this is possible when R_0 for the invading species is greater than unity. However, R_0 must now be calculated at the host population density when it is in equilibrium with the existing community (Box 4.2, eqn 4.4) (Gatto and de Leo 1998).

If a parasite species may also interfere with the reproductive performance of the host, then more complicated population dynamics, including oscillations in host and/or parasite density, are possible (Chapter 3; May and Anderson 1978; Diekmann and Kretzschmar 1991). This is also true when there are multiple species of parasite present.

4.3.2 Stability and community structure

The fundamental questions of parasite community ecology are:

1 Is it possible that a particular community of parasite species can coexist in the host population?
2 Can another species invade an established community?

We address these questions by developing expressions for the steady state (equilibrium) values of host and parasite densities, and examining their stability (see Box 4.3). To illustrate the method, and explain the connection between the mathematics and the ecology, we first develop the argument for the simple case of exploitation competition with no parasite effect on host birth rate, and then introduce interference competition and the effect of parasites on host birth rate.

Exploitation competition

We now assume that each parasite species exploits a niche in the host population independently of the other species, and the only inter-species interaction is via their effects on the population dynamics of the host. The structure of the equilibrium parasite community may then be determined from the model equations (see Box 4.3). In every case the equilibrium solution with parasite species i present is not possible if $R_{0i} < 1$, where R_{0i} is the basic reproduction number for species i, as determined by Box 4.2, eqn 4.3. It is stressed that $R_{0i} > 1$ is not sufficient to guarantee that species a place in the community.

Box 4.3 Stability analysis for parasite communities

If we write the various state variables of a host parasite system (e.g. host population density, mean number of parasites of a species per host) as components of a vector $\mathbf{x}$, then the system dynamics may be expressed by an equation of the form

$$\frac{d\mathbf{x}}{dt} = F(\mathbf{x})\mathbf{x}, \tag{4.5}$$

where $F(\mathbf{x})$ is a matrix-valued, possibly non-linear, function of the components of $\mathbf{x}$. This system has a steady state (equilibrium) when $\mathbf{x}$ remains constant over time, i.e. when $d\mathbf{x}/dt = 0$. We write this equilibrium $\mathbf{x}^*$, and see that if $\mathbf{x}^*$ is not zero (no hosts or parasites) then $F(\mathbf{x}^*) = 0$. If this equation has a solution for which all components of $\mathbf{x}^*$ are positive, the steady-state solution is biologically feasible. We examine the stability of this equilibrium by looking at what happens near equilibrium, where $\mathbf{x} = \mathbf{x}^* + \varepsilon\mathbf{y}$ with ε small. We find that the vector $\mathbf{y}$ changes according to the equation:

$$\frac{d\mathbf{y}}{dt} = J(\mathbf{x}^*)\mathbf{y}, \tag{4.6}$$

where $J(\mathbf{x}^*)$ is the Jacobian matrix of the system, evaluated at the steady state $\mathbf{x}^*$. To construct the Jacobian matrix first find the vector $F(\mathbf{x})\mathbf{x}$, and then calculate the partial derivative of each component with respect to each component of the vector $\mathbf{x}$. The entry in the i, j position of the matrix $J(\mathbf{x})$ is the partial derivative of $(F(\mathbf{x})\mathbf{x})_i$ with respect to x_j. We then substitute $\mathbf{x} = \mathbf{x}^*$.

The solution of our perturbation equation (eqn 4.6) is then:

$$\mathbf{y} = \sum c_i \mathbf{v}_i \exp(w_i t), \tag{4.7}$$

where the w_i are the eigenvalues of $J(\mathbf{x}^*)$, the v_i are the corresponding eigenvectors, and the c_i are constants. Clearly, if one of the eigenvalues w_i is positive, or is a complex number with a positive real part, then $\mathbf{y}$ grows exponentially with time and the vector of system state variables $\mathbf{x}$ moves away from its steady-state value. Alternatively, if all the eigenvalues w_i are either negative or complex numbers with negative real parts, then $\mathbf{y}$ decreases exponentially with time and $\mathbf{x}$ moves towards its steady-state value. We have therefore shown that the steady state $\mathbf{x} = \mathbf{x}^*$ is stable whenever all the eigenvalues of $J(\mathbf{x}^*)$ are negative or have negative real parts, and that it is unstable otherwise.

The stability analysis presented here is concerned with *local stability*. That means that the system will return to equilibrium after it has been moved from it by a small amount. In order to show that the system will return to a particular equilibrium, regardless of its present state, one has to establish *global stability*, which is much more difficult. For further discussions on this topic see books on dynamical systems, e.g. Drazin (1992), Robinson (1994), Sánchez (1968), or Zwillinger (1997).

Stability as indicated by the eigenvalues of $J(\mathbf{x}^*)$ corresponds to the *basic reproduction number* (R_0) being less than one, and instability corresponds to the *basic reproduction number* being greater than one (see Diekmann *et al.* 1990; Heesterbeek and Roberts 1995b; Roberts and Heesterbeek 1995; Diekmann and Heestebeek 2000).

Given that an equilibrium parasite community exists and is feasible, its stability may be determined by examining the eigenvalues of the Jacobian matrix (see Box 4.3). For exploitation competition this matrix has a structure that guarantees stability (Roberts and Dobson 1995). Therefore, if an equilibrium community exists it is stable.

The existence and stability of a parasite community symbolized by $\mathbf{x} = \mathbf{x}^*$ (see Box 4.3) are determined by the argument presented above. The question of whether the community may be invaded by another species is examined by considering the stability of the steady state formed by adding an extra zero component to $\mathbf{x}^*$, to represent the potential invader. It is found that this steady state is unstable when $R_0 > 1$ for the invading species, where R_0 is defined by Box 4.2, eqn 4.4. Therefore the criterion for invasion depends on the epidemiological parameters of the invading species and the host population density determined by the existing community.

We illustrate these results for a community that may include up to three parasite species. Figure 4.1 shows a one-parameter bifurcation diagram constructed by setting $\lambda_i = \lambda\nu_i$ for $i = 1, 2, 3$ and assuming that the ν_i are constant. Hence the ν_i measure the relative ease of transmission of the different parasite species, and λ measures an environmental variable that affects the transmissibility of all three species in the same way. Parameter values, which are for illustration only, are given in the figure caption.

From Fig. 4.1 it can be seen that for $\lambda < 0.88$ no parasite community is possible. In this parameter range $R_{01} < 1$, $R_{02} < 1$, and $R_{03} < 1$. For $0.88 < \lambda < 1.9$ the community consists of just species 1, with mean parasite burden M_1 and host population density N. For $1.9 < \lambda < 2.6$ the stable community consists of two parasite species. It is still feasible that species 1 can exist in isolation, but it is vulnerable to invasion by species 2. For $2.6 < \lambda$ the three-species community is feasible (at least locally), but as λ increases further the parasite species are forced to extinction.

Different aspects of the host and parasite population biology then enter into this relationship in subtle ways, and four key factors act to influence the structure of a parasite community:

1 Short-lived host species with high fecundity will tend to have fewer parasite species and, as we may expect, the species absent are those with the smallest values of R_0.
2 Similar life-history attributes among parasites tend to allow more species to coexist. This produces the counter-intuitive result that diversity decreases as the parasites using the host become dissimilar in their life-history attributes. Similarity in life-history attributes may be an important condition for coexistence.
3 More parasite species exist in a community when the parasite distribution is highly aggregated. More aggregation effectively means that high burdens of two parasite species are unlikely to occur in the same individual host animal, so the two species can coexist largely by living in different subsections of the host population. This life-history attribute is a joint feature of the host and parasite species, since the degree of aggregation reflects physiological, immunological, behavioural, and genetic differences in the susceptibility of individual hosts to infection. The parasite species that are excluded are those with low levels of aggregation, implying that large degrees of aggregation should be apparent in the distributions of parasites in highly diverse parasite communities. In contrast, the structure of parasite communities is likely to be less diverse in habitats that are uniform, or where there is less variation among hosts in their susceptibility to parasites.
4 A high host-density can allow new parasite species to invade, but when parasite density increases with host density this may exclude other species from the community by causing increased host mortality in concomitantly infected hosts. This is likely to be an important effect in agricultural situations, where artificially increasing the density of a host species may lead to one or two parasite species dominating a parasite community and causing the majority of economic losses.

The models imply that parasite communities appear to have a 'core' of species that are always present in the community, and 'satellite' species that enter host populations when these reach higher density. In each case, species that might be classified as 'core' species are simply those with higher

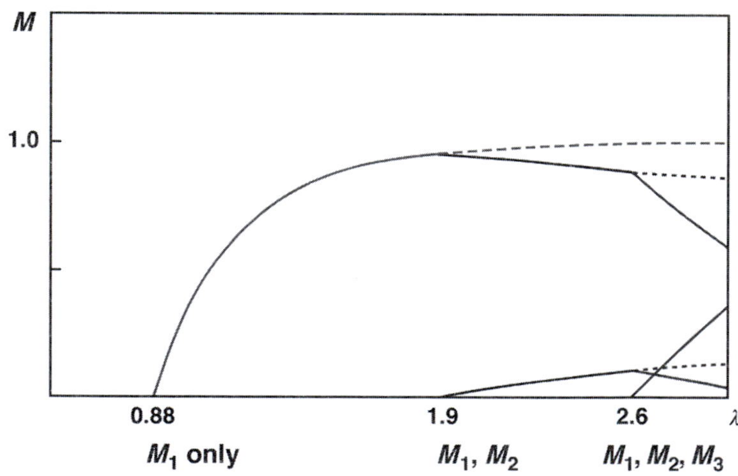

Figure 4.1 Bifurcation diagram showing the structure of parasite communities under different conditions of transmission, the mean number of parasites per host (M) is plotted against the transmission parameter (λ). Parameter values are: $(H_1, H_2, H_3) = (1, 0.6, 0.1)$; $(\nu_1, \nu_2, \nu_3) = (3, 2, 1)$; $(\mu_1, \mu_2, \mu_3) = (1.0, 1.1, 1.5)$; $(\alpha_1, \alpha_2, \alpha_3) = (0.1, 0.1, 0.1)$; and exponents of the negative binomial distributions $(k_1, k_2, k_3) = (0.5, 0.2, 0.3)$; with host birth rate $B(N) = 0.27 - 0.5N^3$ and host death rate $D(N) = 0.27 + 0.5N^3$.

values of R_0, while those that might be classified as 'satellites' are those with the lower fecundity, survival, or transmission rates, leading to lower values of R_0. Species with high values of R_0 invade the host community at lower population densities, and consistently have higher intensities than parasite species that can only establish in higher density host populations.

Interference competition

We now return to the situation where direct competition between parasite species is possible. The structure of the model does not now guarantee a unique equilibrium, and if one does occur, then its stability is not guaranteed. The stability of the parasite community to invasion by another species is determined by the value of R_0 for that species, calculated in the presence of the established community, but the expression is now more complicated than those presented in Box 4.2 (see Roberts and Dobson 1995). The likelihood of invasion is increased if there is a predominantly negative correlation between the distributions of parasite species. Also, if a parasite species in the community reduces the hosts' reproductive performance, then the community is more prone to invasion. The destabilizing effects of parasites that reduce host reproductive performance are well known for single species (May and Anderson 1978), although in some circumstances the opposite is possible (White and Grenfell 1997).

4.4 Competition between tapeworms

As an illustration of the modelling findings shown in the previous section, we shall now explore the relationships between different tapeworm species in sheep in New Zealand. The domestic dog and sheep have been hosts to a community of three species of cestodes in New Zealand. All three, *Echinococcus granulosus*, *Taenia hydatigena*, and *T. ovis*, are tapeworms within the definitive host (dog) and form cysts in the intermediate host (sheep), but only *E. granulosus* poses a threat to human health. Because of this threat, a regime of compulsory dog treatment, combined with an extensive education programme, was introduced in the late 1950s (Roberts *et al*. 1986, Gemmell *et al*. 1986). At that time *E. granulosus* and *T. hydatigena* were common parasites, but *T. ovis* was rare (see Box 4.4), but by 1990, *E. granulosus* had been eradicated through control measures, *T. hydatigena* was less common, and *T. ovis* was widespread.

To explore the different response of these three species to control measures, a model similar in concept to that described in the previous section was applied to the system (Roberts *et al*. 1986, 1987). The key to an understanding of the cestode dynamics lay in the pattern of the age infection curves shown in Box 4.4. The steady increase in the mean numbers of *E. granulosus* with age shows that there was little, if any, acquired immunity regulating this parasite. The value of R_0 was close to 1 so that it only took a minimal control effort to reduce R_0 below 1 and eradicate the parasite. In contrast, the plateau in mean numbers of *T. hydatigena* demonstrated the presence of a rapidly acquired and long-lasting immunity, hence a high degree of population regulation that translated into a strong control effort required for parasite eradication.

There is no reciprocal immunity between *E. granulosus* and *T. hydatigena* (Gemmell *et al*. 1987), hence any interaction between these parasites is via exploitation competition. However, previous exposure to *T. hydatigena* suppresses infection with *T. ovis* through interference competition, hence the emergence of *T. ovis* in response to the control measures. This example demonstrates how the response to changing conditions of one parasite in the community (*T. ovis*) could not be anticipated from the basic reproduction number of that parasite in isolation, it was necessary to consider the community structure and the nature of inter-species competition to deduce the correct response.

4.5 The parasites of alpine ruminants

One of the advantages of the cestode system described in the previous section was that in domestic systems a control programme can be applied that provides experimental data about the species interaction. In natural populations of hosts this is not always possible and the only way we can examine parasite communities is from post-mortem worm counts.

Box 4.4 Competition between tapeworms: a paradox

A national control programme in New Zealand succeeded in eradicating *Echinococcus granulosus* but, as a consequence, *Taenia ovis* emerged as a new problem. The population dynamics of cestodes are regulated by acquired immunity in the intermediate host, in this case the sheep. The basic reproduction number may be estimated from:

$$R_0 = 1 + \frac{D}{A}, \tag{4.8}$$

where A is the age at which an animal becomes immune, and D is the duration of acquired immunity.

In *E. granulosus* there is almost no acquired immunity; the levels of infection continue to increase with age. In contrast, sheep rapidly acquired immunity to *T. hydatigena*. Contemporary estimates of R_0 are presented in Table 4.2.

Although *T. ovis* would have a similar basic reproduction number to *T. hydatigena*, there is reciprocal immunity between the two species, with *T. hydatigena* able to exclude *T. ovis* (Gemmell *et al.* 1987).

Through the compulsory mass chemotherapy of dogs, the New Zealand hydatid control programme eradicated *E. granulosus* between 1958 and 1990. Control reduced parasite transmission and R_0 to 31% of its pre-control value. This was insufficient to eradicate *T. hydatigena*, but was sufficient to reduce its infection pressure on the sheep to a level at which acquired immunity was no longer maintained and this allowed *T. ovis* to invade the cestode community.

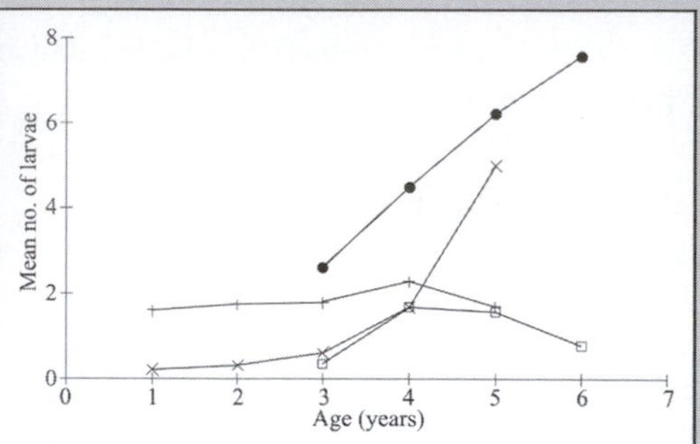

Figure 4.2 The mean numbers of larvae of *Echinococcus granulosus* (×, ●) and *Taenia hydatigena* (+, □) in male and female sheep, respectively, at different ages in New Zealand in 1958 (after Roberts *et al.* 1987).

Table 4.2 Tapeworm infections of sheep and values for R_0

Parasite	*E. granulosus*	*T. hydatigena*	*T. ovis*
Host: male sheep	1.3	2.5	2.5
Host: female sheep	1.6	4.3	4.3

After some decades of low population-size, and even a risk of extinction for species like the alpine ibex, the numbers of several wild ruminants species in the Alps have increased since the 1970s (Perco 1987; Lovari 1988). Most alpine valleys are colonized by at least one of the wild ruminant species. There is territorial overlap between species and domestic ruminants are also present on many of the alpine pastures during the summer, so the ruminants can share parasites. The community ecology of the parasites of the ruminant species was investigated in a part of the Italian Alps by focusing on the abomasal nematodes. These nematodes were selected for the study for two reasons: first, many of these species are potentially pathogenic for both wild and domestic ruminants (e.g. *Haemonchus contortus*, *Ostertagia ostertagi*, *Teladorsagia circumcincta*, *Trichostrongylus* spp.) and, second, the abomasum represents a homogeneous habitat for many closely related, directly transmitted nematode species. The prevalence of the 20 dominant nematode parasites collected in 1984 from the five wild ruminant species and the domestic sheep are shown in Fig. 4.3.

The patterns between host species are clear. Each host species has one or two dominant parasite species: *Ostertagia leptospicularis* and *Spiculopteragia spiculoptera* in cervids, and *Teladorsagia circumcincta* and *Marshallagia marshalli* in bovids. There are some intermediate species, with prevalences of 10–50%, and some rare species. Many nematode species can be found in different hosts, with nine out of 20 observed in five or more species. For each species, mean abundance (Bush *et al.* 1997) and importance indices (Thul *et al.* 1985) were also calculated and these showed similar patterns.

Since wild ruminant populations in the Alps usually live in sympatric conditions, a better under-

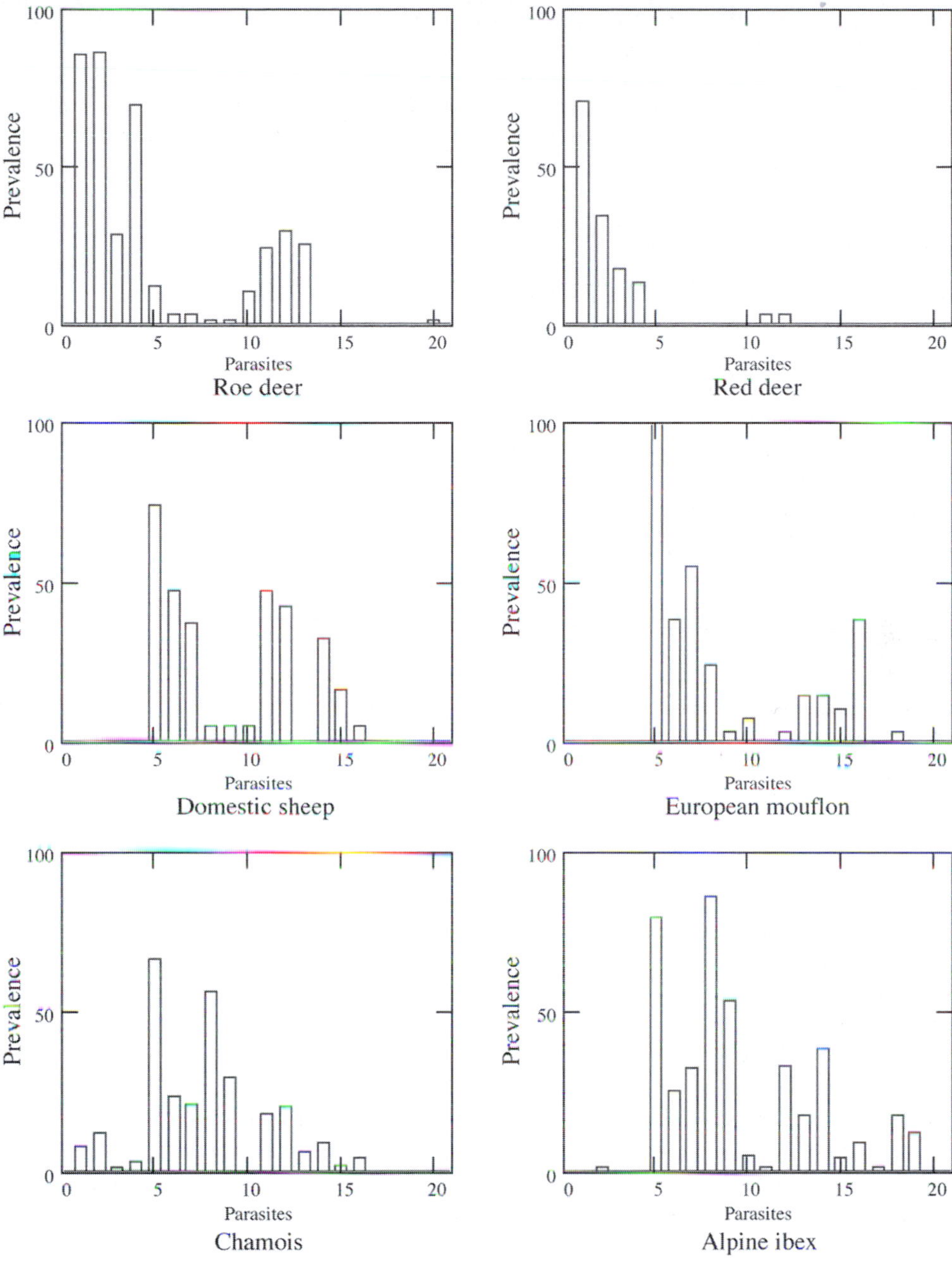

Figure 4.3 Prevalence of common parasites in six species of alpine ruminant. Parasite prevalence in each of six species of ruminant with parasite species indicated as numbers along the horizontal axis. Parasite prevalence in wild hosts was determined from the examination of harvested individuals. Key to parasites: 1 *Spiculopteragia spiculoptera*; 2 *Ostertagia leptospicularis*; 3 *Rinadia mathevossiani*; 4 *Skrjabinagia kolchida*; 5 *Teladorsagia circumcincta*; 6 *Teladorsagia trifurcata*; 7 *Teladorsagia pinnata*; 8 *Marshallagia marshalli*; 9 *Marshallagia occidentalis*; 10 *Ostertagia ostertagi*; 11 *Haemonchus contortus*; 12 *Trichostrongylus axei*; 13 *Trichostrongylus capricola*; 14 *Trichostrongylus vitrinus*; 15 *Trichostrongylus colubriformis*; 16 *Nematodirus filicollis*; 17 *Nematodirus helvetianus*; 18 *Nematodirus abnormalis*; 19 *Nematodirus davtiani alpinus*; 20 *Ostertagia lyrata*. Key to hosts: Roe deer (*Capreolus capreolus*), $N = 280$; Red deer (*Cervus elaphus*), $N = 76$; Domestic sheep (*Ovis aries*), $N = 19$; European mouflon (*Ovis musimon*), $N = 29$; Chamois (*Rupicapra rupicapra*), $N = 101$; Alpine ibex (*Capra ibex ibex*), $N = 155$.

standing of parasite specificity and the patterns of cross-transmission between hosts can be obtained through a discriminant analysis. In Fig. 4.4 the positions of the host animals are plotted on the first two discriminant axis, explaining 90% of the observed variance: axis 1, which accounts for most of the variance, discriminates well between bovids and cervids; and axis 2, which discriminates between host species. Distances between host-species groups were all statistically significant with the exception of domestic sheep and moufflon. The plot of correlations between nematode variables and discriminant axis (Fig. 4.4b) shows which species have more discriminant power, the species that are more specific for a host family. Species with low power are 'generalists', for example *Trichostrongylus axei* and *Haemonchus contortus* that have relatively high prevalence and abundance values in most hosts.

Correlation tests between each pair of helminth species in each host showed no significant negative correlations and a covariance test (Schluter 1984) indicated a significant positive overall association. Hence negative interactions between parasite species are, if present, low and unimportant in determining the structure of the community (Holmes and Price 1986). In one case, even though two helminth species in the ibex alternated in dominance following seasonality, negative correlations between the species were not detected, and when both were present in a host their correlation was significantly positive.

Finally, in another area where the red deer population is increasing, the roe deer is stable and the chamois is slowly increasing, parasite burdens showed the same temporal trend in the three host species despite large differences. Since the three host species have different ecological habits and distinct demography, the fact that their helminth abundances vary in a similar way suggest that abomasal parasitism can be related more to environmental factors common to the three hosts than with specific characteristics of each host population.

In conclusion, the dominant species appear to be 'specialists' for a well-defined taxonomic group of hosts, but some intermediate species are 'generalists'. If the purpose is to investigate the population dynamics of some helminths in wild ruminants, for instance *Haemonchus contortus* that is supposed to be highly pathogenic to roe deer (Zaffaroni *et al.* 1996), we must also consider the presence and the possible role of other host species. Negative interactions between helminth species did not arise for this system, leading to a community structure based on the basic reproduction number, R_0, of each parasite species (Dobson and Roberts 1994). However, it could be important to consider the effect of seasonality on the community composition, and of other abiotic factors as regulators of parasite populations.

4.6 Introductions and the invasion of British fish by helminths

Animal communities continually change as environmental and biological factors provide a greater opportunity for one species to benefit at the expense of another. However, the balance of such communities can be greatly altered by introductions. For example, the introduction of rabbits to Australia caused massive changes to the vegetation and native wildlife. We must take care that such introductions do not cause serious problems for native animals either through direct competition or by introducing parasites that may damage other species. A prerequisite for evaluating the damage caused by invasions is an understanding of the *status quo* before the invaders arrive. For example, Clive Kennedy and his colleagues from Exeter University have recorded changes in the fish and parasite populations in a reservoir and associated river systems in southwest Britain (Kennedy 1994b). This study has shown that while many macroparasite systems exhibit long-term stability, this stability and the community of hosts associated with the parasites can be greatly altered by the introduction of novel parasites.

The range of parasitic fauna of British freshwater species is increasing steadily as alien fish species are accidentally introduced by the pet trade and fish farming (Kennedy 1994a, b). Generally, when these species are introduced they bring parasites into the system, but most invasions fail. The best criterion for predicting whether a parasite will be successful is if it has had success elsewhere. Invading parasites must have the ability to overcome

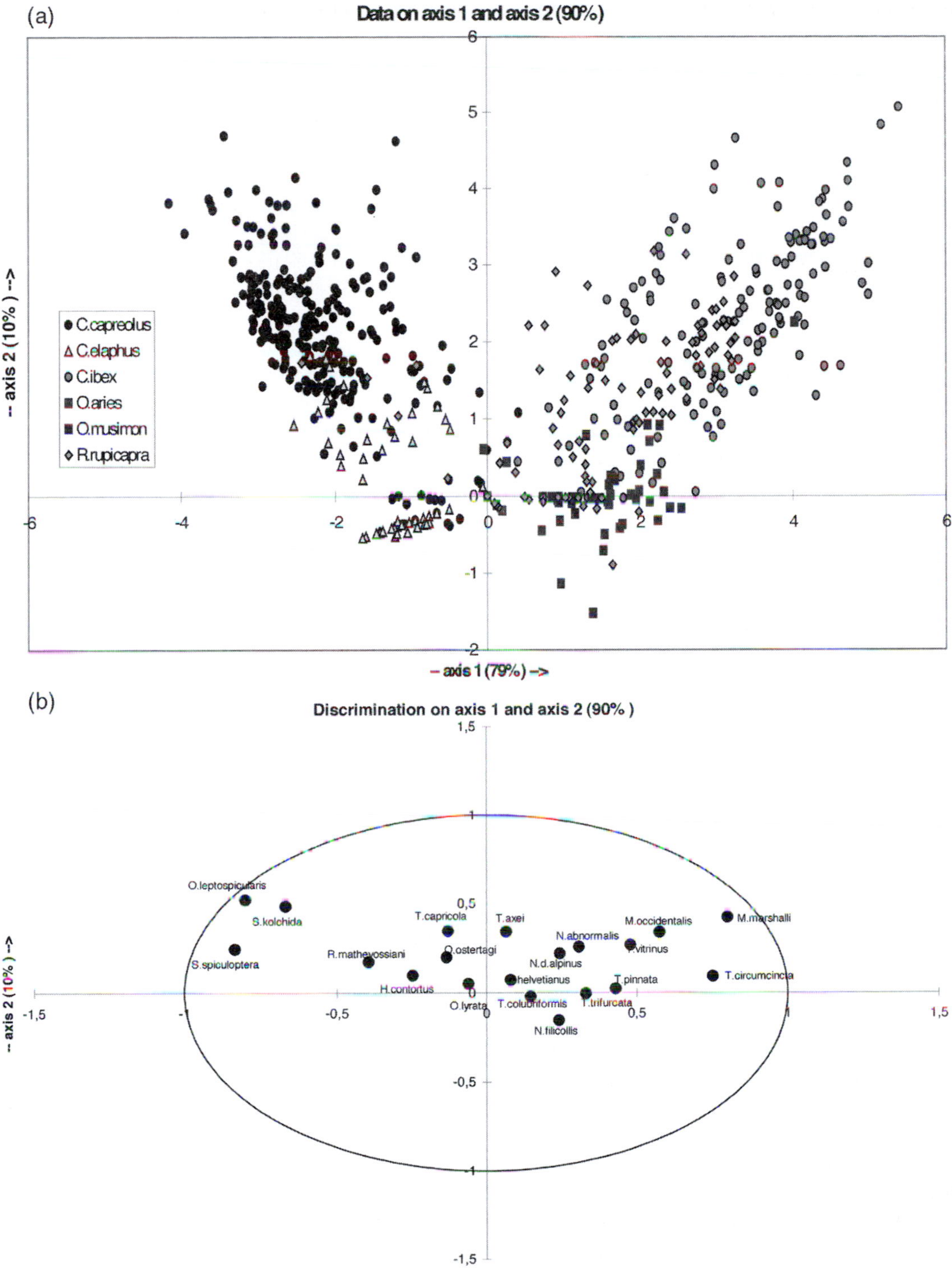

Figure 4.4 Results from a discriminant analysis utilising host species as the grouping factor and log-transformed parasite counts of each species as the discriminant variables: (a) plot of cases on the first two axis; (b) plot of correlations between the first two discriminant variables and initial variables.

environmental barriers in the habitat, biological barriers provided by the hosts, and constraints inherent in their own life cycles. Usually the invaders must enter a locality where the fish community and environmental conditions provide opportunities for transmission similar to those in the habitat in which they developed.

Fourteen species of parasite have successfully colonized British freshwater fish, although several others may be in the process of colonization at present (Kennedy 1994b). The successful colonizers form a heterogeneous group and no clear generalizations can be drawn about their biological attributes. All species have a history of successful introductions to other countries, yet some are still restricted in their distribution in Britain; others are spreading rapidly now that they have become established. Successful colonizers, which have at present a local distribution, are specialists that may have poor natural dispersal abilities. Those species that are distributed more widely tend to be generalists that may have been distributed by movements of commercial fish species, in particular carp, for stocking either gardens or commercial fish farms.

Perhaps the easiest way to accidentally introduce a parasite into a new location is by introducing an infected host. Quarantine or isolation of potentially infected hosts is the best way to prevent accidental introductions of pathogens of humans and domestic livestock. However, the adult stages of many parasitic helminths may live as long as their hosts, so quarantine regulations are often circumvented when fish are moved between different areas. In the past, the slow transportation of fish or fish eggs across an ocean or an overland route provided an indirect means of quarantine that allowed time for a parasite to be lost in transit. Today live fish may be flown into production farms within 48 h and parasites can become rapidly established. This rapid movement of parasites around the globe has led to an increase in the parasite fauna of fish communities in many countries, particularly those where fish farming is widespread.

4.7 Possum parasites in New Zealand

Hosts that invade new habitats can also be used to examine some key questions about parasite community structure. Populations of hosts that colonize, or are introduced to, new habitats tend to harbour a sub-set of their normal parasite fauna. This occurs in a way that is directly analogous to the founder effect in population genetics: the small population of hosts that colonizes a new environment contains a sub-sample of the total variety of parasites species (and genes) present in the original population (Dobson 1986, 1988b). All of the above discussion suggests that it is more likely that 'core' species of parasites will colonize with the host, with less common species likely to be absent. Nevertheless, chance may also cause other species to be absent, particularly when the population size of the initial introduction is low. Multiple introductions of the same host to different areas permit the opportunity to examine how the structure of parasite communities alter when one or more species are absent. Possums in New Zealand present a particularly good example since they are a significant pest species.

The population of up to 80 million common brushtail possums (*Trichosurus vulpecula*) that inhabits New Zealand derives from less than 300 that were introduced from Australia in the last century. The possum is now a major pest in New Zealand, where it destroys native vegetation and other flora, and is a reservoir for bovine tuberculosis (*Mycobacterium bovis*) (Cowan 1990; Roberts 1996). Recently, comprehensive surveys of the parasites of possums in New Zealand and Australia have been carried out, as part of an investigation of possible methods by which the New Zealand population may be controlled. A sample from the results is presented in Table 4.3.

Given this community, the interesting questions to be asked are:

1 Is *Paraustrostrongylus trichosuri* (*Pu*) (or *Parastrongyloides trichosuri, Pa*) uncommon in (or absent from) the South Island due to founder effects, i.e. because it was not present in the possums that were originally introduced, or are there other factors that make the habitat unsuitable?

2 If *Paraustrostrongylus trichosuri* or *Parastrongyloides trichosuri* were introduced to the South Island, would it invade the parasite communities, and what would its effect be on the population dynamics of the possum?

Box 4.5 Possum parasites in New Zealand and Australia

The major gastro-intestinal parasites of possums are[1,2]:

Tc *Trichostrongylus* spp.: *T. colubriformis* and *T. retortaeformis*, probably derived from sheep and rabbit parasites respectively. Found in possums throughout New Zealand, except for Stewart Island. Also reported (*T. retortaeformis*) from Victoria, Tasmania and New South Wales, but not Queensland or South Australia.

Pu *Paraustrostrongylus trichosuri*: found in possums from various locations in the North Island, but only from one location in the South Island of New Zealand. Uncommon in Australia, but found in 7/7 possums from New South Wales.

Pa *Parastrongyloides trichosuri*: found in possums from various locations in the North Island of New Zealand, but not in the South Island. Uncommon in Australia, but found in 6/7 possums from New South Wales.

Be *Bertiella trichosuri*: a cestode, found in some locations in New Zealand, and in all locations in Australia.

Ad *Adelonema trichosuri*: not found in New Zealand, but found in all locations in Australia.

[1] New Zealand: 15 locations approximating the original sites of introduction (Stankiewicz *et al.* 1996).

[2] Australia: possums sampled from Victoria, Tasmania, South Australia, and Queensland, plus seven possums from New South Wales (Coman unpublished report).

Table 4.3 Sample data from New Zealand farmland: prevalence (%), mean number of parasites per host and an estimate of the exponent of the fitted negative binomial distribution (*k*)

Location (possums)	Species	Prevalence	Mean	*k*
Paraparaumu (26)	*Tc*	19.2	38.85	0.175
	Pu	69.2	234.19	0.441
	Pa	65.4	82.42	0.181
Wanganui (69)	*Tc*	39.1	47.54	0.035
	Pu	02.9	0.87	0.021
	Pa	63.8	76.0	0.427
Banks Peninsular (56)	*Tc* only	62.5	566.3	0.365
Stewart Island (72)	No parasites			

3 If *Adelonema trichosuri* were introduced from Australia, would it invade the parasite communities of New Zealand, and what would its effect be on the population dynamics of the possum?

The answers to these questions are not obvious, largely due to the heterogeneity of the data. At each location in New Zealand possums were collected from one or more of seven different habitat types. In addition, climatic conditions are varied, with mean annual rainfall ranging from 819 mm to 2824 mm, and mean annual temperature ranging from 9.7 °C to 15.7 °C. One consistent feature however is the positive covariance of the Tc/Pa interaction, and the negative one for Tc/Pu. Hence communities consisting of just *Tc* would be more vulnerable to invasion by *Pu* than by *Pa*.

Table 4.4 Sample data from New Zealand Farmland: inter-species parasite correlation coefficients

Location (possums)	*Tc/Pu*	*Tc/Pa*	*Pu/Pa*
Paraparaumu (26)	−0.157	0.629	−0.048
Wanganui (69)	−0.024	0.321	−0.055

This suggests that *Pu* could be a suitable vector to be used as an agent of biological control.

4.8 Parameter estimation

Models provide a powerful tool for exploring the structure and dynamics of parasite communities. The results from models are often robust to wide variations in parameter values, but even so it can be difficult to estimate some of these parameters for parasites of wild animals. In this section we consider some methods that can be used to estimate the likely magnitude of model parameters in the absence of data from field trials. Although these methods only provide an indication of the magnitude of a particular parameter, this is often sufficient to develop a qualitative description of the system.

4.8.1 Determinants of abundance

Comparative studies of nematode populations of mammals have revealed two major determinants of abundance in parasite communities. First, parasite abundance depends on host population density. Looking across comparable host species, parasite abundance increases with increasing host density (Arneberg *et al.* 1998b). Theoretically this link is generated by host densities positively affecting parasite transmission rates: as host densities increase, each parasite egg or larva enjoys an increased probability of contacting a host (Anderson and May 1978; May and Anderson 1978). This implies that transmission rates constrain parasite population growth, at least in some host species. It also supports the interpretation of parasite community structure as being a consequence of 'supply-side' ecology (Lewin 1986): you have what you are exposed to (Bush 1990); with host density being an important determinant of exposure rates, particularly in parasite communities inhabiting low-density host populations. Second, there is a trend for large-bodied parasite species to live at lower infection densities than more small-bodied species (Fig. 4.5). This relationship implies that negative density-dependent processes operating within host individuals often influence the dynamics of nematode populations. Thus, there is evidence that both between-host processes and within-host processes are important in determining abundance in parasite communities, which immediately raises the question of what might determine the relative importance of the two seemingly opposing processes. The question is not new (Esch *et al.* 1990a), but knowing that within-host processes are related to parasite body size and between-host processes to host population density could enhance our ability to answer it.

4.8.2 Determinants of life-history traits

Life-history strategies are a combination of demographic traits that have been favoured by selection. They may change in time, under the pressure exerted by natural selection, but the array of possible modifications is bounded by physiological, physical, and phylogenetic constraints. A common way of analysing life-history strategies is to search for general patterns in demographic traits over a wide range of body sizes. In fact, of all characteristics of a species, body size is one of the most

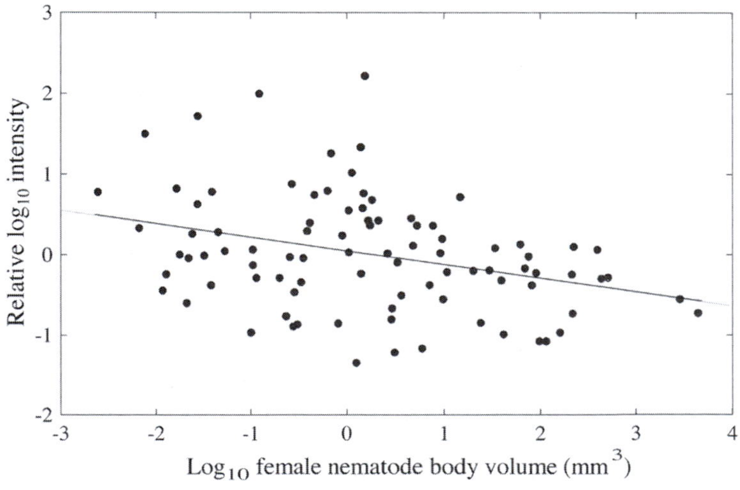

Figure 4.5 The relationship between $\log_{10}$ parasite body size (female volume) and relative $\log_{10}$ intensity across 92 mammalian nematode species. Values of relative intensity are residuals from the regression between $\log_{10}$ host body mass and $\log_{10}$ intensity (mean number of parasites per infected host) (after Arneberg *et al.* 1998a).

influential on its biological and behavioural characteristics, including host density, fertility, and longevity (Calder 1987; Charnov 1993; Peters 1983; Silva and Downing 1995). Hosts having large body size are expected to live at lower population densities and to have long life-expectancies and low reproductive-rates. Conversely, higher population densities, shorter life-expectancies, and high birth-rates characterize small host species.

Parasitic nematodes vary enormously in size, and this leads to systematic changes in development rates, longevity, and reproductive patterns. In general, pre-patency period increases with body size (Morand 1996; Skorping *et al.* 1991) as do reproductive lifespan (though less clearly) and rate of egg production. The major pattern that emerges from a study by Skorping *et al.* (1991) is a continuum from small nematodes with short developmental periods, low fecundity, and short reproductive period (such as trichostrongyles), to larger species (like the ascarid nematodes) with long developmental periods, high fecundity, and a long reproductive-lifespan. Egg size does not seem to be associated with any other life-history trait, although a correlation between fecundity and the thickness of the eggshell and the protection it gives to the larvae cannot be ruled out (Poulin 1995). This suggests that selection may operate almost independently on the free-living and adult stages of a parasitic nematode. These patterns are very different from those found in comparative analyses of life-history variation in other taxa (Skorping *et al.* 1991). Mammalian species in particular can be arranged along a fast–slow continuum, from small early maturing species with high fecundity, small offspring, and short lifespans, to species with the opposite traits. In nematodes the benefits of larger size and prolonged development appear primarily as increased fecundity, while in mammals they are manifested as fewer, higher quality offspring that are better able to compete for resources and mates. The contrast between the two patterns may reflect the energy-rich environment exploited by the intestinal nematodes and their need to continually locate new susceptible patches of habitat (hosts). The energy-rich environment should favour larger body size, because larger parasites live longer and have a higher fecundity.

The space available for parasite occupation in the host (Poulin 1995), or the energy available for growth and reproduction, may constrain the evolution of increasingly larger parasite body size. The energy processed by an organism is strictly related to its metabolic rate, which in turn scales allometrically with body size for both the host and the parasite. If the energy budget provided by the host can be a limiting factor for parasite body size, an association between parasite body size (which determines metabolic rate) and host body size should be detectable from experimental data. Larger hosts (which process more energy and live longer) are expected to harbour larger parasites with larger reproductive output. A comparative phylogenetic analysis showed that the evolution of body size in female pinworms of Old World primates appears to depend only on the body size of their hosts (Harvey and Keymer 1991). Another study showed that parasite body size is positively correlated with host body size for mammals and small invertebrate hosts (Morand *et al.* 1996). Here we have discarded the data corresponding to the smallest non-mammalian hosts (<4 cm) and then re-estimated the allometric coefficient using major-axis regression. The log–log regression between host and parasites body size now yields a coefficient that approximates unity, which means that the relationship between host and parasite body sizes is isometric.

The link between host and parasite body sizes is provided by the relationship between the fertility and pathogenicity of the parasite: an increase in egg production leads to an increase in the energy sequestered by the parasite to its host, and therefore to an increase in host mortality. Such a trend has been detected in both the human and snail hosts of schistosomiasis, where the degree of illness and snail mortality is related to parasite fecundity (Anderson and Crombie 1984). Other experimental studies have provided evidence that parasitic helminths have a significant impact on the energy budgets of their hosts (e.g. Booth *et al.* 1993; Hudson and Dobson 1995). Chronic infections often lead to increased metabolic rate and reduced body mass, which may have a long-term impact on host fitness. Negative interactions between the rate of parasite reproduction and host pathology are also predicted by theoretical analysis and observations in microparasites (Anderson and May 1982; Dwyer *et al.* 1990; Fenner 1983).

To test the hypothesis that host body size constrains life-history traits of macroparasites by limiting their body size, De Leo and Gatto (1996) have recast a standard epidemiological model of macroparasites by setting model parameters as simple allometric functions of parasite and host body size. The parasite damage to the host has been assumed proportional to the *relative* amount of energy that each parasite sequesters to its host. Since the energy processed by the host and the energy absorbed by the parasite are functions of host and parasite body size, respectively, the parasite-induced mortality α has been set proportional to the ratio between parasite metabolic rate and host metabolic rate, $\alpha \sim \Omega^{0.75}/W^{0.75}$. Here Ω is the parasite body size, W is the host body size and 0.75 is the classic allometric coefficient for metabolic rate. The proportionality coefficient was estimated as that which gave the best fit to available data.

The standard threshold theorem for parasite invasion has been used to determine the range of body sizes for the parasite to establish in a mammalian host population (De Leo and Gatto 1996). Recall that a parasitic species can establish in the host population if its basic reproduction number (R_0) is greater than 1. From Box 4.2, eqn 4.3, we see that R_0 depends upon the following parameters: population density of the host at the disease-free equilibrium (K), the parasite induced host mortality (α), a transmission coefficient ($\Lambda = s\lambda$) that depends upon the fertility of parasites in the patency stage and the survival of parasites from pre-patency to patency stage, a saturation constant (H), which has been set so that R_0 attains realistic values (i.e. smaller than 10), the mortality of the parasites in the patency stage (μ), and the host mortality (d). Host population density is then assumed to depend on host body size W(kg) as $K = 16.2W^{-0.70}$ (No. km^{-2}), and host mortality rate computed as $d = 0.4W^{-0.26}$ (yr^{-1}). Parasite mortality has been set equal to $\mu = 2.9W^{-0.25}$ (yr^{-1}) and the parasite fertility rate to $\lambda = 218500W^{0.5}$ (number of eggs parasite^{-1}yr^{-1}). The fractions of worms surviving to maturity may be derived from considerations of slow-fast dynamics of a more general model explicitly including the prepatency stage. It follows that $s = \sigma/(\sigma + \mu + \alpha + b)$, where $\sigma = 30\Omega^{-0.4}$ is the rate at which pre-reproductive worms are recruited to the adult stage (Ω is measured in mm^3; Skorping *et al.* 1991).

For any given host body size there is a corresponding range of parasite species able to invade the host (Fig. 4.6). Parasite species having a size exceeding the upper bound cannot establish in the population, because they will on average kill their host before producing secondary infections. Parasite species having a size smaller than the lower

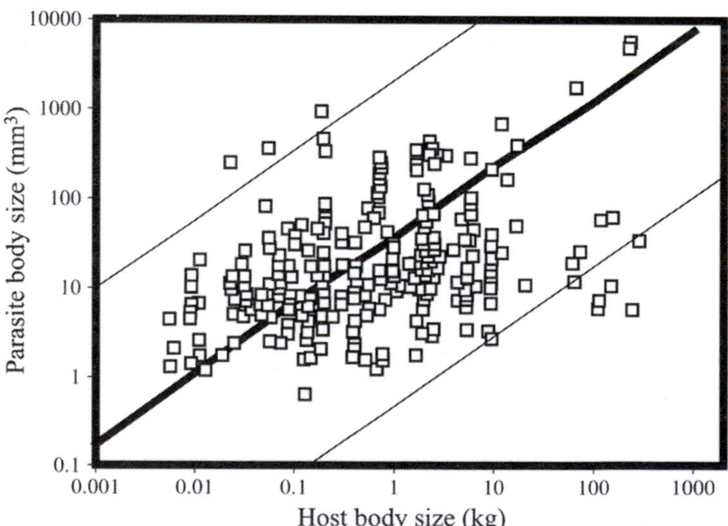

Figure 4.6 Optimal parasite size as a function of host body size. The *lower* and *upper lines* indicate the estimated range of parasite body size in the community as a function of host size. The *squares* represent the data observed by Morand *et al.* (1996).

bound cannot establish either: their reproductive output is too small to guarantee the infection of new hosts. Both the upper and lower bounds scale with host body size, thus demonstrating that smaller hosts can harbour only short-living parasites with small body-sizes and small reproductive-output, while larger hosts favour larger parasites. The allometric slope resulting from the theoretical analysis is equal to 1, as supported by the empirical data presented by Morand *et al.* (1996). Therefore, by increasing host size by one order of magnitude, parasite body size is expected to increase by one order of magnitude. Predictions from theoretical argument thus fit field data reported in Morand *et al.* (1996), showing that the energetic determinant of parasite pathogenicity constrains the size of the parasites able to invade a host.

4.9 Synthesis

In essence this chapter has found that parasite species that wish to invade an established community must do so under the conditions imposed by the community itself, hence the order of invasion by species is important. If the distributions of two parasite species tend to be negatively correlated, then they are more able to coexist in a population but the community is more prone to invasion. If one or more parasite species reduces the reproductive potential of the host, then the community is more prone to invasion.

The case studies have demonstrated several key features of the model. For example, the existence of 'core species' such as *Teladorsagia circumcincta* in all alpine ruminants except red deer and the *Trichostrongylus* species that were present in possum populations throughout New Zealand. The structure of a parasite community is determined not only by the life-history traits of the parasite species and the demography of the host species, but also by the interactions between parasite species as mediated by the host. A good example of this was the behaviour of the community of three larval cestodes in sheep, where the balance was changed by tapeworm control in dogs reducing the infection pressure for all three parasites.

The primary determinant of a parasite species' place in a community is its basic reproduction number, R_0. Traditionally R_0 is calculated in the absence of all parasites and while it is true that $R_0 < 1$ under these circumstances ensures that the species cannot persist, it is not necessarily true that $R_0 > 1$ implies that it can. A parasite species can persist in a community if $R_0 > 1$ when R_0 is calculated in the absence of the species under review, but in the presence of all other parasite species in the community. Hence, in the tapeworm example the basic reproduction number of *Taenia ovis* was greater than 1 in isolation, but less than 1 in the presence of *T. hydatigena*. Parasite control reduced R_0 for both parasites in isolation, but increased R_0 for *T. ovis* in the presence of the other parasite, enabling it to invade the community.

There are many ways in which the detail of these models may be increased. For example, acquired immunity could be important in the dynamics of some parasite species of wild animals, seasonally dependent host-population dynamics or parasite transmission could occur, or the age structure of the host may be relevant. It has been shown that even the most benign parasite-population dynamics arising in farmed animals can be transformed into something quite complicated by the interaction between seasonal host dynamics and acquired immunity (Roberts and Heesterbeek 1998). Conversely, complex host dynamics due to seasonal reproduction can be stabilized by host–macroparasite interactions (White and Grenfell 1997).

Models are useful for asking 'What if?' questions about population dynamics. There have been too few studies of parasite communities of wild animals that have reported quantitative results, and the studies that have been carried out show considerable heterogeneity. Hence it is useful to have methods that relate parasite abundance and body size (hence fecundity) to host population density, and parasite life-history traits to host and parasite body sizes. These techniques provide useful parameter ranges for use in conceptual studies.

Host–parasite population dynamics provide an ideal environment in which to study the dynamics of metapopulations (Earn *et al.* 1998; Nee *et al.* 1997). The parasite provides the metapopulation, and each individual host provides a habitat patch. An infected host corresponds to an occupied patch, and inter-host parasite transmission corresponds to inter-patch migration. Hence, not only do parasite communities provide a convenient model system for controlled experiments in metapopulation dynamics, but the burgeoning field of metapopulation studies in ecology should provide new insights to the dynamics of parasite communities. These ideas are explored further in Chapter 6.

CHAPTER 5

Microparasite transmission and persistence

J. Swinton, M. E. J. Woolhouse, M. E. Begon, A. P. Dobson, E. Ferroglio, B. T. Grenfell, V. Guberti, R. S. Hails, J. A. P. Heesterbeek, A. Lavazza, M. G. Roberts, P. J. White, and K. Wilson

The transmission dynamics of microparasites are an interplay between the spread of infection through a naïve population and the way in which the resulting post-epidemic population is returned to susceptibility through demographic processes.

Here we address questions like: What are the characteristics of the deterministic invasion threshold of infection? How do these differ from the stochastic fadeouts? How does host population density and transmission routes affect infection levels? What mechanisms do directly transmitted pathogens use to persist in small populations?

5.1 Background

Microparasites are those pathogens whose epidemiology can be usefully modelled with a classification of hosts into susceptible, immune, or recovered individuals. They generally include the viruses, bacteria, and protozoa. The epidemiology of microparasites of wildlife naturally has much in common with the better-established studies of microparasite dynamics in human and domestic animals. While much of the basic theory was established in the context of humans, there are a number of features of wildlife epidemiology that give it a particular flavour, including variability in host demography, such as seasonal variation in birth rates, and the significance of social structuring, such as territoriality versus group living.

Theory has been largely conducted in terms of two important concepts: threshold theory and fade out theory. Threshold theory relies on the notion of the basic reproductive number, R_0 and how changes in host demography change R_0 for the pathogen; it defines the necessary conditions for an epidemic to occur (see Box 3.4). By contrast, fadeout theory is about what happens in the aftermath of an epidemic and whether the pathogen has been so infectious as to run out of Susceptibles and thus become extinct.

5.2 The fundamental theory of microparasite transmission

The observed epidemiological patterns of microparasites have been reviewed at some length by Dobson (1995), and the mathematical models constructed to deal with them by Heesterbeek and Roberts (1995a). In this section we present only the simplest outline of the theory and the reader should refer either to these papers or Scott and Smith (1994). In particular, this chapter concentrates on how host densities affect parasite ecology, rather than the equally important and intertwined question of how parasites affect host abundance (Chapter 3; Grenfell and Dobson 1995).

A simple model of microparasite epidemiology, capable of demonstrating three important infection patterns, can be constructed by assuming that a host population is divided into: Susceptibles, S, who have never experienced infection; Infecteds, I, who are currently infectious; and Recovereds, R, who experienced infection in the past and are now immune. In addition, we assume that infection itself causes no mortality but that all individual hosts have a fixed mortality risk from other causes and that the birth rate is sufficient to maintain the total population size. Such a model can be summarized in the flow diagram shown in Fig. 5.1 and the basic mathematics are presented in Box 5.1.

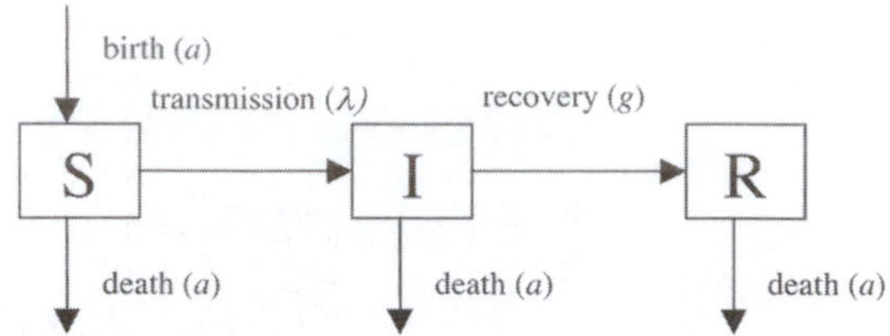

Figure 5.1 Flow diagram representing the basic SIR (Susceptible, Infected, and Recovered) model. Hosts are classified into one of three classes, young are born into the susceptible cohort, come into contact with an Infected then pass into this category and then either die from the infection or recover, usually with the development of life-long acquired immunity.

Three possible outcomes from this SIR model are illustrated in Fig. 5.2. The simplest infection pattern happens in a population when there are no births during the period of the epidemic (i.e. $a = 0$, Fig. 5.2(a)). This is the characteristic epidemic curve with an initial rapid epidemic growth, as the pathogen passes through the population, infecting and removing Susceptibles. The declines in the number of available Susceptibles leads to a decline in the growth rate, and then to a decline in the numbers infected, as the remaining Infecteds recover from infection without successfully infecting any Susceptibles.

When the host population has a turnover of births and deaths (i.e. $a > 0$, Fig. 5.2(c)), then a similar *initial* epidemic is observed in the wholly susceptible population. However, the births of new Susceptibles mean that not all of the later infections are dead-ends so the chain of transmission is maintained and the epidemic process can be repeated once Susceptibles are no longer too thinned out by the presence of Immunes (Fig. 5.2(c)). Eventually the infection settles down to an endemic equilibrium. An intermediate situation can also occur when birth rates are small but not quite negligible. We see the same oscillatory dynamics as in

Box 5.1 Mathematical formulation of the SIR microparasite model

A precise formulation of the SIR model for a population of size N can be created using differential equations for the sizes in each class S, I, and R. By assuming that the total birth rate aN is proportional to the total population size; that the death rates of Susceptibles, Infecteds, and Recovereds are aS, aI, and aR, respectively (so that the population size N is unchanging); that hosts recover from infection at rate gI; and that the rate of infection per susceptible given by the 'force of infection' $\lambda(I, N)$ we can produce the following linked differential equations:

$$\frac{dS}{dt} = aN - \lambda(I, N)S - aS \tag{5.1}$$

$$\frac{dI}{dt} = \lambda(I, N)S - gI - aI \tag{5.2}$$

$$\frac{dR}{dt} = gI - aR \tag{5.3}$$

A common assumption (discussed in Box 5.3), is that the force of infection λ has the form $\lambda(I, N) = bI/N$. It is then mathematically straightforward to show that this model can only support sustained infection when $R_0 > 1$, where R_0 is by definition the parameter combination $b/(a + g)$. Biologically R_0 has a natural interpretation as the rate at which new cases are produced by a single infective ($b\,S/N$) multiplied by the average infectious period ($1/(a + g)$), when transmission is optimal for the parasite and almost the entire population is susceptible ($S = N$). It is the number of secondary cases produced by a single primary case in a wholly susceptible population. Other assumptions about the force of infection lead to different mathematical forms for R_0, but this biological interpretation remains. For further details see Heesterbeek and Roberts (1995a) and for a broad overview see Diekmann and Heeseterbeek (2000) or Anderson and May (1991).

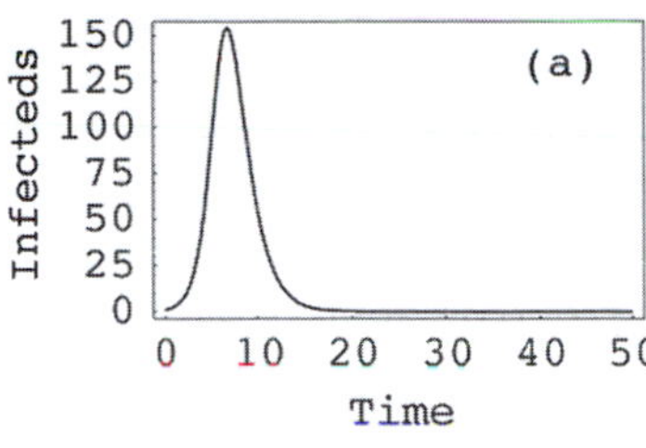

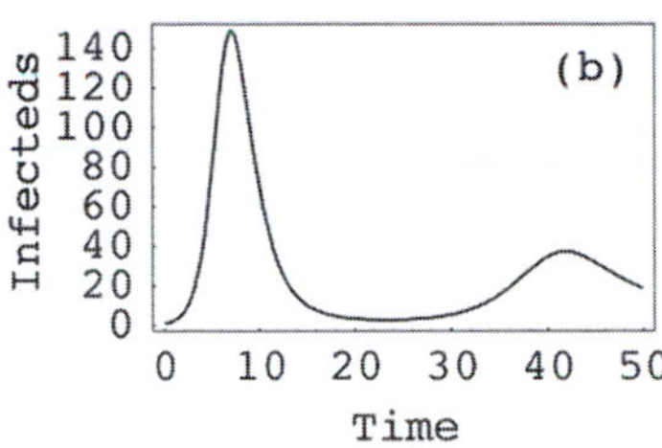

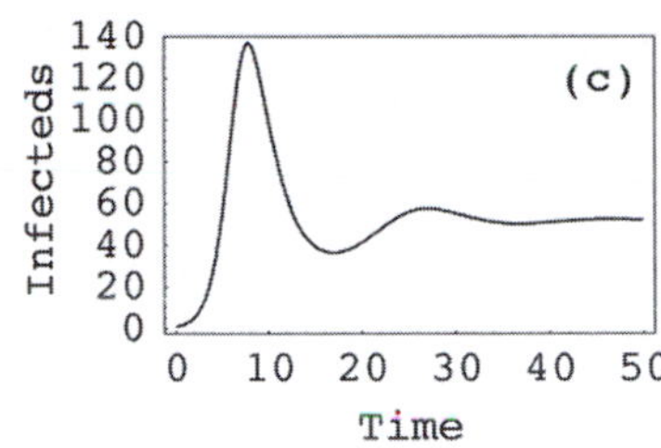

Figure 5.2 Three types of epidemic. Epidemic curves produced from the SIR model presented in Box 5.1, illustrating three scenarios of varying host birth rates. When there are no host births during the period of the epidemic curve (a), the number of Infecteds rises and falls as Susceptibles are removed from the population until the pathogen is finally eliminated and there are no Infecteds remaining. When host birth rate is high (c), then the birth of Susceptibles maintains the infection endemically and the pathogen persists. In the intermediate situation, with low host birth rates (b), the epidemic curve falls to a low level and it then becomes a matter of chance if the infection fades out or persists to generate a further oscillation.

Box 5.2 Phocine distemper virus (PDV) and harbour seals

During just a few months of 1988, about 20 000 harbour seals (*Phoca vitulina*) and a smaller number of grey seals (*Halichoerus grypus*) died in the North Sea following infection with phocine distemper virus (PDV) in one of the best-characterized epidemics in marine mammals (Heide-Jørgensen *et al*. 1992a, b; Heide-Jørgensen and Härkönen 1992; Barrett *et al*. 1995). PDV is a morbillivirus, closely related to canine distemper virus (CDV) and one of a family of viruses including measles (Barrett 1995). It takes several days for infectiousness to develop. Disease mortality is significant, although those animals that do not die recover within about a week and then appear immune for life (Harder *et al*. 1992b). In the PDV epidemic, transmission occurred at the seal haulouts, where the harbour seals congregate and come into close contact allowing direct transmission to occur.

Within each of the single geographic regions, infection typically followed the epidemic pattern type shown in Fig. 5.2(a). Since seals only pup once a year, the infection had died out before new Susceptibles arrived, so there were no new Susceptibles born into the population.

Figure 5.3 (redrawn from Grenfell *et al*. 1994) shows the weekly seal mortality observed in a single bay, The Wash, off the East coast of England in the 4 months of the epidemic and compares this with the expected number from the SIR model. The model captures well the characteristic rise and fall of the morbillivirus in a small population but fails to capture the finer scale details observed. See also Box 3.3 and Box 6.4.

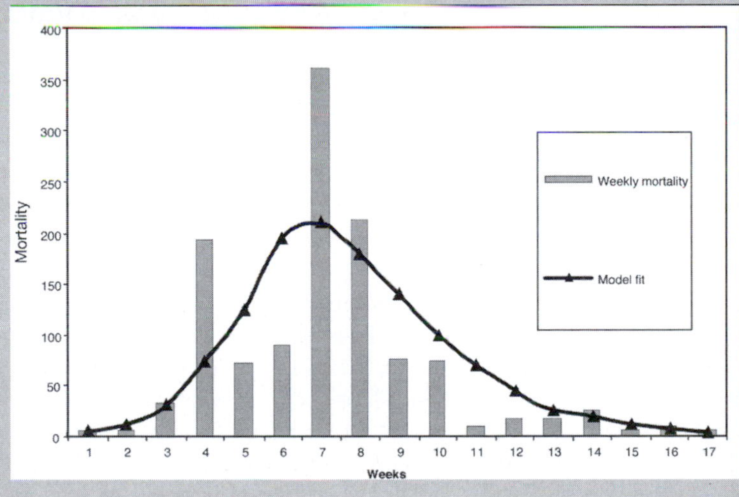

Figure 5.3 Course of PDV infection in seals in the Wash (UK) illustrated as the number of dead seals recovered by weeks after 1 August 1988, together with the fit of a simple SIR model (after Grenfell *et al*. 1992b).

Fig. 5.2(c), but the depth of the trough of infection is now much deeper (Fig. 5.2(b)). At this stage the model is in danger of breaking down because infection densities have dropped to a low level. Populations consist, in fact, of discrete individuals and the assumption that we can measure them by continuous densities, implicit in the use of the differential equation, is violated. In fact, what is likely to happen is that there will be one last infected individual who will recover without infecting any more newborns, and infection will *fadeout*. The subsequent pattern of infection will then be very different to that in the graph, and highly dependent on the pattern of subsequent re-introduction from outside the system (a so-called recurrent epidemic). For a detailed treatment of the intricacies involved in the three situations sketched here, see Diekmann and Heesterbeek (2000).

The simple SIR model outlined in Box 5.1 is enough to reproduce a number of observed epidemic behaviours, some of which have been observed in free-living populations. A nice example is the pattern of infection observed in the 1988 phocine distemper virus (PDV) seal epidemic illustrated in Box 5.2, with further details in Box 6.4.

5.3 Transmission of microparasites

A necessary condition for microparasites to persist for more than a few generations is that, under optimal conditions, the average infected host infects more than one Susceptible during the infectious period. In the context of the *SIR* model, in which infection is constrained only by the lack of Susceptibles, optimal conditions for parasite transmission requires a population composed almost entirely of Susceptibles. It is for this reason that the basic reproductive number, R_0, is defined in terms of what happens when a pathogen enters a new population (Box 5.1).

5.3.1 Threshold theory and host demography

The basic reproductive number, R_0, defines a threshold ($R_0 > 1$) for a pathogen to invade. If R_0 increases with host density or population size, it follows immediately that there must be some threshold host density or population size below which infection cannot persist. For typical *SIR*-type models, R_0 contains a factor describing the rate of contact between Susceptibles and Infecteds, and it is this factor that we expect to respond to host population density. Understanding the nature of this response is the key to understanding threshold theory. For many wildlife populations that extend continuously over habitat ranges (such as elm trees subject to Dutch elm disease), the natural unit of both measurement and population ecology is the number of individuals per unit area. For others (such as the harbour seal populations described in the PDV case study, Box 5.2), it is the total numbers observed at particular haulouts that form the relevant epidemiological and measurement unit. Once the relevant population variable has been identified it is then possible to consider how it influences the contact rate. There are a number of different paradigms in use for the influence of host demography on contact rate that are discussed in Box 5.3.

Which scaling actually occurs for a given system is best identified empirically. An analysis of the cowpox–rodent system carried out by Begon *et al.* (1998) is shown in Box 5.4. Given knowledge of the number of Susceptibles and Infecteds at each point in time, they use the numbers infected at the next time-point to derive the numbers infected in a time period. Using a regression analysis with this as response and numbers susceptible and infected as predictors, they demonstrated that explaining infection dynamics did require a non-linear interaction term and, more equivocally, that mass action was to be preferred over pseudo mass action in this context.

Another observation of infection-intensity scaling with population size comes from experiments with the gypsy moth *Lymantria dispar* and its nuclear polyhedrosis viruses (a baculovirus; Box 5.5). Here, the transmission parameter *declined* with increasing host and pathogen density (D'Amico *et al.* 1996). However, this pattern has proved difficult to interpret, although defoliation or behavioural changes have been suggested as possible explanations (D'Amico *et al.* 1996). Heterogeneities in susceptibility to infection may lead to a reduction in the transmission parameter with increasing pathogen density (§5.3.4; Dwyer *et al.* 1997), and density-dependent increases in patho-

Box 5.3 R_0 and the 'law' of mass-action

Suppose that the number of infectious individuals is I within a population of size N. The per capita rate at which Susceptibles become infected is the force of infection (λ). In Box 5.1 this was represented as a function of I and N. De Jong *et al.* (1995) named and contrasted two particular representations: first pseudo mass-action, with a force of infection $\lambda = \beta I$; and second, mass-action, with a force of infection of the form $\lambda = \beta I/N$. Pseudo mass-action can be interpreted as saying that a given Susceptible makes contacts at random with a fixed *fraction* of the individuals in the population, while mass-action can be interpreted as saying that given Susceptible makes contacts at random with a fixed *number* of individuals from the population, but only a fraction I/N of these can lead to a new case (Heesterbeek and Roberts 1995). The 'pseudo' prefix does not imply that the assumption is necessarily inappropriate; the same contrasts have also been labelled as those appropriate to vector-transmitted and directly-transmitted infection (Anderson and May 1991), or to sexually-transmitted and directly-transmitted infection (Thrall and Antonovics 1997). Yet another usage has been to label these contrasts as density-dependent and frequency-dependent transmission (Thrall *et al.* 1998). Moreover, these are just two possible scalings from many that may be relevant to different modelling settings.

One of the reasons that this distinction is important is that mass-action implies that the basic reproductive number R_0 (as defined in Box 5.1, where mass-action is assumed) is independent of N, while pseudo mass-action implies that R_0 is proportional to N. Since R_0 has a threshold for infection to be able to invade (i.e. $R_0 > 1$), it follows that mass-action implies that N has no effect on invasibility, while pseudo mass-action implies that there is a threshold N value below which infections cannot invade.

How should the modeller choose a plausible representation of the scaling of transmission? It is important to be careful about the definition of population 'size', distinguishing population numbers and population densities as variables. The choice of variable is likely to be dictated by the problem at hand. While a choice of scaling in either case is best empirically justified, the *interpretation* of mass-action in terms of a fixed contact acquisition rate is only really applicable when population size is used.

gen resistance may lead to reduction in the transmission parameter at high host densities (Wilson and Reeson 1998).

Transmission in the moth *Plodia interpunctella*, by contrast, occurs mainly through cannibalism of infectious cadavers, and so an increase in cannibalism at high host-densities should lead to an increase in the rate of contact between host and pathogen. Indeed, laboratory experiments using *P. interpunctella* and *Bacillus thuringiensis* showed that transmission increased with host density and decreased with pathogen density (Knell *et al.* 1996).

There is some evidence that stresses associated with high host-density can increase the host's susceptibility to infection. However, there is also evidence from Lepidoptera and Coleoptera that some insects may direct *more* resources into pathogen resistance when at high densities than at low (Kunimi and Yamada 1990; Goulson and Cory 1995; Wilson and Reeson 1998; Reeson *et al.* 1998; Barnes and Siva-Jothy 2000). These insects typically experience wide fluctuations in population density both within and between generations, and this is reflected in a phenotypically plastic life-history that responds to changes in host density. The high-density morphs of species like the African armyworm (*Spodoptera exempta*), exhibit a range of adaptations to the threats that accompany life at high densities; threats that include increased competition for food, increased predation risk, and, most importantly in the present context, increased risk of contracting disease (Wilson and Reeson 1998). These phenotypic changes may have a marked impact of pathogen transmission in the field and on the dynamics of the host-pathogen interaction (White and Wilson 1999).

Increasing host population size or density will make per capita contact rates increase under two broad circumstances (Heesterbeek and Metz 1993; Diekmann *et al.* 1996). First, when individuals contact other conspecifics within a given socio-spatial arena and crowding increases the number of individuals within that arena. For example, if badger population densities are correlated with social group size then overall transmission of tuberculosis may increase (White *et al.* 1997). Second, when the

Box 5.4 Cowpox in rodents: mass action in operation

Cowpox virus is a member of the genus *Orthopoxvirus*, and is endemic in Europe and some western states of the former USSR (Baxby and Bennett 1997). Although natural infection and disease occurs in cattle, man, and domestic cats, such cases are relatively uncommon, and the reservoir hosts are generally accepted to be wild rodents (reviewed by Bennett and Baxby 1996). Antibody and virus have been detected in wild ground squirrels (yellow suslicks) (*Citellus fulvus*) and gerbils (*Rhombomys opimus*, *Meriones libicus*, and *Meriones meridianus*) in Turkmenistan and Georgia (Marennikova *et al*. 1984; Tsanava *et al*. 1989), from root voles (*Microtus oeconomus*) on the Kolskiy Peninsula in northern Russia (Lvov *et al*. 1988), and evidence of infection has been obtained by PCR from various rodents in Norway (Tryland *et al*. 1998). In Great Britain, a high prevalence of cowpox virus antibody has been detected in wild bank voles, *Clethrionomys glareolus*, field voles, *Microtus agrestis*, and wood mice, *Apodemus sylvaticus* (Crouch *et al*. 1995) and cowpox virus-specific DNA in bank voles and wood mice detected by PCR (unpublished data).

Begon *et al*. (1998) studied cowpox dynamics in two field sites in north-west England and found annual cycles in the number of susceptible bank voles $S(t)$ (reflecting the annual arrival of newborn hosts) with corresponding but delayed cycles in the numbers seropositive for cowpox, $I(t)$ (Fig. 5.4).

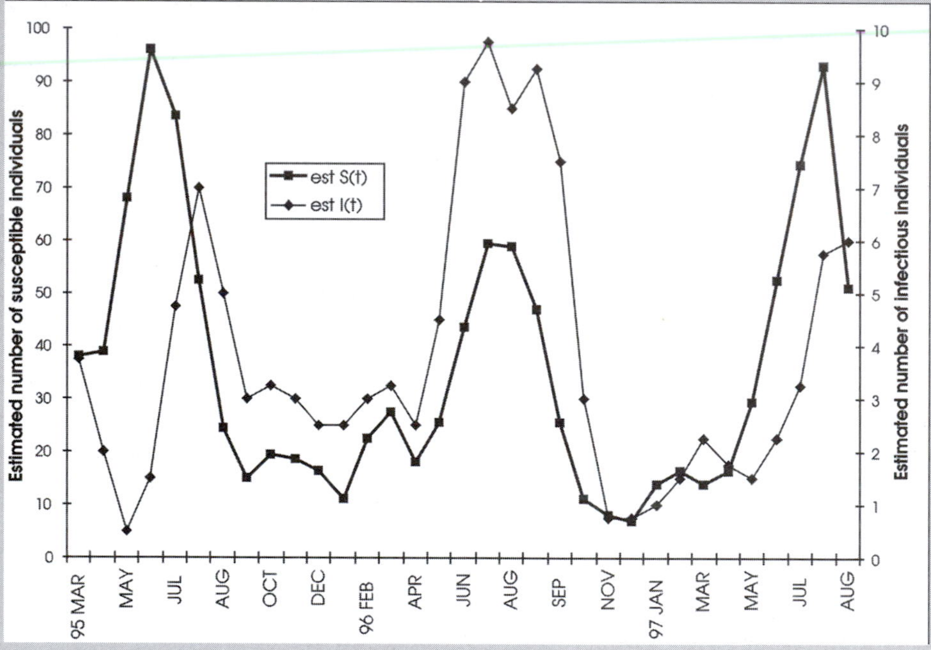

Figure 5.4 The estimated number of bank voles infected with, or susceptible to, cowpox virus in 4-weekly samples between March 1995 and September 1997 at Manor Wood, England (after Begon *et al*. 1998).

social structure of the population allows the contact with most of the other hosts over the lifetime of infection. For example, harbour seals with daily regrouping at the haulout might do this over the week of PDV infectiousness; pathogens with longer infectious periods would require less rapid mixing. In either case, population increases will lead to increased contact rate, and pseudo-mass action models may be applicable. By contrast, if population increases are associated with an increased number of groups over an expanded range (e.g. possums in New Zealand: Ryan *et al*. 1998), population increase at that scale will not significantly affect contact rates. As such, it is important to recognize that pathogen life-history determines the relevant features of host social structure. More-

Box 5.5 Insect baculoviruses

Baculoviruses, from the family Baculoviridae, are a large group of double-stranded DNA viruses that are restricted to arthropods, most having been isolated from lepidopteran insects. The two main groups are the nuclear polyhedrosis viruses (NPVs), which produce large structures called polyhedra containing many virus particles (virions), and the granulosis viruses (GVs), which produce smaller structures called granules, normally containing a single virion. Both groups infect their lepidopteran hosts when the larva ingests polyhedra or granules while feeding on their host plants. In the alkaline gut of the insect, the proteinaceous coat surrounding the virions breaks down, and they pass into the midgut epithelium. In most Lepidoptera there is a transient phase of infection in the midgut and from here the virus spreads to most larval tissues via the haemolymph, or via the tracheoles (Volkman 1997). Several days later, the insect host will be little more than a flaccid bag of virus (Fig. 5.5). Frequently, an infected larva climbs to the tips of shoots shortly before death (Vasconcelos *et al*. 1996). The body tissues then disintegrate, liberating the occlusion bodies (polyhedra or granules) over the foliage beneath, so distributing the virus particles in a suitable place to infect new susceptible hosts. This elevation-seeking behaviour of the host before death is just one of the many ways in which the baculovirus manipulates its host to facilitate transmission and persistence.

over the route of transmission can be very important and it clearly matters whether a pathogen is sexually transmitted or food-borne.

5.3.2 Threshold theory and disease management

If one believed that pseudo mass-action was appropriate for a particular pathogen system, then simple models will suggest that the population threshold exists and (in the simplest case) is the observed size of the susceptible population at the endemic equilibrium (e.g. Anderson *et al.* 1981; Lyles and Dobson 1993; Barlow 1995; Fromont *et al.* 1998a, b). For many infections this is a rather small fraction of the population (e.g. possums: Barlow 1993, 1996) so this may predict that disease eradication can be achieved through substantial populations reduction. However, this conclusion should be treated with caution. Trivially, mass-action and pseudo mass-action make identical predictions when host population sizes are unchanging. So, in this situation, empirical comparisons of any scaling laws are impossible. This is particularly serious when it is proposed to make a radical change in population density and when the system has not been previously observed at a wide range of densities. In addition to the uncertainties about population scaling of transmission, it has also been suggested that the social perturbation arising from such

Figure 5.5 The pine beauty moth, *Panolis flammea*, in the advanced stages of a nucleopolyhedrovirus (NPV) infection. Note that the larva has climbed to the top of the plant before releasing the virus occlusion bodies that fall onto vegetation being eaten by other larva. (Courtesy of Jenny Cory.)

major intervention may itself promote dispersal and enhance transmission, thus working against the desired effect, e.g. bovine tuberculosis in badgers (Box 5.8; Swinton *et al*. 1997).

Control analysis does not necessarily require good answers to these difficult scaling questions. In particular, when considering the impact of vaccination, a good model assumption is that contact structures remain completely unchanged, except that the proportion of infectious contacts is reduced. Indeed, one advantage of vaccination is that, whatever its impact is relative to other forms of disease control, such as culling or immuno-contraception, the impact of vaccination is likely to be more predictable in advance, since it is less sensitive to the uncertainties about the population scaling of transmission.

5.3.3 Ecological complexities in disease management

Complexities such as density-dependent processes that act on the birth and death rate of the host will also influence the efficacy of the different control methods (Barlow 1996). An unusual example of this could occur in the control of swine fever in wild boars (Box 5.6). In some areas, hunting selectively removes the older, immune boars and this mortality may be rapidly balanced by compensatory birth rates that lead to an increase in the number of young susceptible animals that in turn promote transmission. An intriguing theoretical possibility is that a pathogen may even benefit from enhancing its own virulence, so as to increase its access to Susceptibles in this way. This idea was examined in the phocine distemper case study (Boxes 5.2, 6.4) but the relatively slow intrinsic growth rates of seal populations prevents sufficient compensation for re-infection. Nevertheless, the idea could be tested further in a fast breeding host with density dependent birth rate.

When disease management involves the control of infection in a wildlife population with the objective of reducing the risk of infection in an economically important domestic population (e.g. swine fever in wild boar and pigs, tuberculosis in badgers and cattle) then it is important to recognize that disease risk is likely to be related, not only to the prevalence in the wild population, but also to the total numbers infected. A crude analysis that ignored the epidemiology may assume that killing half the wildlife population will halve the risk of infection but the non-linearities discussed above will often modify this conclusion in significant ways. When the goal is disease eradication, rather than mere control, then a clear understanding of these complexities becomes more important.

Such considerations are also relevant when one attempts to map disease risk through GIS-based prediction of host densities from satellite derived environmental data (Box 6.1). For example, Boone *et al*. (1998) found that hantavirus infection of deer mice (*Peromyscus maniculatus*) was less likely where rodent densities were below a threshold value, but that density had little effect on prevalence above threshold.

5.3.4 Host heterogeneity

Heterogeneities in host susceptibility can also significantly affect transmission (Woolhouse *et al*. 1997). Dwyer *et al*. (1997) has shown in insect baculoviruses that incorporating variation between hosts in their susceptibility to pathogens into mathematical models of *L. dispar*-NPV improved the fit of the models to data from field experiments. Such variation can act in a density-dependent manner. Another heterogeneity that has been found to be important in a number of human disease is age-structured mixing, where transmission is more likely to be between similarly aged people than between age groups (Anderson and May 1991). Another heterogeneity of particular importance to conservation biology is the relationship between host genetic variability and small population viability in the presence of parasites (Lyles and Dobson 1993).

5.4 Persistence of microparasites

The previous section concentrated on the transmission of microparasite infection after a pathogen invaded a population of susceptible hosts and investigated this using the concept of the basic reproductive number, R_0. However, such conditions are not sufficient for pathogen persistence. For example, each sub-figure of Fig. 5.2 represents an epidemic with $R_0 > 1$; yet only one corresponds to the long-term persistence of an infection. This section now examines the mechanisms that directly-transmitted microparasites use in order to persist.

Pathogens that develop the means of persisting, after most susceptible hosts have been lost from the host population, will clearly be at a selective advantage. Some pathogens, such as feline immunodeficiency virus (FIV) persists in feral populations of domestic cats *Felis catus* simply by inducing lifelong infectivity (Courchamp *et al*. 1995). Others remain in the host species but develop a 'carrier state' amongst some of the recovered individuals, where certain individuals continue to shed virus after recovery. In some long-lived hosts, the infection may have a long incubation period before the full-blown infection develops, such as in the transmissible spongiform encephalopathies (§5.4.9). A more common strategy is to use a biological reservoir of one kind or another. Some use vectors (Chapter 7) or a 'reservoir' host species, as in the case of tuberculosis in cattle and badgers (§5.4.7) or canine diseases in Ethiopian wolves and domestic dogs (§5.4.5; §8.2). Directly transmitted macroparasites often have long-lived free-living stages and while the microparasites do not have special infective stages, viruses such as the baculoviruses can be sustained in an environmental reservoir (§5.4.8).

Spatial structuring of the host population may also allow pathogen persistence (§5.4.4). This may be simply a consequence of the way the hosts are socially structured, such as brucellosis in ruminants (§5.4.6), or how the population is divided into sub-populations. Here, the pathogen invades a host sub-population and then jumps to another susceptible population, leaving the original population to recover through the immigration and birth of Susceptibles, thus making it a suitable habitat for future invasion. This may provide the key to understanding both the non-persistence of phocine distemper virus in harbour seals (Boxes 5.2, 3.3, 6.4) and the apparent persistence of caliciviruses (Box 5.7).

5.4.1 Fadeout theory and host demography

In those highly infectious pathogens with a short duration of infectiousness and a lack of specialized persistence mechanisms, the dynamics of microparasites are dominated by the deterministic consequences of invasion and the stochastic conditions that lead to fadeout. The probability of fadeout is influenced predominantly by the critical community size.

For pathogens like the morbilliviruses, a reasonable first approximation of the epidemic curve is that almost every member of a population will get infected and then become immune. This leads to a post-epidemic trough, from which infection can only recover through the input of new Susceptibles, usually through birth. Initially, all of these new Susceptibles will be diluted amongst the Immunes, and the chance of new transmission will remain small. Not until the number of Susceptibles has increased sufficiently can a subsequent epidemic re-occur. The input of Susceptibles depends directly on the birth rate and thus, via host ecology, on population size. However the risk of fadeout during this period is largely independent of population size and is controlled by the behaviour of the small number of non-immune individuals. The combination of these two factors, one population-independent and one population-dependent, leads once more to a population threshold. This is an effect that cannot be simply captured in continuously based models, since they allow arbitrarily low numbers of Infectives in the troughs which can then reseed the infection, and is much harder to analyse mathematically beyond the initial work of Bartlett (1957) (see Nåsell 1999a, b).

One example of infection fadeout in wildlife is the loss of phocine distemper virus (PDV) in the North Sea seal epidemic (Boxes 5.2, 3.3, 6.4). Although the infection had no difficulty in spreading ($R_0 > 1$) it could probably not persist because the epidemic ceased before sufficient new Susceptibles were introduced at the annual pupping season. Serological evidence for the continuing transmission of PDV after 1988 in North Sea harbour seals is equivocal. Seropositive samples were reported from Dutch seals born after 1988 (Visser *et al*. 1993), but none from Germany (Harder *et al*. 1993), the Wash (Hughes *et al*. 1992) or the Moray Firth (Thompson *et al*. 1992), There are no reports of continued excess mortality at any of these sites. As we saw in Figure 5.3, observed mass mortality in Wash harbour seals ceased after a few months, and the same pattern was repeated around the North Sea (Box 3.3). It appears that, even if transmission did continue beyond early 1989, it was of a strain of greatly reduced virulence and different transmission characteristics. Thus it seems likely that, if there is a critical community size for PDV in North Sea harbour seals, it will be greater than 50 000 individuals.

5.4.2 Fadeout and persistence: classical swine fever

The range of epidemic patterns, from fadeout to persistence of infection, and how this pattern is influenced by host demography, is nicely illustrated by the case of classical swine fever (Box 5.6) in the wild boars of Italy. This also provides a possible example of the role of compensatory reproduction in enhancing disease persistence.

The introduction of swine fever into the wild boar populations of Italy has come about through three routes: contact with infected captive wild boars, illegal restocking with infected wild boars, and shared habitat with infected free-ranging pigs (Lowings *et al.* 1999). Three different types of infection dynamics have followed these outbreaks. Infection in small populations (Lunigiana, 800 animals/304 km^2; Piacenza: 400 animals/75 km^2) has resulted in rapid fadeout from 45% seroprevalence to 4.5% within 12 months. In large populations (Maremma, 8000 head/3800 km^2) of which about 35% of animals are culled each year, virus was detected for 5 years and seropositive animals for 8 years (Fig. 5.6(a)). A third type of dynamics has been observed in eastern Sardinia, where infection is endemic and has persisted for at least 17 years with a low seroprevalence (Fig. 5.6(b)). The age-stratified serological data suggest a small endemic area out of which the infection spreads unpredictably from year to year and this spread is influenced by wild boar dynamics. In order to eradicate infection in this area, strong hunting pressure has been encouraged so that 45% of the population has been killed each year.

Application of a simple SIR model has identified how the dynamics of the infected population could influence the maintenance of the infection in the eastern Sardinian population (Guberti *et al.* 1998). This showed that the large number of young animals could form a reservoir of Susceptibles. Increased hunting of the wild boar population increased the number and the survival of the newborn Susceptibles, so the virus had enough susceptible animals to be maintained in the wild. Four main factors determine the persistence of the infection in the wild:

1 the size of the susceptible wild boar population;
2 the presence of domestic herds of pig or wild herds in the area;
3 depopulation or very high hunting pressure promoting persistence by increased population turnover through compensatory reproduction;
4 the presence or the selection of low virulence strains (Guberti *et al.* 1998).

5.4.3 Persistence of highly virulent pathogens: caliciviruses

The newly emerging lagomorph caliciviruses, European brown hare syndrome virus (EBHSV) and

Box 5.6 Classical swine fever

Classical swine fever is an RNA pestivirus in the Togaviridae. Infection typically results in high morbidity and mortality, though infections with low-virulence virus do exist. Suidae are the sole natural hosts, and direct contact between infected and susceptible animals is the principal means of viral transmission—though in domestic pigs, swill feed has also been identified as a source of infection. Infected animals shed large amounts of virus for up to 40 days, sometimes intermittently. The virus is excreted in oronasal and lacrimal secretions, urine, and faeces. Recovered animals show specific antibody and lifelong immunity. European Union pig protection programmes have shown that swine fever outbreaks in wild boars cause severe economic losses to domestic pigs. Oral vaccination does not induce durable protection and depopulation of wild pigs is considered the only suitable approach.

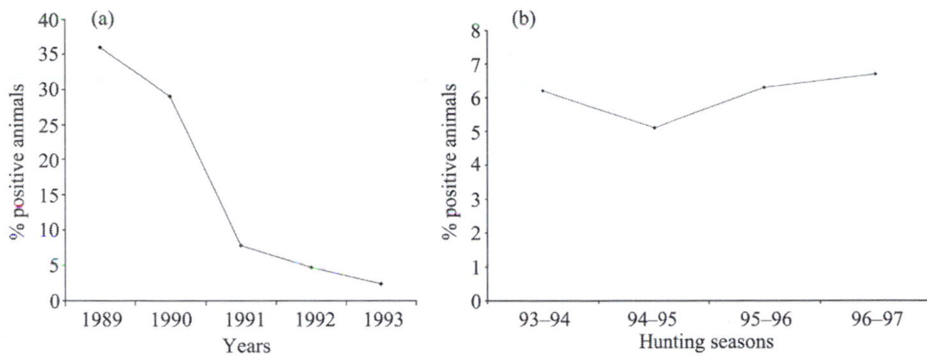

Figure 5.6 Longitudinal serology to classical swine fever in wild boars showing (a) slow decline of seroprevalence at Maremma (Tuscany) from initial infection in June 1983 and (b) long-term persistence in Eastern Sardinia, Italy where infection has been present since 1981 in this population (after Lowings *et al.* 1998).

rabbit haemorrhagic disease virus (RHDV) (see Box 5.7), are highly pathogenic infectious pathogens that could be of great significance both ecologically and economically. Their impact is very variable, and probably highly sensitive to host demography.

Host population density affects the rate of transmission and the host age structure determines the size of the susceptible and refractory juvenile populations, which vary greatly over the year. In a study of hares in the Modena region of Italy, Lavazza *et al.* (1997) found that mortality due to EBHSV was negligible at high densities (>15 hares/ha) due to rapid transmission, meaning that most hares were exposed whilst still very young. At lower densities (<8/ha) juveniles were more likely to become exposed after their refractory period and hence to

Box 5.7 Lagomorph caliciviruses

Rabbit haemorrhagic disease virus (RHDV) and European brown hare syndrome virus (EBHSV) constitute a novel and distinct subgroup of *Caliciviridae* (Wirblich *et al.* 1994). They are both host-specific, the former infecting only European rabbits (*Oryctolagus cuniculus*) (Lenghaus *et al.* 1994), and the latter infecting both brown hares (*Lepus europaeus*) and mountain hares (*Lepus timidus*). Both viruses are highly infectious and have case-fatality rates of up to 97% for RHD and over 50% for EBHS, with death occurring within 2 days. An interesting feature of EBHS is the 'juvenile refractory period': hares less than 8 weeks old are infectious and become immune, but do not suffer symptoms (Lenghaus *et al.* 1994, Lavazza *et al.* 1997). Additionally, EBHSV is highly robust and has been found to remain infectious for 3–4 months in the field in Italy.

First identified in China in 1984, RHDV spread rapidly throughout Europe and North Africa in the late 1980s and early 1990s. It was imported into Australia and New Zealand for use as a biological control agent (Coman 1997) because of its host-specificity, coupled with its rapid and, relative to myxomatosis, humane killing ability. But these lagomorph caliciviruses also pose a threat for conservation of high-quality, rabbit-grazed swards, and the loss of an important food source for threatened predators.

RHDV's epidemic dynamics vary greatly, in terms of seasonal timing, speed of spread, mortality, and persistence. In arid areas of Australia, epidemics tend to be very intense, peaking 2–3 days after introduction and lasting for a few weeks (Barlow and Kean 1998). Mortality is high and local fadeout of infection is likely. In more temperate areas, epidemics are usually less intense, last for months, and mortality is much lower. In Spain, RHDV may have become endemic, limiting the rabbit population to around a third of its former level and removing important food sources for predators such as kites and the endangered Spanish lynx.

A further interesting factor is the discovery of a non-pathogenic strain of RHDV, which does not cause symptoms but confers immunity to pathogenic RHDV. This strain may be widespread and highly prevalent in parts of the UK (Trout *et al.* 1997) and Europe.

succumb to infection. At low density, host extinction is a possibility, and it was recommended that populations be maintained above 8–15/ha to avoid this. Barlow and Kean (1998) developed a simple homogeneous-mixing model for RHDV, and found that the juvenile refractory period reduced the epidemic intensity and enhanced subsequent population recovery. In areas with a long breeding season the population may have a chance to recover. Thus the local population dynamics in terms of the length, timing, and intensity of the breeding season are important.

Given its propensity for intense epidemics with fast mortality rates, how does a pathogen such as RHDV persist? There is no evidence of a reservoir species and experimental data from Australia could not identify a carrier state (Lenghaus *et al.* 1994). Barlow and Kean (1998) considered that local persistence is unlikely to occur and that after fadeout, subsequent outbreaks are caused by re-infection. However, epidemics can exhibit very fine spatial heterogeneity, killing all rabbits in a warren, whilst leaving those in a neighbouring one unaffected; an unbroken 'epidemic wave' probably does not occur often (B. Richardson, personal communication). In the UK, data from Ramsey Island also support this view (R. Trout, personal communication). If RHDV is as robust as EBHSV (Box 5.7), then some burrows could act as an environmental reservoirs triggering another outbreak.

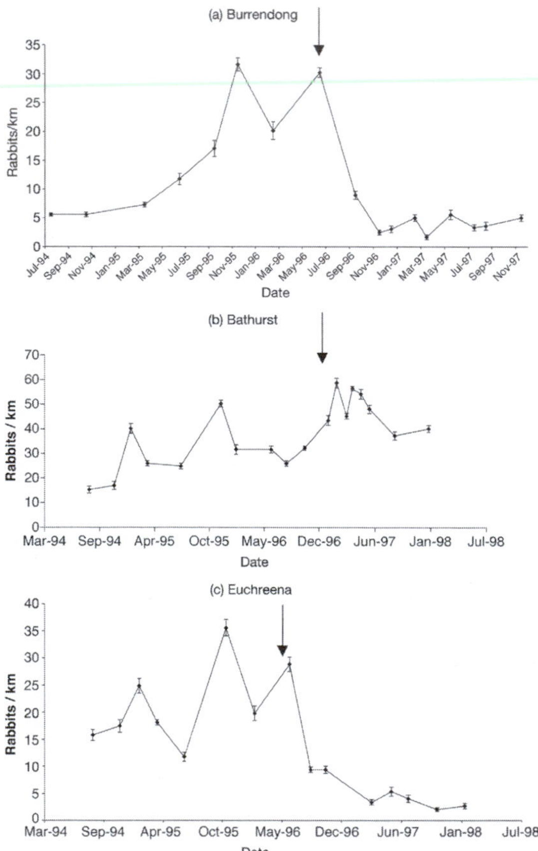

Figure 5.7 The impact of RHD on rabbit abundance. Figures show numbers of rabbits at three sites in central-western New South Wales, Australia in 1996–97. *Arrows* indicate the arrival of RHD. Infection appears to have little impact in site (b), the most densely populated: the difference may be due to the presence of juveniles (refractory to symptoms) at the time of the outbreak (after Saunders *et al.* 1999).

Persistence through re-infection from neighbouring populations will be facilitated by the high natural mortality of rabbits (typically 60% per annum in adults) leading to rapid replacement of immune individuals by Susceptibles. Additionally, maternal antibodies, which last for longer than the juvenile refractory period, may inhibit the development of immunity to RHDV, thus depriving the offspring of surviving females of the opportunity to become immune whilst refractory to symptoms, and so enhance the supply of new Susceptibles following an epidemic.

The robustness of the viruses means that they may be transmitted via a number of routes and apparently over long distances. Whilst experimental conditions have shown that it can be transmitted by close contact, indirect transmission probably predominates in the field. Due to rabbits' restricted home range, this is likely to be mainly short-range transmission. Short-range transmission could be vector-borne transmission via the flea, *Spilopsyllus cuniculi*, as well as direct transmission from cadavers and through the oro-faecal route. Vector transmission by mosquitoes is another potential transmission route that would operate over longer distances. However, since mortality is so rapid there is only a short window of opportunity prior to death during which the host is viraemic and so direct transmission from cadavers and via carrion-feeding animals may be more important. Infectious virus has been detected in the droppings of some predators.

5.4.4 Spatial structure and critical meta-population distributions

Host populations that are structured into sub-populations promote persistence of microparasites by allowing epidemics to occur asynchronously in the various sub-populations and avoiding deep global troughs (Bolker and Grenfell 1996). However, if the sub-populations are small and isolated, then infection to fadeout will occur and some form of intermediate coupling is required for persistence (Rohani *et al.* 1996). The notion of the critical community size, in fact, needs to be generalized to the critical meta-population distribution to accommodate the conditions when there are a number of different host patches of different sizes and connectivities. Indeed, this is particularly important for infections of wild-mammal populations that are frequently at low and variable densities, with low reproductive rates.

This approach may be particularly important to conservation problems (Chapter 6 and 8), such as canine distemper virus in black-footed ferrets (Forrest *et al.* 1988) or Ethiopan wolves (*Canis simensis*) in the Bali mountains (Laurenson *et al.* 1998) or rabies in the Serengeti (Gascoyne *et al.* 1993)—see Chapter 8. All these are situations in which generalist pathogens are freshly re-introduced to susceptible populations by chance or through changing species distributions (McCallum and Dobson 1995; Woodroffe 1998). Models, and their theoretical and empirical underpinnings, clearly need to be developed for these settings in which a single pathogen can affect several different species in many different patches.

5.4.5 Reservoir hosts and inter-species transmission

There are a number of key examples in which infection has spread from one host species within which it appears to be endemic to another in which it cannot persist, though there have been few systematic tests of such hypotheses. Cleaveland and Dye (1995) suggested three conditions to test the hypothesis that a reservoir host is acting as a source of infection to other species:

1 that reservoir host populations should show evidence of persistent infection;
2 that cases should occur in the reservoir host in the absence of cases in the other species;
3 that outbreaks in the other species should follow cases in the reservoir host.

On this basis, they demonstrated that domestic dogs were the likely reservoir of rabies infection in the Serengeti. Similarly, Rhodes *et al.* (1998) suggested, on the basis of a density model, that side-striped jackal (*Canis adustus*) populations in Zimbabwe were unable to sustain endemic infection with rabies and, by implication, that domestic dogs were re-introducing infection into the population.

Further examples in which inter-species transmission have been studied are brucellosis, bovine tuberculosis, and the transmissible spongiform encephalopathies. We now consider each of these pathogens.

5.4.6 Inter-species transmission: brucellosis

An example of inter-species transmission is provided by brucellosis, which is present in a number of well-studied wildlife and livestock systems, including bison in Yellowstone, USA, and chamois in the Italian Alps.

Brucella abortus has been present in the Yellowstone bison herd since its introduction by infected cattle in the early years of this century (Dobson and Meagher 1996; Meyer and Meagher 1997). Ironically, it is now perceived to be a threat to the local cattle industry (Baskin 1998). Determining the risk of transmission of *B. abortus* from wildlife to cattle requires a quantitative understanding of the transmission dynamics of pathogens within the host population. Three routes of transmission between bison have been described (Fig. 5.8), these are vertical (mother to calf), horizontal as a sexually transmitted disease, and what Heesterbeek has called 'diagonal' when an infected bison cow aborts and contaminates the environment (Williams *et al.* 1997). It is only this latter route that could transmit *B. abortus* to cattle but, in contrast to cattle, brucellosis-induced abortion is rare in bison (Meyer and Meagher 1997). In addition, the vertical and horizontal routes of transmission are frequency-dependent, hence systematic culling of the Yellowstone bison herd will not lead to the eradication of the disease.

By contrast with the Yellowstone experience, the primary issue with brucellosis in the Western Italian Alps is the risk from livestock to wildlife populations. The species of concern are the chamois (*Rupicapra rupicapra*) that suffered an outbreak between 1994 and 1996 in the Western Italian Alps. Since the first case in Switzerland in 1950, *B. abortus* infection has been reported from only six other animals, and it was suspected that infection arose from infected bovine herds. No other cases were recorded until 1988 when *B. melitensis* was isolated in the French Alps in an area of relatively high (10%) prevalence of brucellosis in sheep (Garin-Bastuji *et al.* 1990; Garin-Bastuji 1993). Since then a further ten cases have been recorded. The sporadic nature of brucellosis in chamois is confirmed by serological surveys that show infection is either sporadic (Corti *et al.* 1984) or absent (Tolari *et al.* 1987; Gauthier 1991; Gennero 1993)

Circumstantial evidence for inter-species transmission comes from the Susa valley where *Brucella abortus* was first recorded in chamois in December 1994. This followed an outbreak in a bovine herd in the valley and chamois were presumably exposed when they grazed on the contaminated pasture (Ferroglio 1998). Prevalence in chamois subsequently decreased from 28% in 1995, to 18% in 1996, and 8% in 1997. All the infected animals were adult and exhibited classic clinical signs of chronic disease. More males were positive than females, probably because males tend to use low pastures more than females and were more exposed to the infected pasture. Confirmation that chamois are just a spill-over host comes from the observation that when brucellosis is eradicated from livestock it also disappears from wildlife (Rementzova 1964; Leon-Vizcaino 1991).

The high mortality of the disease in chamois could be why chamois epidemics appear self-limiting (Bouvier and Burgisser 1958). However it is interesting to note that the infection also fades out in the solitary moose (Corner and Connell 1958; Forbes *et al.* 1996) but is endemic in many of the gregarious ruminants such as bison (*Bison bison*), elk (*Cervus canadiensis*), and caribou (*Rangifer tarandus*). This may be partly due to the higher rate of induced mortality but may also be because of increased transmission from aborted foetuses in the herding species. Indeed diagonal transmission could be the way to explain persistence in species where abortion occurs, such as elk and caribou (Thorne and Morton 1978; Tessaro and Forbes 1986).

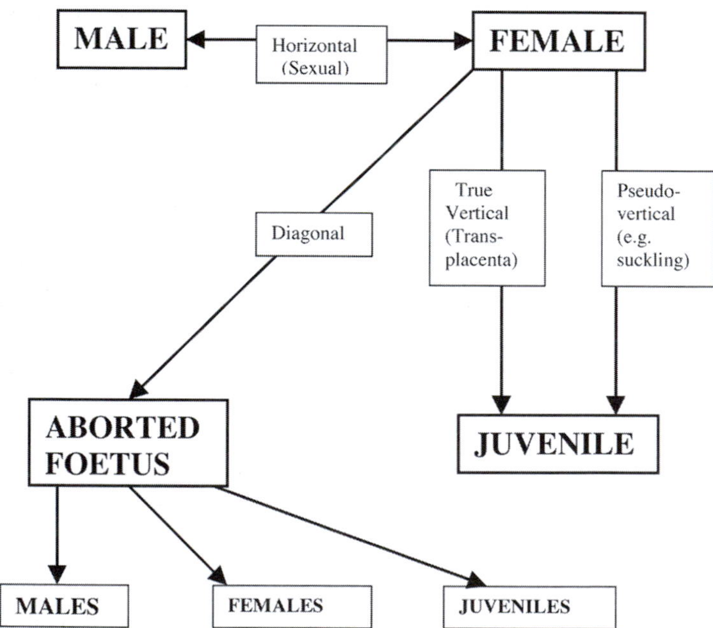

Figure 5.8 Transmission routes of *Brucella abortus* in bison. Note that cross-infection to cattle can not occur through the vertical or horizontal routes, and can only occur through diagonal transmission via abortion. Interestingly, the prevalence of abortion in bison is rare.

Brucellosis also provides an example of density-dependent transmission. Where elk receive supplementary feed, herd size may reach several thousand and the prevalence of brucellosis is high (37%, Herriges *et al*. 1992); while in areas where elk herd size is not artificially increased, the observed prevalence is less than 3% (Merrell and Wright 1978; Rhyan *et al*. 1997). For comparison, in Yellowstone, bison brucellosis is present whenever herd sizes exceed about 200 animals (Dobson and Meagher 1996).

5.4.7 Inter-species transmission and reservoirs: bovine tuberculosis

Another pathogen found in both wildlife and cattle is bovine tuberculosis. As tuberculosis control in cattle has become increasingly effective, areas of control failure have been attributed to wildlife reservoirs of infection, most notably in possums in New Zealand and badgers in the south-west of England (Box 5.8). One lesson that epidemiologists have learned in the past decades is how difficult it has been to identify any one 'reservoir' species with certainty: authors disagree, for example, on whether wild deer *Cervus elaphus* also function as a reservoir in New Zealand (Hickling 1995; Wobeser 1995). Emerging molecular techniques, have the potential to offer much in this area (Clifton-Hadley 1998), especially when combined with the rigorous approach of Cleaveland and Dye (1995) discussed in §5.4.5.

5.4.8 Persistence and environmental reservoirs

Baculoviruses persist in the natural environment between epidemic outbreaks in their invertebrate hosts, although the actual mechanism of persistence still remains a mystery (Box 5.9). New isolates of baculovirus are usually found when the host population is at high density, but how they persist during periods of low host-density is unknown. One possibility is that they may be present as sublethal or latent infections. There is limited evidence that a very slow, replicating infection persists in a laboratory culture of *Mamestra brassicae* (Hughes *et al*. 1993, 1997), though similar infections are yet to be found in field populations. Alternatively, they may frequently go locally extinct, but persist globally through the action of long-distance dispersers, such as birds (Entwistle *et al*. 1993). However, whilst it has been demonstrated that birds have the potential to disperse baculoviruses, few data are available to enable an assessment of how frequently this occurs. Even so, there is considerable empirical evidence that baculoviruses can persist locally in sheltered microhabitats, such as the soil or crevices on the bark of trees, for weeks, months, and even years (McLeod *et al*. 1982), they die quite quickly when exposed to sunlight. As such, local persistence is very much dependent upon microhabitat.

5.4.9 Long incubation period: transmissible spongiform encephalopathies

In contrast to environmental reservoirs, there are a wide range of biological reservoirs. Microparasites use their hosts as reservoirs in different ways. For example, an infection may be sustained simply as a chronic infection within a host, a strategy associated with sexually transmitted diseases (Smith and Dobson 1992), such as chlamydia in koalas *Phascolarctos cinereus* (Augustine 1998). A further strategy is to develop a long incubation period, a particularly good strategy in hosts with low mortality rate and in pathogens where no alternative biological or environmental reservoir is used. One group that has a long incubation period are the transmissible spongiform encephalopathies that are examined in this section.

Transmissible spongiform encephalopathies (TSEs) are neurological diseases affecting a wide range of mammals, including humans (Table 5.1). Those of major importance include: scrapie in sheep; bovine spongiform encephalopathy (BSE) in cattle; and Creutzfeld–Jakob disease (CJD), especially the 'new variant' form (vCJD) in humans. Transmissible mink encephalopathy (TME) is known as a disease problem on fur farms. The only known TSE in wildlife is chronic wasting disease (CWD) in mule, deer, elk, and white-tailed deer in Colorado and Wyoming in the USA. However, TSEs have been widely reported from zoo animals in the UK in recent years. These cases were mostly due to BSE transmitted via contaminated meat or commercial feedstuffs. BSE is

Box 5.8 Badgers, possums, and tuberculosis

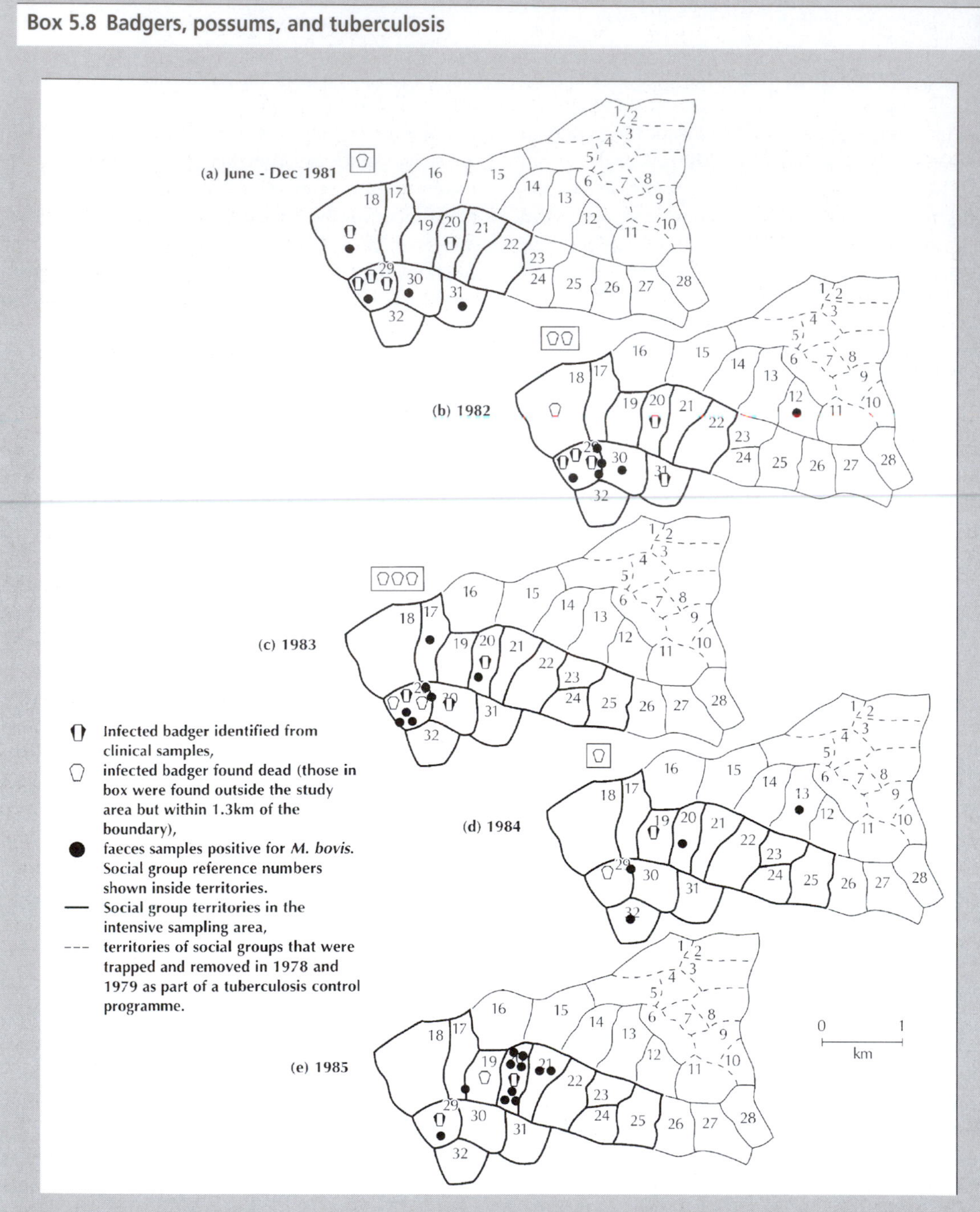

Figure 5.9 Spatial pattern of bovine tuberculosis at Woodchester Park, showing long-term infection localized within a few social groups (after Cheeseman *et al*. 1988a, b).

continued

Bovine tuberculosis (Tb) is caused by the bacterium *Mycobacterium bovis* and while the pathogen is of little threat to humans there is an international effort in Europe to try and eradicate the pathogen from domestic cattle herds. Overall, cases are relatively uncommon in cattle in Great Britain, although most cases are concentrated in the south-west of England and coincide with high densities of badgers *Meles meles*. More recently, the disease has spread spatially and moved into new parts of central and western Britain, causing increased concern amongst farmers. Cattle in other countries, notably New Zealand, the Republic of Ireland, Italy, and Spain also have *M. bovis*. In Britain, badgers have been implicated as a significant source of infection in cattle (Clifton-Hadley *et al*. 1995). Badger removal has been followed by reduced rates of herd breakdown and while the evidence is largely circumstantial and controversial, a committee of scientists found it compelling (Krebs 1997). Direct contact between badgers and cattle appears to be unusual, and pasture contamination with faeces and urine has been proposed as the main route of infection, simply because badgers shed large numbers of bacilli in their urine (Krebs 1997). While it is generally considered that the badgers are a source of infection, there is little evidence to quantify the extent to which badgers receive spill-over infection from cattle. Simple husbandry methods to separate badgers and cattle could have a significant role in reducing risk of infection to either or both species. However, other wildlife species may also carry the pathogen and be an important cause of initial breakdown. Interestingly, some badger social groups remain chronically infected for many years, while neighbouring groups remain uninfected (Smith *et al*. 1995). This pattern of temporally continuous but spatially patchy infection is thought to be caracteristic of maintenance wildlife hosts (Morris and Pfeiffer 1995).

In New Zealand the possum is the major reservoir, with deer, ferrets, and pigs. The transmission of tuberculosis between possums is believed to be largely due to den-sharing, fighting, and mating (Roberts 1996; Kao and Roberts 1999). As such, other mechanisms are necessary to explain transmission to cattle. In one experiment, a possum was semi-anaesthetized and introduced into a field of cattle (Sauter and Morris 1995). The cattle approached the possum and investigated it by nuzzling and licking, so it is likely that if the possum had Tb, the cattle would have become exposed. When the experiment was repeated with a semi-anaesthetized ferret, the cattle avoided it. These observations demonstrate a potential route of infection and imply that ferrets might be less important as a reservoir of infection.

particularly adept at jumping species. For example, there was an epidemic of feline spongiform encephalopathy in domestic cats in the United Kingdom associated with the BSE epidemic in cattle. However, there have been no reports of BSE infection in wildlife to date.

TSEs have long incubation periods and progressive pathologies and are invariably fatal once clinical signs have developed. There appears to be no effective immune response, there is no cure, and diagnostic tests for pre-clinical infections are still under development. The nature of the infectious agent remains controversial but the leading proposal is the 'prion' hypothesis that suggests that TSEs are caused by the introduction of an abnormal form of the prion protein (PrP), which spreads by recruiting normal protein molecules to the abnormal state. Certainly, the late pathology of TSEs is characterized by accumulating plaques of abnormal PrP in the tissues, especially the central nervous system. TSEs are transmitted by many and varying routes: some, such as scrapie in sheep, are clearly horizontally transmitted, although the precise route is uncertain and scrapie may also persist in the environment. BSE is transmitted via ingestion of infected tissues or of commercial feedstuffs prepared from infected tissues. Some TSEs might be vertically transmitted from mother to offspring and several TSEs have been transmitted iatrogenically (i.e. during medical treatments involving infected tissues).

The epidemiology of chronic wasting disease (CWD) is described in detail in Williams and Miller (2000) and Spraker *et al*. (1997). CWD was first recognized in 1967 in captive animals and subsequent outbreaks have affected up to 90% of adult animals at some sites. Forty-nine cases had been reported in wild animals by 1995; annual incidences are increasing but this may simply reflect improved reporting. Cases are confined to adults more than one-and-a-half years old. Transmission routes are uncertain but the high incidences in captive animals are suggestive of horizontal transmission—there is no evidence of exposure to contaminated feed and no evidence of a genetic component to susceptibility. Vertical transmission is possible but hard to distinguish from high rates of horizontal

Box 5.9 Modelling baculovirus persistence

Hochberg (1989) incorporated heterogeneity in baculovirus persistence in a model where the occlusion bodies (OBs) in the environment were divided between those on the leaf surfaces exposed to sunlight and hence with high mortality, yet available to infect susceptible hosts, and those in a protected microhabitat, which were not available to Susceptibles. The relative importance of the reservoir to persistence depended on the rate of movement of OBs from leaves to soil by rainfall or invertebrates, the input of OBs directly into the soil, for example, by infected insects falling off the plant, the loss of OBs through decay and then the translocation of OBs from the soil back up to transmissible surfaces. The presence of a reservoir was irrelevant if either the rate of translocation was so high that the OBs returned from the reservoir as soon as they entered it, or so low that the soil acted as a sink. However, at intermediate rates of translocation, the reservoir acted as a buffer. At times of high virus-density on the leaves, there was net movement into the protected habitat, so enhancing the persistence of the virus. Conversely, at times of low virus-density on the leaves, there was net movement out of the reservoir back to the leaf surfaces. This heterogeneity in the pathogen population stabilized the pathogen host system to produce more constant host abundance and persistence of the pathogen. This raises the question of whether the movement of pathogen particles out of the protected microhabitats back to transmissible surfaces is sufficient for the presence of a reservoir to have any impact on the dynamics. Manipulative experiments have provided estimates for the 'translocation rate' and suggested that under some conditions reservoirs can be ecologically important (Hails, unpublished). The study in question involved wild-type baculoviruses in agricultural habitats where the habitats are considered ephemeral and cultivation destroyed the reservoir each year. Even so, such virus movement could act to buffer the within season dynamics, and could be of greater importance in more stable habitats.

transmission. Transmission between captive and wild animals may have occurred at boundary fences. Consistent with the minimum age of cases, the incubation period is at least 1.5 years, based on the occurrence of cases after the introduction of new stock into a CWD-free population. Efforts to eradicate CWD from captive populations by culling and decontamination have failed and the need to limit the spread of CWD and have led to movement restrictions. It is not known whether CWD can persist endemically in wild populations.

Since the real impact of TSEs on individuals is so unclear, it is not surprising that we do not know whether they have a significant impact on population dynamics. Outbreaks of scrapie in domestic sheep can cause mortalities exceeding 25% per year (Elsen *et al.* 1999; Woolhouse *et al.* 1999). However, such high mortalities due to TSEs are unusual because the incubation period is typically long in relation to the host's life expectancy. Most infected animals, therefore, die before showing clinical signs. A crucial question is whether there is mortality associated with pre-clinical infection? Diagnostic tests for pre-clinical TSE infection are only now being developed.

In general, risk factors for the possible occurrence of TSEs in wildlife will reflect the transmission routes of these diseases. Cannibalism is clearly a potential risk factor. Also, feeding on carcasses and ingestion of placental tissue (which may apply to herbivores in nutritionally poor habitats) may facilitate TSE transmission. Cross-species transmission is obviously most likely to be possible from prey to their predators.

5.5 Synthesis

This chapter has attempted to survey a wide range of patterns of microparasite dynamics under the dual approach of transmission and persistence. We have argued that each can be associated with a threshold phenomenon: one for invasion, with deterministic R_0 based-analyses, and the second for fadeout, with stochastic analyses of the critical community size. These two kinds of threshold phenomena are undoubtedly both important but, as with human infections, invasion phenomena are easier to characterize theoretically. The two are alternative sides of the coin of the violent recurrent epidemic: probably rare in 'established' coevolved infections (with exceptions such as the morbilliviruses), but important in some novel pathogens.

Understanding these density-dependent infection processes is essential for a rational approach

Table 5.1 Species that have been naturally or experimentally infected with transmissible spongiform encephalopathies

Human	Cattle	Mule deer
Chimpanzee	Sheep	Elk
Macaque	Goat	White-tailed deer
	Moufflon	Nyala
Puma		Gemsbok
Cheetah	Rat	Arabian oryx
Ocelot	Mouse	Scimitar-horned oryx
Tiger	Hamster	Greater Kudu
Cat		
Mink		

to the practice of ecologically managing disease. For example, under what circumstances will infection be promoted rather than reduced by host culls because of compensatory reproduction as suggested in the classical swine fever example? Can the same kind of transmission scaling (approximately mass-action in population numbers) seen in the cowpox study also be expected for the insect baculoviruses that have no recovered or immune class and are indirectly transmitted? Developing this approach depends on improvement in biological understanding. For some pathogens like TSEs, many of the appropriate experiments are under way, but results may take years to obtain. Mathematical models can, therefore, be extremely valuable in identifying, or at least confirming, research priorities.

Another theme that emerged from the case studies is the difficulty of assessing the significance of inter-species transmission assigning species as 'reservoirs' of infection. There has been much recent progress in understanding the dynamics of deterministic models of multi-species transmission (Hudson and Greenman 1998) and there are many theoretical challenges in extending these insights to the stochastic models needed to consider the dynamics of fadeout and persistence.

CHAPTER 6

Spatial aspects of disease dynamics

G. R. Hess, S. E. Randolph, P. Arneberg, C. Chemini, C. Furlanello, J. Harwood, M. G. Roberts, and J. Swinton

Most host species are spatially distributed in patches that will influence the way in which disease spreads and how pathogens persist. How can landscape epidemiology improve our ability to predict, prevent, and respond to disease outbreaks? How has metapopulation theory enhanced our understanding of disease dynamics? How can landscape epidemiology and metapopulation theory be combined to increase understanding of the spatial aspects of disease dynamics? This chapter brings these spatial issues together and explores them in detail.

6.1 Background

Landscape epidemiology focuses on the biotic and abiotic factors that influence the manner in which disease agents spread spatially between sub-populations of hosts (Pavlovsky 1966; Meade *et al.* 1988). With the advent of advanced technology for gathering and analysing extensive data on the structure of the landscape and the way host populations use this landscape, coupled with a realization that spatial structuring is important, landscape epidemiology has blossomed as a field of study.

Advances in remote sensing and geographic information systems (Box 6.1) allow researchers to explain current distributions of disease and map sites of potential disease outbreak across large areas (e.g. Beck *et al.* 1994; Hay *et al.* 2000). In fact, mapping the spatial patterns of disease occurrence has a long history in epidemiology (Meade *et al.* 1988). One of the most famous historical examples is John Snow's 1854 map of cholera cases in relation to water pumps in London (Howe 1972). The clustering of cases around a particular water pump supported Snow's belief that cholera was a water-borne pathogen. His insistence that the handle of the pump be removed probably represented the first recorded instance of a biologically appropriate, and therefore effective, intervention to halt the transmission of a water-borne infection.

More recently, recognition of a fundamental biological association between disease and environmental variables has provided the key to understanding the distribution of disease agents, allowing public health practitioners to move beyond retrospective fire-fighting to prospective, preventative action. For example, recent research has led to an increased understanding of cholera on a global scale (Colwell 1998; Pascual *et al.* 2000). *Vibrio cholera* preferentially attaches to chitinous marine plankton, such as copepods, and can be detected in zooplankton in regions in which cholera is endemic. As Colwell put it, 'ocean currents sweeping along coastal areas thereby translocate plankton and their bacterial passengers'. By tracing the environmental factors responsible for copepod production, and establishing the remotely sensed correlates of these factors (Beck *et al.* 1995), ecological and epidemiological information have been brought together to allow predictive explanations and advance public health measures (Colwell 1998).

In conservation biology, metapopulation theory has been influential in the study of populations whose dynamics are influenced by their spatial structure (Hanski and Simberloff 1997). The coupling of many locally unstable populations reduces the risk of metapopulation extinction, because dispersal movements introduce asynchronies among the populations (Allen *et al.* 1993; Ruxton 1994). When a population becomes extinct, individuals dispersing from other populations can re-colonize the site it occupied. Similarly, some studies have highlighted how local dispersal between a sufficiently large number of non-persistent populations can result in metapopulation persistence (Hassell *et al.* 1991; Comins *et al.* 1992; Rohani and Miramontes 1995). Spatial structure has also been shown to enhance the coexistence of many competing species by creating stable spatial segregation that might, at extremes, give rise to small 'islands' within the habitat (Hanski 1983; Hassell *et al.* 1994; Tilman 1994). Noting that hosts provide patches for pathogens, mathematical epidemiologists have applied metapopulation theory to the analysis of disease transmission in spatially structured populations (Lawton *et al.* 1994; May and Nowak 1994; Grenfell and Harwood 1997; Nee *et al.* 1997).

In this chapter, we examine developments in landscape epidemiology that might allow epidemiologists to improve their prediction and response to disease outbreaks in wildlife populations. We also review the use of metapopulation theory to advance the understanding of the spatial dynamics of disease transmission and persistence. Finally, we consider how linking concepts from landscape epidemiology and metapopulation theory might further understanding of the spatial dynamics of infectious wildlife diseases.

6.2 Landscape epidemiology

Landscape epidemiology is a way of linking spatially structured wildlife populations with the transmission dynamics of the parasites and pathogens infecting them. The survival, reproduction, and dispersal of wildlife species is largely determined by the spatially variable, often fragmented, physical (abiotic) and biological (biotic) attributes of their habitat. The same is true for parasites, for which the individual hosts form an important element of their biotic environment. It is because of this close link between environmental conditions and the transmission dynamics of many parasites and pathogens that the term 'landscape epidemiology' has been coined, developing the idea that the landscape can be mapped in terms of spatially variable infection risk factors and disease incidence.

Any feature of the landscape that can affect the distribution and abundance of disease agents can be geo-referenced and manipulated within a computerized geographical information system (GIS) (Box 6.1). In addition to spatial variation, many factors vary temporally on differing scales: seasonal patterns, cycles over several years, random variation from year to year, and longer term trends. Static maps, therefore, are not enough. Definition of the relevant landscape characteristics must present temporal dynamics in ways amenable to analysis. This is particularly true of climate—the single most significant extrinsic factor driving the dynamics of directly transmitted wildlife disease systems.

Data on complex temporal weather cycles are too unwieldy in their raw state to form a useful layer in a GIS, but they can be reduced to biologically meaningful, uncorrupted summaries by methods such as temporal Fourier analysis (Chatfield 1980; Rogers and Williams 1994; Rogers *et al.* 1996). This procedure provides a habitat fingerprint in terms of the mean level, amplitude, and timing of the principal cycles of each abiotic factor (e.g. annual, biannual, triannual). Relationships between such factors can also be summarized within the GIS and used to characterize the landscape. For example, different vegetation systems show varying degrees of delay between rainfall and maximum photosynthetic activity. Typically, longer delays occur in farming systems (where farmers react to rainfall by sowing seeds) than in natural systems (where existing vegetation or seeds respond with new growth). This approach might also capture more subtle elements of fragmented landscapes.

Unfortunately, meteorological records are rarely available in sufficient detail, or at sufficiently fine temporal or spatial resolutions, for satisfactory use in landscape epidemiology. Satellites, on the other hand, provide complete cover of areas on continental scales in real time, recording reflected or emitted radiation from the earth's surface at various electromagnetic wavelengths (Box 6.1). A major advance

Box 6.1 Geographic information systems and remote sensing

The analysis of spatial patterns in ecology, including the distribution of disease, has been enhanced by the development of geographic information systems (GIS) (Haines-Young *et al*. 1993; Kitron 1998; Hay *et al*. 2000). These are computer-based systems for storing, manipulating, and displaying mapped data. A variety of data layers can be included: abiotic, such as climate, soil type, vegetation; biotic, such as the presence of an organism and its population size; and natural and anthropogenic features, such as mountains, rivers, roads, and housing (Hay *et al*. 2000).

Global positioning systems allow the location of each datum to be fixed precisely, so that the different data layers can be overlaid to detect relational patterns. For example, the habitat of a vector species and its relationship with human settlements can be mapped, thereby displaying the potential risk for disease transmission on a range of spatial scales from local through regional to continental (Fig. 6.1).

Once the data have been mapped and entered into a GIS, locational information can be extracted and submitted to statistical spatial analysis, and the processed data returned to a GIS for re-mapping. The patterns generated from these procedures often yield insight into the underlying biological mechanisms, which can inform more rigorous and versatile process-based analysis. For example, multivariate discriminant analysis shows repeatedly that temperature and moisture conditions are the most significant determinants of arthropod distributions. This directs attention towards these environmental variables as drivers of arthropod demographic processes.

Considerable effort has been directed at augmenting or replacing ground-collected climatic information with remotely sensed data for use in GISs (Rogers and Randolph 1991; Washino and Wood 1994; Kitron *et al*. 1996; Hay *et al*. 1996, 2000; Hay and Lennon 1999). Sensors mounted on an aircraft or satellite measure the electromagnetic energy radiated from a source or reflected from an object. Calibrating and interpreting remotely sensed data against corresponding ground-based observations have established clear relationships between biophysical variables and electromagnetic signals. A wide range of biophysical variables can be measured remotely, including elevation, chlorophyll absorption characteristics, biomass, temperature, surface texture, and moisture content.

With careful analysis, remotely sensed signals can be related more closely to the integrated conditions within the microhabitat by using measurements taken from standard weather stations. This can be important in epidemiological studies when climate influences transmission. Satellite imagery has the potential to provide far more detailed, real-time measures of a wide variety of environmental conditions on a regional or continental scale than can climatic data derived from widely scattered meteorological stations. For example, large spatial variability of rainfall across short distances makes remote sensing a more accurate predictor of local rainfall than spatial interpolation of data from meteorological stations (Hay and Lennon 1999).

Different types of satellites provide data with different uses and limitations (reviewed by Hay *et al*. 1996). Images are stored and transmitted as digital data, each value being a picture element, or pixel, that refers to a small area in the satellite sensor's field of view. Any given pixel represents the aggregate spectral signal from all sources within the pixel bounds. Satellite data volume is limited by on-board image storage facilities and by opportunities for transmission of data to receiving stations. As a consequence of these limitations, remotely sensed images have either high temporal or high spatial resolution, but not both.

The Landsat thematic-mapper (TM) and the *satellite pour l'observation de la terre* (SPOT) satellites have high spatial-resolution and low temporal-resolution. TM images have a pixel size of 30 m with a repeat-time of approximately 16 days; SPOT produces 10 m pixels with a repeat time of about 26 days. Frequent cloud contamination of the images results in few clear images of this sort each year. The absence of seasonal information, together with high cost, generally limits the biological application of high spatial-resolution imagery to the production of habitat maps for relatively small areas.

In contrast, the National Oceanographic and Atmospheric Administration (NOAA) series of meteorological satellites yield two global-coverage images per 24-h period, but with a maximum spatial resolution of 1 km. A Moderate Resolution Spectroradiometer (MODIS) sensor, with a spatial resolution of 0.25 km, was placed into orbit in 2000. The high-temporal resolution imagery from this sensor allows seasonal changes to be monitored in real time at a relatively fine spatial scale, offering huge potential for a number of biological applications, particularly those that explore dynamical biological processes.

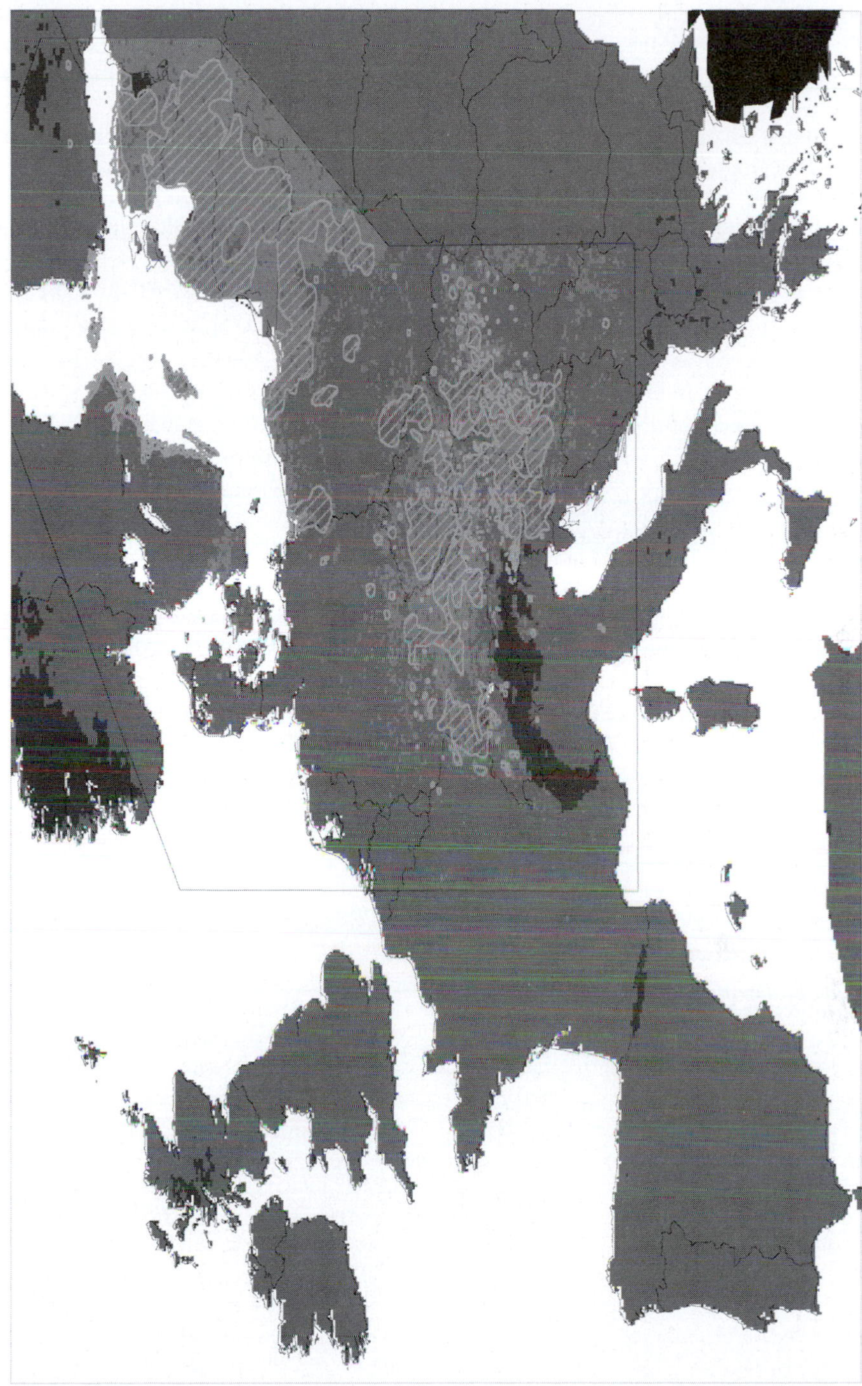

Figure 6.1 Predicted (red) and observed (yellow hatched) pan-European distributions of foci of tick-borne encephalitis virus based on analysis of remotely sensed environmental variables and elevation within the outlined area. The virus occurs extensively to the east of this area, but is not yet mapped in any detail. Frequent cloud contamination in high mountain areas (darker green) prevents analysis there (after Randolph 2000). See also Plate 1.

has been the quantification of correlations among data from meteorological satellites, ground measurements of environmental factors (e.g. saturation deficit, rainfall, and temperature), mortality rates, and population densities of arthropod vectors of parasites (Chapter 7; Rogers and Randolph 1991; Randolph 1994; Hay *et al.* 1996, 1997), and incidence of disease (Rogers 1991). This allows researchers to augment, or even replace, climatic data with satellite data, and to make epidemiological predictions across large geographical areas and into regions where appropriately monitored data on climate are unavailable. Predictions made on the basis of past conditions can be tested and updated as those conditions change.

6.2.1 Statistical and biological approaches

The colourful maps resulting from geographic information systems, whether or not they contain layers of remotely-sensed data, are often dismissed as no more than pretty pictures. This is unfortunate, because they contain vast quantities of biologically meaningful, digitized information. There are two ways in which these data can be harnessed for epidemiological purposes: statistical and biological. Statistical approaches seek correlations between environmental conditions and the distribution of disease agents; biological approaches identify causal biological processes that underlie observed patterns.

These complementary approaches are best illustrated with vector-borne diseases. Vectors, the crucial link in the chain of transmission, are subject to population processes that are well understood and quantifiable using methods from the historically and economically significant field of entomology (Varley *et al.* 1973). With a few notable exceptions (e.g. Colwell 1998), vector-borne disease systems are amongst the best-worked examples. Using statistical patternmatching, researchers have attempted to define the environmental characteristics of sites where vectors are known to occur and other sites where they do not occur (e.g. Rogers and Randolph 1993; Beck *et al.* 1994; Merler *et al.* 1996; Rogers *et al.* 1996; Estrada-Peña 1997). On the basis of multivariate analyses of randomly selected subsets of points throughout the known distribution of vectors, and approximately equal numbers of points from areas outside these distributions, the probability of vector occurrence can be predicted for large geographical regions.

Relatively simple multivariate techniques, such as discriminant analysis, have provided acceptably

Box 6.2 Classification-tree analysis of tick distributions

A distribution map for the tick *Ixodes ricinus* in Trentino, Italy, illustrates the landscape-epidemiology approach. Researchers from the ITC-irst and The Center for Alpine Ecology developed a risk-assessment system based on predictive classification tree models (Merler *et al.* 1996; Furlanello *et al.* 1997). This system has been used to predict areas likely to be infested by ticks based on a number of environmental factors: altitude, soil substratum, degree of exposure, vegetation, roe deer density and distribution. Tree-based classifiers have been used in medical decision-making for automated medical diagnosis, and can also be used in a predictive sense (Hand 1997). This is an important aspect of GIS-based epidemiology, in which predictions for extensive landscapes are extrapolated from intensive site-sampling.

The model for Trentino indicates that altitude is the strongest explanatory variable, with the presence of ticks decreasing above approximately 1020 m (see Fig. 6.2). Geological substratum (limestone preferred) and roe deer density are the next most important factors, confirming previous findings from a different data set (Merler *et al.* 1996; Furlanello *et al.* 1997). Altitude and geological substratum have a direct impact on conditions in the tick's microhabitat and therefore on rates of tick development and survival when they are not on a host. Deer are the major host for the adult stage of the tick and are essential for tick population maintenance. This model correctly classified 65% of sites where ticks were absent and 81% of those where ticks were present, with prediction at the spatial resolution of 50×50 m. New methods for statistical machine learning can be applied to GIS-based landscape epidemiology. Bagging (§2.5) and boosting techniques have resulted in classification accuracy of up to 81% for uninfested sites, and 81–89% at infested sites (Furlanello and Merler 2000).

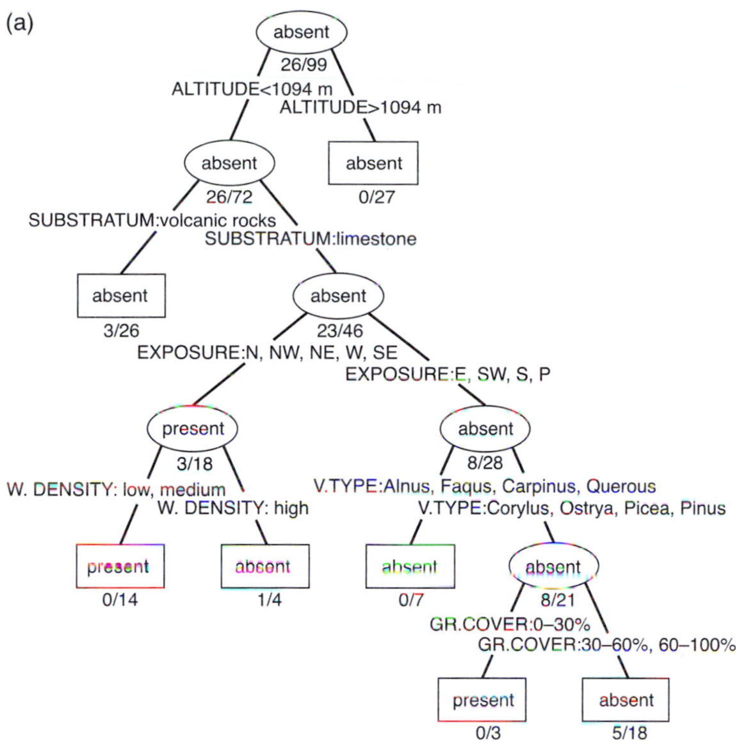

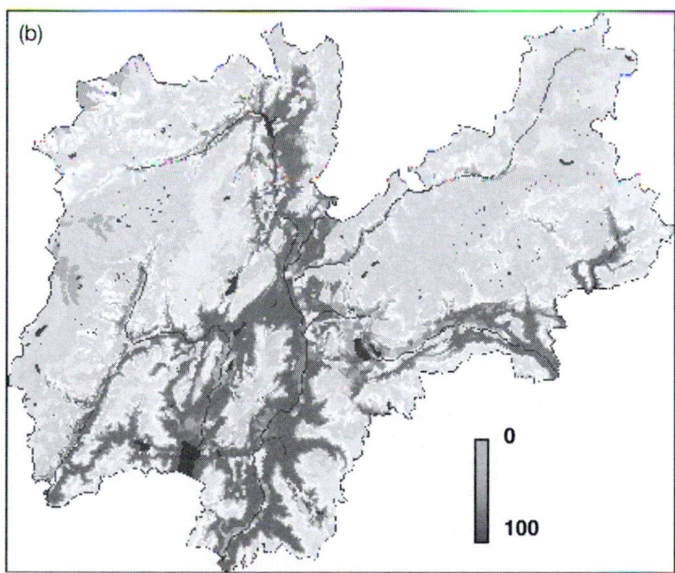

Figure 6.2 (a) Tree model of the ecological and environmental factors determining the distribution of ticks within Trentino; and (b) the final spatial GIS model produced using these data with an aggregation of 100 tree models. Colours from green to red indicate increasing probability of tick infestation; blue represents water. The risk map was computed at 50×50 m resolution in a window of 116×97 km. See also Plate 2.

accurate results. For example, model predictions were more than 80% accurate when compared to recorded distributions of ticks and tsetse flies across Africa at coarse spatial resolutions of 1×1 and 8×8 km (Rogers and Randolph 1993; Rogers and Williams 1993, 1994; Rogers *et al.* 1996). Using an alternative approach, classification tree models, Merler *et al.* (1996) and Furlanello *et al.* (1997) have predicted areas likely to be infested by the ticks *Ixodes ricinus* at the finer spatial resolution of 50×50 m, but over smaller areas (Box 6.2). These statistical methods can help identify the relative importance of a number of factors in determining distribution patterns and this can in turn help biologists identify process-based explanations.

6.2.2 The need for a biological approach

Statistical pattern-matching is a valuable tool in the first search for the factors that influence the spatial variation in risk of disease, but it is always limited by two fundamental problems. First, it offers little intrinsic understanding of the processes that generate the observed patterns. Second, such methods cannot give us measures of changing vector, pathogen, or parasite abundance.

At best, a predicted high probability of occurrence reflects high suitability of environmental conditions; so high-occurrence probability might correlate with high abundance. It is gratifying how such simple models capture the distribution patterns. On the other hand, it seems unlikely that statistical methods alone can capture the biological complexity that intervenes between the presence and abundance of certain vectors, and the prevalence of transmitted parasites and resultant disease (see Chapter 7). For example, the risk of infection by a tick-borne parasite is rarely a simple reflection of the presence and abundance of the vector in the way that it is for insect-borne parasites (Dietz 1982; Rogers 1985; Hudson *et al.* 1995).

The outcome of interactions between factors whose rates are changing simultaneously, frequently is unpredictable and sometimes counterintuitive. Unexpected correlations can develop between environmental variables and distribution patterns of vectors, parasites, and disease. Biological, process-based models address directly the demographic processes and their rates, which can then be built into the final pattern according to their interactions. They are also more versatile, allowing precise analyses and predictions to be matched to the questions posed.

6.3 Metapopulation modeling in epidemiology

Host animals provide habitat patches for their parasites and a number of authors (e.g. Lawton *et al.* 1994; May and Nowak 1994; Nee 1994) have recognized that epidemiological theory has addressed exactly the same issues as metapopulation theory. Metapopulation theory (Box 6.3) can provide insight into epidemiological processes at two spatial scales (Fig. 6.4; Grenfell and Harwood 1997). First, one may consider individual hosts as patches to be exploited by pathogens, so that the focus is on the movement of pathogens among individual hosts. Second, one may consider populations of hosts as patches. In this case, the focus of metapopulation theory is on the movement of infected hosts among populations (patches) and the consequent transmission of pathogens among hosts.

6.4 Individual hosts as patches

Lawton *et al.* (1994) noted that, if the species of interest is a parasite, each host organism may be considered as a habitat patch containing a local population of parasites. Infection is equivalent to colonization, and death or recovery of the host is equivalent to local extinction. Hence the similarity between the Levins (1969) equations used in metapopulation dynamics and many of the descendants of the logistic equation that are used in epidemiology for systems in which hosts can reacquire infection after recovery (Box 6.3; Grenfell and Harwood 1997; Nee *et al.* 1997).

A susceptible animal is an empty patch from the parasite's perspective; and the community of parasites in the host is the community that occupies that particular patch. Correspondence between metapopulation and parasitology terminology can be used to translate models between the disciplines (Table 6.1). There is also a correspondence between

Box 6.3 What is a metapopulation?

A metapopulation is a set of populations, distributed over a number of distinct patches that are connected to varying degrees by dispersal (Levins 1969, 1970; Hanski and Gilpin 1991; Hanski 1991; Hanski and Simberloff 1997). The dynamics of a metapopulation are a function of both within-population dynamics and among-population movement. At any point in time, a particular patch can be occupied or empty. The number of occupied patches in a system at equilibrium is a balance between patch population extinction and recolonisation of empty patches.

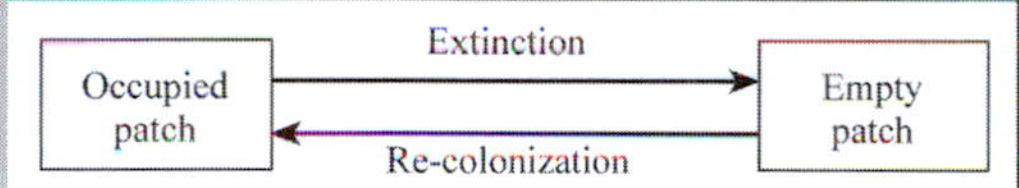

Figure 6.3

The rate at which the number of occupied patches changes is the difference between the recolonization rate and the extinction rate. In its most familiar form, Levins' (1969) model is:

$$\frac{dp}{dt} = mp(1-p) - xp, \tag{6.1}$$

where p represents the proportion of available patches that are occupied, t is time, m is the immigration rate, and x is the rate of local extinction. This equation has a single stable, non-zero equilibrium given by:

$$p^* = 1 - (x/m), \tag{6.2}$$

if $m > x$. All patches ultimately go extinct, if the local extinction rate exceeds the immigration rate. Levins' model depends on the following assumptions:

1 all patches have identical characteristics, identical within-patch population dynamics (which are ignored), and experience identical environmental conditions;
2 the number of patches is large enough that a deterministic model is appropriate;
3 all patches are equally accessible from all other patches;
4 patches that have undergone local extinction are available for immediate recolonization.

The structural equivalence between Levins' model and the logistic equation can be seen by rewriting the metapopulation equation as:

$$\frac{dp}{dt} = (m-x)p\left[1 - \frac{p}{1-(x/m)}\right]. \tag{6.3}$$

This is the familiar logistic form, with $m - x$ equivalent to r, the rate of increase when p is small, and $(1 - x/m)$ equivalent to K, the carrying capacity. Similarly, Anderson and May (1991) demonstrated the equivalence of their simple Susceptible–Infected–Removed (SIR) model and the logistic equation. It is this latter case that provides a potential linkage to a broad scale, landscape epidemiological approach.

Caveats on metapopulation theory

Metapopulation theory has replaced island biogeography as the central paradigm in conservation biology (Hanski and Simberloff 1997). In the process there has been a tendency to consider any population that is spatially structured as a metapopulation. Hanski and Simberloff (1997) have suggested that 'meta-population' be confined to populations that are spatially structured into assemblages of local breeding populations, and for which migration among the local populations has some effect on long-term dynamics, including the possibility of population re-establishment following extinction. The latter phenomenon is usually referred to as population turnover.

A number of authors (Harrison 1994; Thomas 1994; Harrison and Taylor 1997) have suggested that very few populations actually conform to the classical metapopulation ideal. They suggest that many so-called metapopulations are actually:

1 patchy populations, where migration is so high that there is little or no chance of extinction;
2 mainland-island metapopulations, where one or more patches contains a population that never goes extinct and acts as the main source for recolonization of vacant patches;
3 artificial metapopulations, which have been created by anthropogenic fragmentation of previously continuous habitat;
4 local populations, which track ephemeral patches of habitat where extinction coincides with the destruction of the habitat patch, so that there is no possibility of recolonization.

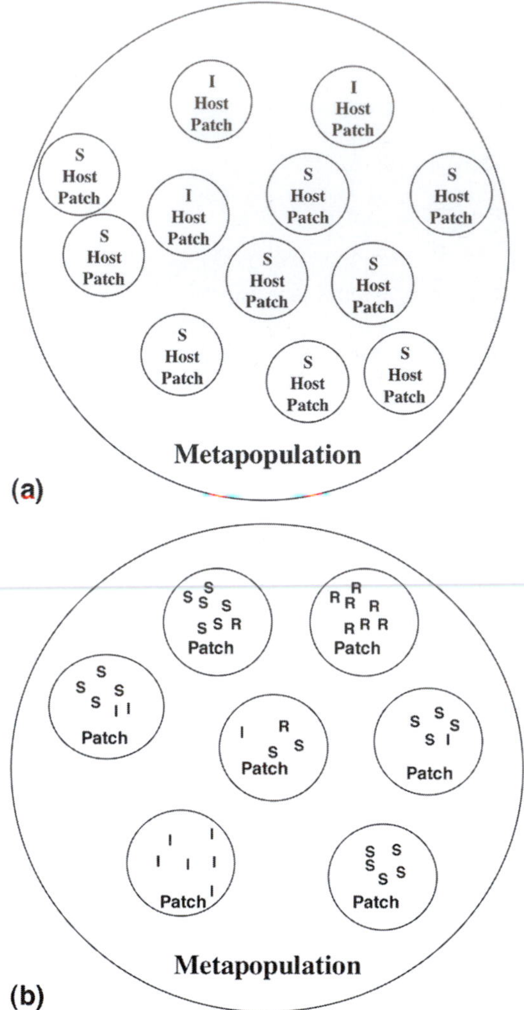

Figure 6.4 Metapopulation theory can provide insight into epidemic processes at two spatial scales. (a) Individual hosts as patches for populations of pathogens. Here, each host is occupied by a population of pathogens (I, for infected) or is pathogen-free (S, for susceptible). The metapopulation is the set of hosts. Colonization is the appearance of a pathogen population on a susceptible host. Extinction is the disappearance of the pathogen population from an infected host. (b) Populations of hosts as patches. Each patch is occupied by a population of susceptible (S), infected (I), and recovered and immune (R) hosts. The metapopulation is the set of host populations. Colonization is the appearance of infected individuals in a population. Extinction is the recovery of all infected individuals in a population. A population of recovered and immune hosts cannot be recolonized (re-infected) until more susceptible hosts are born or immigrate into the population.

the dynamics of different metapopulation models and the dynamics of various disease systems. For

Table 6.1 Correspondence between terms from metapopulation theory and epidemiology when considering individual hosts as patches

Metapopulation dynamics	Parasite population dynamics
Habitat patch	Host animal; infra-population (Sousa 1994)
Empty patch	Susceptible animal
Occupied patch	Infected animal
Colonization	Infection
Extinction	Recovery or death
Migration	Transmission
Metapopulation	Host population; metapopulation (Sousa 1994)

example, Gotelli (1991) examined four simple metapopulation models from the perspective of the origin of colonists (i.e. from within or outside the metapopulation) and rescue effect (Table 6.2). Here, the term 'rescue effect' is used to describe immigration of individuals into a patch on the brink of extinction (i.e. with a small population; Brown and Kodrick-Brown 1977). Gotelli's (1991) propagule rain models, in which colonists originate from outside the metapopulation, are analogous to a disease system with reservoir hosts. Further, the derivation of the formula for the minimum amount of suitable habitat (Hanski *et al.* 1996) required for metapopulation persistence is identical to that used to determine the proportion of a host population that must be vaccinated in order to eradicate a parasite (Lawton *et al.* 1994; Nee 1994).

6.4.1 Patch occupancy and patch density

Patterns of patch occupancy exhibit similarities between parasites and free-living animals in ways that suggest the relationship between metapopulation theory and epidemiology might be meaningful. For example, host population density is an important determinant of worm intensity of infection in mammals (§4.8.1). There is a positive relationship between host density and parasite prevalence (patch occupancy), because colonization of patches (transmission) increases with increasing host density (§4.8.1; Arneberg *et al.* 1997). Parallel patterns have been found in metapopulations of free-living

Table 6.2 Correspondence between metapopulation and disease dynamics when considering individual hosts as patches. Four models examined by Gotelli (1991) are presented. In all models, p is the proportion of available patches that are occupied, t is time, m is the immigration rate, and x is the rate of local extinction

Model	Equilibria	Stable when	Metapopulation dynamics	Disease dynamics
$dp/dt = mp(1-p) - xp$	$p^* = 0$ $p^* = 1 - (x/m)$	$x > m$ $x < m$	• Recolonization from inside metapopulation • No rescue effect—populations go extinct (temporarily) even if many patches are occupied	• Hosts only infected by pathogens from within host population • Pathogen populations go extinct on hosts (temporarily), even if pathogen prevalence is high
$dp/dt = mp(1-p) - xp(1-p)$	$p^* = 0$ $p^* = 1$	$x > m$ $x < m$	• Recolonization from inside metapopulation • Rescue effect—immigrants can save populations from extinction	• Hosts only infected by pathogens from within host population • Immigrating pathogens keep hosts infected, if many hosts are infected
$dp/dt = m(1-p) - xp$	$p^* = m/(m+x)$	$m,\ x > 0$	• Recolonization from outside metapopulation (mainland source) • No rescue effect—populations go extinct (temporarily) even if many patches are occupied	• Hosts infected by pathogens coming from outside the metapopulation (e.g. reservoir hosts) • Pathogen populations go extinct on hosts (temporarily), even if pathogen prevalence is high
$dp/dt = m(1-p) - xp(1-p)$	$p^* = m/x$ $p^* = 1$	$x > m$ $x < m$	• Recolonization from outside metapopulation (mainland source) • Rescue effect—immigrants can save populations from extinction	• Hosts infected by pathogens coming from outside the metapopulation (e.g. reservoir hosts) • Immigrating pathogens keep hosts infected, if many hosts are infected

animals. Among such animals, the fraction of patches occupied (prevalence) often increases with decreasing patch isolation (increasing host density) (e.g. Smith 1980; Harrison *et al.* 1988; Thomas and Jones 1993). These patterns have been found for a wide range of organisms, including insects and mammals. The biological mechanism appears similar to variation in parasite transmission rates: distantly located (low density) patches are less frequently occupied because they are harder to colonize (Hanski 1994; Hanski and Gilpin 1997). Such similarities between free-living animals and parasites indicate broad generalizations about the role of patch (host) density in determining population dynamics of animals (parasites) exploiting fragmented habitats (Nee 1994; Nee *et al.* 1997).

6.4.2 Parasite load distributions explained by metapopulation dynamics

An element that is sometimes incorporated in metapopulation models, but not in epidemiology, is the 'rescue effect' (Brown and Kodrick-Brown 1977). The rescue effect is a decreasing rate of extinction of sub-populations in patches with an increasing fraction of patches occupied. This mechanism has profound effects on metapopulation dynamics, producing a strong propensity for a species to occupy either most or very few of the patches. Equivalently, when looking at parasites and considering each individual host as a patch, parasite species tend to occur with either high or low prevalence (Nee *et al.* 1991; Hanski and Gyllenberg 1993). Bimodal distributions of prevalence have been documented within samples of large taxonomic groups and within local macroparasite communities (Fig. 6.5; Stock 1985; Bush and Holmes 1986; Stock and Holmes 1988; Esch *et al.* 1990b).

Additionally, when no single patch within the metapopulation produces sufficient emigrants to recolonize empty patches, species occupying few patches are predicted to go extinct unless they are saved by immigration from outside the metapopulation (Hanski and Gyllenberg 1993). This is potentially the situation for parasites, because the hosts (i.e. patches) eventually die. An examination of the prevalence of 89 nematode populations in mammalian hosts reveals a bimodal distribution (Fig. 6.5). Parasite populations with either high or low prevalence are observed significantly more frequently than populations with intermediate values.

Two other explanations have been proposed for bimodal distributions of parasite prevalence. One is that bimodal distributions are sampling artifacts (Nee *et al.* 1991), another is that they are generated by species having different niches (Brown 1995). However, neither of these explanations predicts that low prevalence populations should disappear when communities are isolated from other communities. Determining which mechanisms generate these patterns is a key question in epidemiology. If bimodal distributions of patch occupancy are

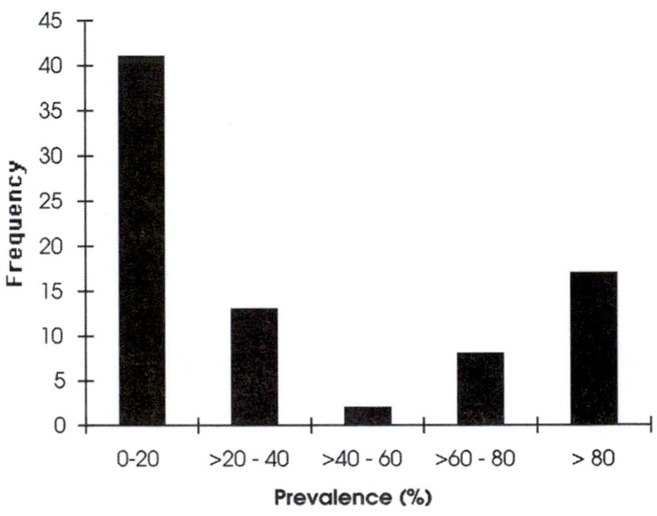

Figure 6.5 Frequency distribution of nematode prevalence in species of mammals. The distribution can be considered bimodal, because there are more observations in the lowest and highest classes than expected. The area across which hosts are sampled plays a critical role in uncovering this bimodal distribution. Among mammalian nematodes, the bimodal distribution of prevalence is seen only when using data from sampling areas that are small relative to the host's home range size. Data from larger sampling areas provide no evidence of bimodality, consistent with predictions from metapopulation theory (Hanski and Gyllenberg 1993). All data represent a complete examination of nematodes in the gastrointestinal tract of 12 host species, each sampled with at least 30 individuals.

generated by metapopulation dynamics among parasites, parasite populations occupying a low proportion of patches are predicted to be extinction prone. If so, two factors can be identified as important determinants of persistence in parasite populations. First, the biological mechanisms generating the rescue effect will affect whether a population occupies a low fraction of the patches and thus render it extinction prone. Second, the nature of spatial dynamics will affect whether extinction prone populations are saved by immigration from elsewhere.

6.5 Populations of hosts as patches

If we consider groups of hosts to be patches, the metapopulation analogy applies very well to parasites that invoke host immunity. For example, the spatial organization of human hosts into well-defined social units—families, villages, towns, and cities—produces well-defined patches of resource for their microparasites (Grenfell and Harwood 1997). Once a population recovers from a disease, it cannot be re-infected until more susceptible hosts enter by birth or immigration. Grenfell and Harwood (1997) have used measles in groups of human hosts as an example of the application of this perspective in epidemiology. The spread of phocine distemper virus among mammals of the order Pinnipedia is also amenable to a metapopulation approach (Boxes 6.4 and 5.2).

6.5.1 Movement among host populations

Metapopulation theory has become the basis for a number of widely used and promoted approaches in wildlife conservation that increase movement among populations. For example, the design of conservation corridors: strips of suitable habitat between sub-populations. Other conservation efforts that increase the movement of individuals among populations include captive breeding and release programs (Ballou 1993), and translocation of individuals among populations (see Table 8.3; Griffith *et al.* 1993). Although almost all metapopulation models suggest that increased movement among populations reduces the probability of metapopulation extinction (e.g. Hanski 1985; several papers in Saunders and Hobbs 1991), few have examined the possibility of any negative effects with increased movement (Henien and Merriam 1990; Hess 1996a, b). Since animals moving among patches carry pathogens with them, increasing colonization rates might also have a cost. Incorporating some negative effects of movement among populations can alter predictions from metapopulation models under certain circumstances (Hess 1996a).

The potential risk of spreading disease among populations depends upon a number of factors, including the nature of the pathogen, mechanisms by which infection is maintained the relative frequencies of transmission between and within popu-

Box 6.4 Metapopulation modelling of phocine distemper virus epidemics

The phocine distemper epidemic in the harbour seal (*Phoca vitulina*) population of the North Sea (see also Box 5.2) provides an example that can be modelled in a metapopulation framework in which populations of hosts are considered patches. Suitable habitat patches ('haulouts') are discrete and easily defined, and bear some relationship to breeding groups. They are surrounded by habitat that is unsuitable for breeding and disease transmission (ocean and human-occupied coastline). Colonies go extinct from time to time, but sites can be re-established. Each breeding group of seals constitutes a metapopulation patch that can be colonized by the virus. The availability of mortality data from a range of locations around the North Sea makes it possible to parameterize a metapopulation model (see Fig. 6.6). Swinton *et al.* (1998) investigated the metapopulation dynamics of the phocine distemper virus within such harbour seal communities. They concluded that a very large population—much larger than the known population—would be required for phocine distemper to persist in the North Sea, even with a metapopulation structure for the host. The interplay between the rapid fadeout of infection, as captured by a stochastic model, and the slow birth rate for new Susceptibles, makes persistence unlikely. This example illustrates the possibility of combining metapopulation dynamic models with those allowing the representation of stochastic fadeout (Chapter 5) to generalize the idea of a critical community size to that of a critical metapopulation distribution.

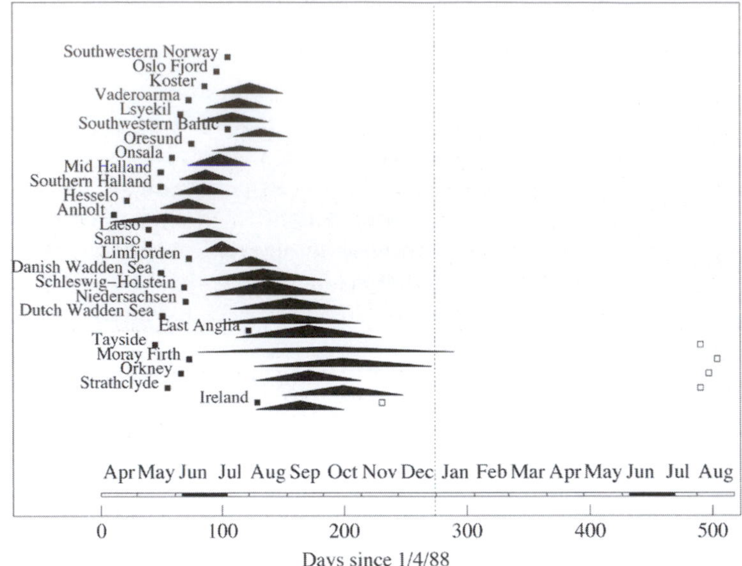

Figure 6.6 Spatial spread of phocine distemper virus showing both local and global fadeout. The graph shows reported mortality from the 1988 North Sea PDV epidemic (Box 5.2): *closed box*, first recorded case in each location; *triangle*, centred on peak reporting time, with width equal to length of period in which mid-90% of cases recorded and height proportional to logarithm of total number of cases; *open box*, marks last recorded case in those location where known (in other locations cases are those up to December 1988). Shaded period of month bar is approximate pupping season (after Swinton *et al.* 1998).

lations, and the frequency of dispersal events. The introduction of emerging diseases into a metapopulation setting, in which infected individuals can move freely among populations, could affect negatively the entire metapopulation. Of less concern is the spread of a generalist pathogen maintained in a widely distributed reservoir population in which the frequency of transmission from a reservoir to endangered populations is higher than transmission among populations comprising the metapopulation (e.g. rabies in domestic dog/wild dogs).

Hess (1994, 1996a) incorporated elements of mathematical epidemiology into theoretical metapopulation models by considering two types of occupied patches: those containing only susceptible individuals; and those containing infected individuals (Box 6.5). He used both analytic and simulation approaches to determine the qualitative conditions under which disease might threaten a metapopulation, to determine if the spatial configuration of a metapopulation affects disease dynamics, and to examine the effect of a quarantine on the impact of disease. The models are appropriate to a host-specific disease in a metapopulation setting. They do not reflect the dynamics of diseases maintained in reservoir hosts, although they could be modified easily to do so, e.g. by using Gotelli's (1991) propagule rain term for recolonisation.

Results from the models developed by Hess (1994, 1996a) suggest that highly contagious diseases of moderate severity spread widely, increasing the probability of metapopulation extinction. The simulation model demonstrated that disease might be difficult to control, if infected individuals can move freely among many populations. However, a centralized, single-population quarantine yielded dramatic reductions in metapopulation extinction probability, if movement among patches was more constrained, such as in a hub-and-spoke or stepping stone arrangement (Fig. 6.8).

Yeh (1998) added host genetic-resistance to Hess' (1996a) simulation model using a simple one locus, two allele, genetic system in which the recessive genotype was resistant to the disease. Each individual was considered bisexual, but no self-mating was allowed. The initial gene frequencies in the metapopulation were in Hardy–Weinberg equilibrium and each patch had the same number of resistant individuals. Since the population size in each patch was small, the effect of genetic drift was accounted for using the formulas derived by Tyvand (1993) to calculate the genotype frequencies of each new generation. Results from this model indicated that host resistance to disease can reduce the probability of metapopulation extinction,

Box 6.5 Incorporating disease dynamics into a simple metapopulation model

Hess (1994, 1996a) incorporated disease dynamics into Levins' (1969) metapopulation model by considering two types of occupied patches: those containing only susceptible individuals and those containing infected individuals. This allowed examination of the infection and recovery process.

The disease was represented by having a higher extinction rate for infected patches than for susceptible patches. Because two types of occupied patches are considered, two equations result:

$$\frac{dS}{dt} = mS(1 - I - S) - Sx_S - m\delta IS \qquad (6.4)$$

$$\frac{dI}{dt} = mI(1 - I - S) - Ix_I + m\delta IS. \qquad (6.5)$$

Here, S is the proportion of susceptible patches, I the proportion of infected patches, x_S is the extinction rate for susceptible populations, and x_I is the extinction rates for infected populations. A susceptible patch becomes infected, with probability δ if it receives immigrants from an infected patch. In this set of equations, infected populations do not recover. Note that $1 - I - S$ is the proportion of empty patches. The assumptions associated with this model are reasonable when the transmission process within each population is fast relative to transmission among populations.

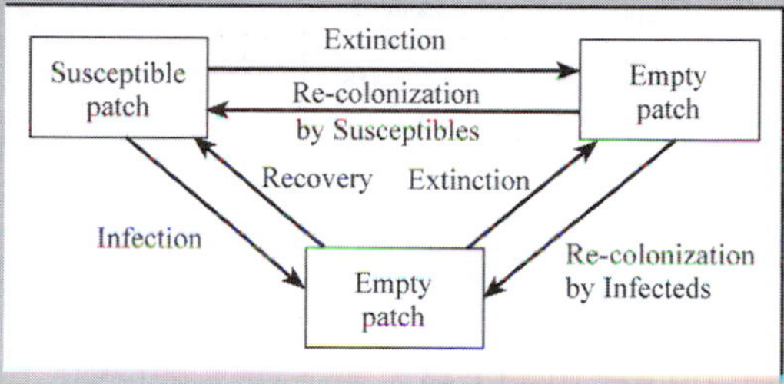

Figure 6.7

if the resistant gene is present above a threshold level.

With further research, metapopulation approaches such as those highlighted here can be used to increase understanding of spatial dynamics of wildlife diseases. For example, how does connectivity among populations affect the spread of disease maintained in reservoir hosts? What happens when the rate at which infections are introduced from the reservoir hosts varies among the connected host populations? The answers to these questions can help direct control strategies when host populations are geographically disjoint, but connected by occasional movement among them.

Metapopulation models can also be used to examine how different inter-patch connection patterns affect global properties, such as the fadeout of infection (Chapter 5), and how different sampling strategies for observing such connection patterns bias predictions for global properties. Such an examination would be a useful theoretical complement to empirical mark–release studies. For example, transmission models characterized by the 'small-world' property—that despite strong local aggregation, most individuals are only a few transmission steps from any other—have rather different transmission dynamics than models using more regular lattice structures (Watts and Strogatz 1998). It is not clear at present how to tell from an empirically realistic sampling strategy whether or not a network possesses the 'small-world' property.

6.6 Synthesis

Many wildlife populations live in spatially heterogeneous landscapes; some can be described as metapopulations, although there are some important caveats (Box 6.3). The dynamic structure of these landscapes, particularly as a result of fragmentation (e.g. Saunders *et al.* 1991) and other anthropogenic changes, impacts the ability of wildlife populations to survive and move in the landscape. Landscape changes, including changes in the location and size of wildlife populations, also affect the distribution of pathogens. Thus, there is an important link between the dynamics of host and pathogen populations and the dynamics of landscape change. The dynamics of this link are not captured in either landscape epidemiology or metapopulation approaches alone. This deficiency might be addressed by combining the two approaches, and using landscape and global

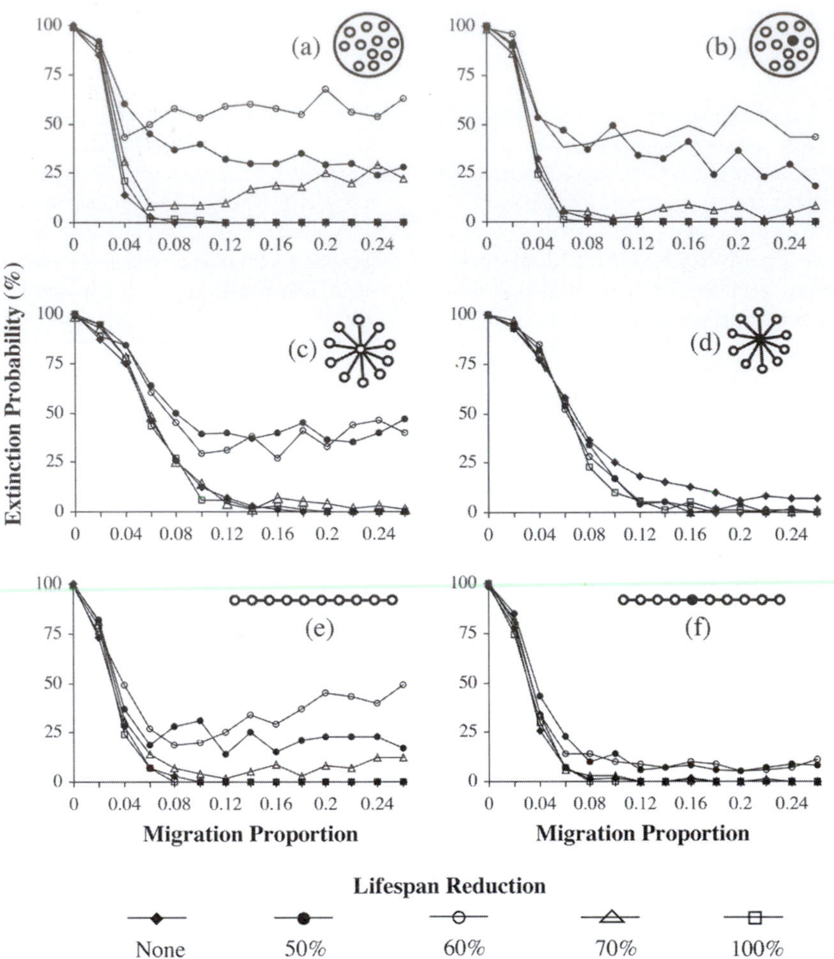

Figure 6.8 Probability of metapopulation extinction within 150 time-units plotted against the proportion of individuals moving during each time step. Three spatial configurations are shown: island (a, b), hub-and-spoke (c, d), and stepping-stone (e, f). Each configuration is shown without (a, c, e) and with (b, d, f) a quarantine patch (in *black* in the inset figures in b, d, f). A range of life-span reductions caused by the disease is plotted; life-span reductions <50% and >70% are not plotted, because they cluster on the line for no lifespan reduction (after Hess 1996).

change models to drive landscape dynamics (Fig. 6.9).

Landscape epidemiology provides a methodology for mapping the spatial distribution of conditions favourable to infection, based on biotic and abiotic parameters relevant to pathogen presence and abundance. The data for these maps increasingly are produced using remote sensing technology. To a large degree, the same remotely sensed data, with the application of different classification techniques, are used to map wildlife habitat. Wildlife habitat maps can form the basis of spatially explicit metapopulation models (e.g. Bowers and Harris 1994; Dunning *et al.* 1995).

Both metapopulation and epidemiological theories are concerned with how extinction and recolonization determine the persistence of patchily distributed populations. Metapopulation theory provides a framework for linking the dynamics of spatially separated populations of both pathogens and hosts. It may be applied to epidemiological processes hierarchically—one can examine the movement of pathogens among populations of hosts at a large spatial scale and the movement of

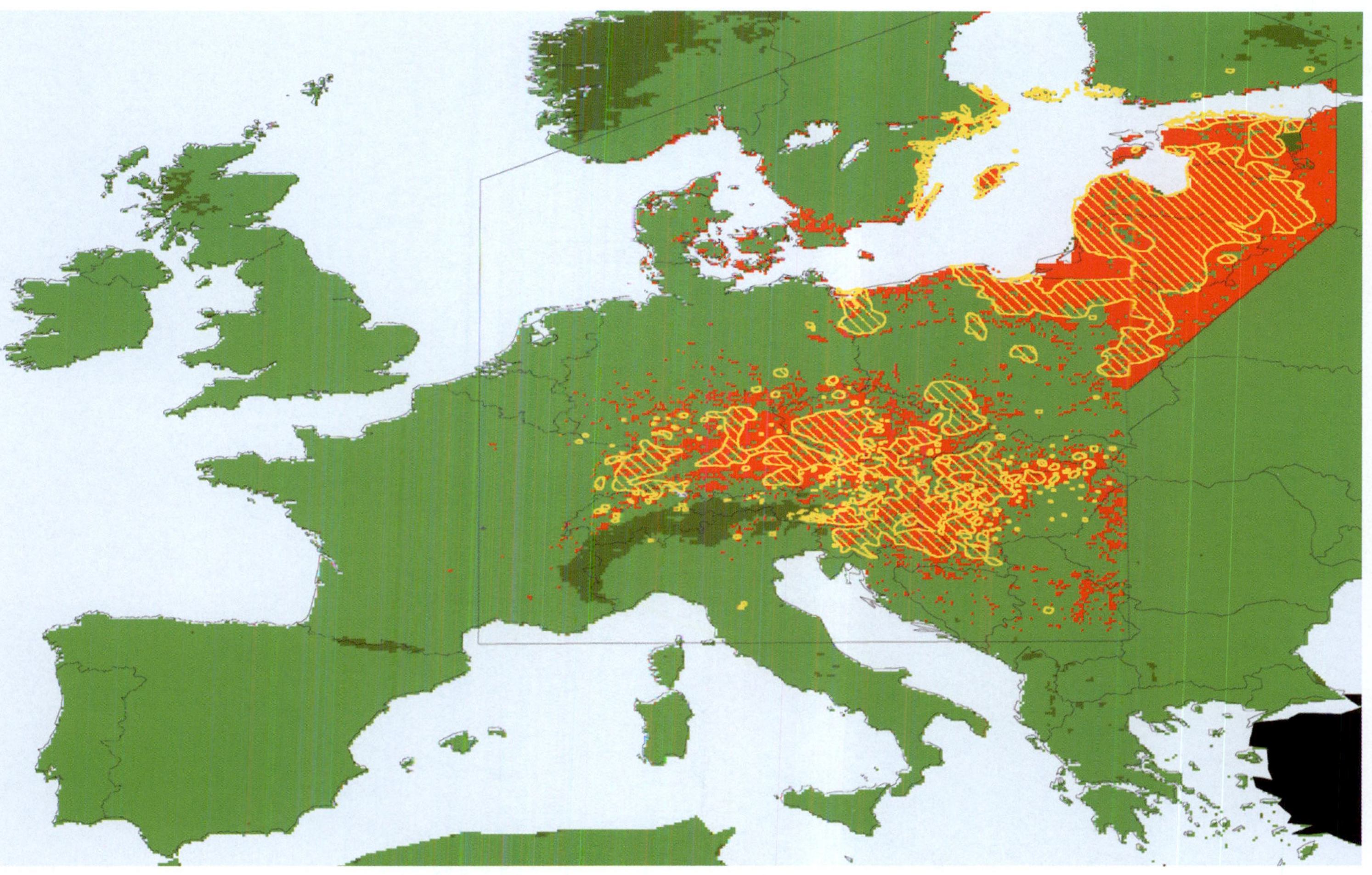

Plate 1 Predicted (red) and observed (yellow hatched) pan-European distributions of foci of tick-borne encephalitis virus based on analysis of remotely sensed environmental variables and elevation within the outlined area. The virus occurs extensively to the east of this area, but is not yet mapped in any detail. Frequent cloud contamination in high mountain areas (darker green) prevents analysis there (after Randolph 2000). See also Figure 6.1.

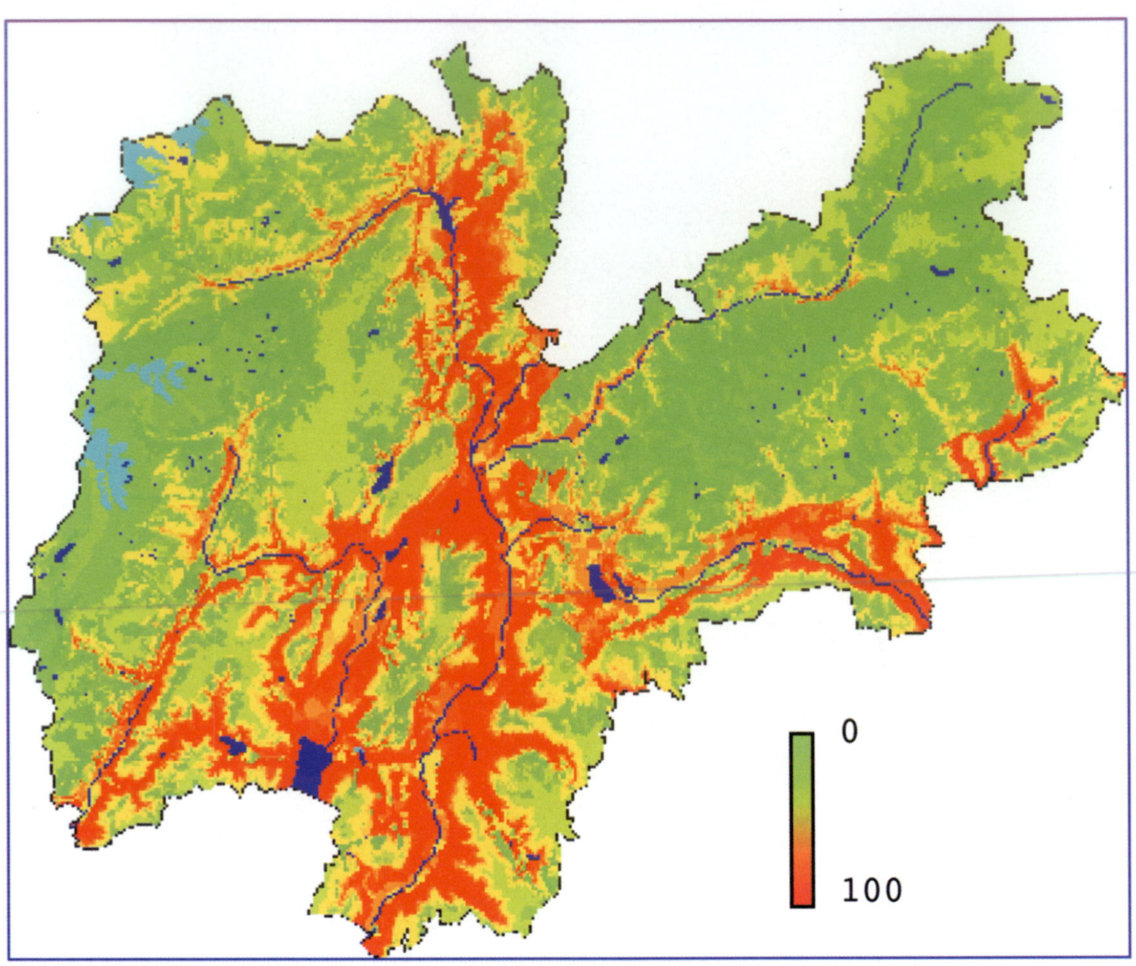

Plate 2 The final spatial GIS model produced using these data with an aggregation of 100 tree models. Colours from green to red indicate increasing probability of tick infestation; blue represents water. The risk map was computed at 50 × 50 m resolution in a window of 116 × 97 km. See also Figure 6.2(b).

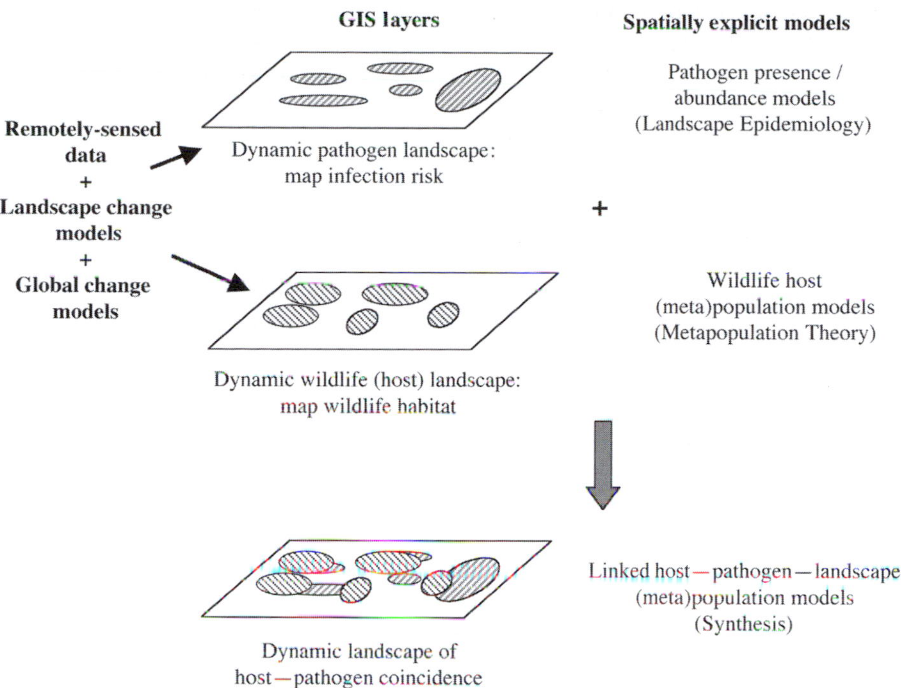

Figure 6.9 A conceptual framework for linking landscape epidemiology and metapopulation theory. Remotely sensed data are used to create geographic information system layers of pathogen and host habitat landscapes. Landscape and global change models drive landscape dynamics, and metapopulation-based epidemiological models are used to simulate host–pathogen dynamics.

pathogens among individual hosts within a single population (i.e. Fig. 6.4). Various forms of the recolonization and extinction terms can be used to examine the effects of transmission dynamics on the spread and persistence of diseases (Table 6.2).

Landscape and global change modelling are increasingly vital fields, as researchers seek to understand the causes and ecological implications of changing conditions (e.g. Neilson *et al.* 1992; Baker *et al.* 1992; He *et al.* 1999). Since changes in both climatic conditions and landscape structure will affect populations of pathogens and their hosts, any evaluation of the long-term effects of pathogens on wildlife populations should include consideration of these processes. Landscape and global change models can be used to simulate changes in the environment in which pathogens and hosts interact.

At present, statistical patternmatching methods are the most reliable and inclusive way of identifying the likely impact of present and future multivariate climatic conditions on the potential geographical range of parasites (Rogers and Randolph 2000; Randolph and Rogers 2000). The ideal, however, is to develop process-based population models, which are in any case a necessary foundation of disease transmission models. For example, population models that predict patterns of tick seasonal dynamics on the basis of geographically variable climatic conditions, address both the spatial and temporal aspects of infection risk (Chapter 7; Randolph and Rogers 1997). As long as they are designed to capture the interactions of pathogens with their hosts and their abiotic environment, process-based models will capture the spatially variable patterns of pathogen seasonal dynamics across extensive areas.

The spread of disease among wildlife populations across broad landscapes can be modelled by integrating such single-population models into a spatially explicit, metapopulation construct, based on maps produced by landscape-epidemiological analyses and maps of wildlife habitat. With landscape and global change models driving changes in the landscape through time, in ways that are meaningful to pathogen and host population growth, it should be possible to develop dynamic, linked host–pathogen–landscape metapopulation models (Fig. 6.9). Such models would allow analysis of the spatial dynamics of disease within the context of spatially dynamic landscapes, potentially bringing more realism to the analysis of control strategies for infectious diseases of wildlife.

CHAPTER 7

The ecology of tick-borne infections in wildlife reservoirs

S. E. Randolph, C. Chemini, C. Furlanello, C. Genchi, R. S. Hails, P. J. Hudson, L. D. Jones, G. Medley, R. A. Norman, A. P. Rizzoli, G. Smith, and M. E. J. Woolhouse

Ticks feed on a wide variety of vertebrate hosts and transmit a range of parasites that cause mortality and suffering to both livestock and wildlife. Here we address questions such as, How do ticks differ from insect vectors? What are the essential biological processes governing the patterns of infection and persistence of tick-borne diseases in wildlife populations?

7.1 Background: the challenge

Blood-sucking arthropods can provide efficient transport of infectious agents from one host to another. The large variety of combinations of host, vector, and pathogen locked into eternal triangles is testimony to the advantage to parasites of utilizing this form of transmission. We consider in detail those infections transmitted by ticks, since these are almost all zoonoses, with wildlife species playing a central role in maintaining endemic cycles of infection. Ticks are relatively long-lived and many species are catholic in their blood-sucking habits, so they can act as the principal disease reservoir and often carry a community of pathogens. Humans or their livestock are frequently infected, often with fatal consequences, as they accidentally intrude on these natural cycles.

Dye and Williams (1995) have challenged students of vector-borne infections by asking 'does having a vector make a difference?' They were not, of course, denying that insects and ticks are instrumental in carrying many pathogens from host to host. They were inquiring whether there was anything special that happens within the vector to make the essence of that transmission process any different from direct transmission. If the answer is 'No', the whole of the vector element can be collapsed out of models of transmission. For the answer to be 'Yes', there must be some kind of non-linearity operating as the pathogen passes through the vector, some sort of density-dependence, so that the transmission potential changes with infection prevalence or intensity. This could arise if, for example, the pathogen had an effect, either beneficial or adverse, on vector survival or host contact rates.

Insect-borne transmission is a relatively rapid process, being achieved over periods of a few days according to the insect's feeding interval and the incubation period of the pathogen within the insect. High mobility and survival by the vector are key features that improve transmission rate, so there will be strong selection for the pathogen to avoid jeopardizing either. Ticks differ biologically from insects in ways that directly affect their performance as vectors (Randolph 1998), opening up many more opportunities for non-linear phenomena during the very much slower transmission process.

7.2 Tick biology: what is special about ticks as vectors?

Ticks are wingless and relatively immobile. Nest-dwelling soft ticks, the Argasidae, live in semi-permanent, or seasonally repeated, close association with their hosts, which makes them vulnerable to specialized protective host responses. So they hide in natural cracks. Most hard ticks, the Ixodidae, do not live in nests. Typically they climb to some vantage point on the vegetation from where they contact a passing host, exposing themselves to a greater range of climatic factors. Unlike the insects, ticks minimize the costs of achieving contact with their hosts by taking very few but very large meals. This is taken to extreme by the Ixodid ticks, which feed only once per life stage, as larvae, nymphs, and adults, and reproduce only once, after the adult meal. Some species take all three, or two of the three, meals from the same individual host, but the majority of species drop to the ground between meals, where they undergo development to the next stage before feeding again. This development takes weeks or months to complete and its rate is temperature-dependent, so the non-parasitic inter-stadial period varies seasonally with the climate.

To exploit such a vector, a pathogen must survive trans-stadially (Fig. 7.1). It is acquired from an infected host by a tick of one stage, maintained through the tick's development and moulting processes, and transmitted horizontally to a new host by the following tick stage. Many pathogens are passed trans-ovarially from females via the eggs to larvae of the next generation, thereby exploiting the tick's fecundity to achieve considerable potential for vertical amplification of prevalence in addition to any horizontal amplification. Thus the parasite's transmission cycles are determined by the tick's rates of development, survival, and reproduction, any of which may be non-linearly related to infection intensity. Interestingly, evidence to date of such interactions refers to the larger protozoal parasites. The fecundity of *Babesia*-infected ticks was reduced (Davey 1981; Gray 1982), but the feeding success and survival of ticks improved, as they fed on hosts with higher parasitaemias of *Babesia microti* (Randolph 1991). Both the feeding success and

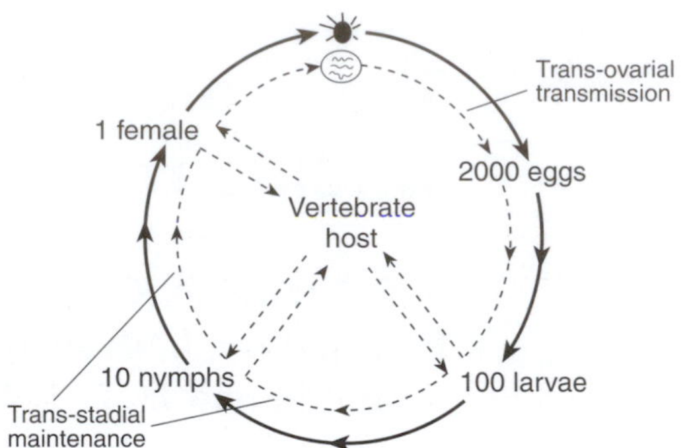

Figure 7.1 The transmission cycle of a tick-borne parasite (*dashed circle*) depends on host relationships, life cycle, survival, and reproductive rates of the tick (*solid circle*). The parasite is trans-stadially maintained from one tick stage to the next within a generation, or may be transmitted trans-ovarially via the eggs to larvae of the next generation. Successive tick stages are less numerous according to the mean inter-stadial mortality suffered—the example shown in the figure would allow population equilibrium. Each tick stage feeds on a different host individual, then drops to the ground to develop to the next stage.

reproductive output of *Theileria parva*-infected ticks were impaired, and there was a non-linear relationship between piroplasm ingestion by nymphs and subsequent salivary gland infection levels in adult ticks (Watt and Walker 2000). Intriguingly, there is also preliminary evidence that *Babesia*-infected rodents are more active, precisely at the time when ticks must attach if they are to become infected (Hughes 1998).

7.3 Factors in the transmission dynamics of tick-borne infections

This section concentrates on the transmission dynamics of tick-borne infectious disease agents, attempting to draw generalizations from examples of the few specific systems that have hitherto received attention. Before examining various models in detail (§7.4), we ask what factors of tick-borne infections should be included within models of transmission dynamics and what is their likely impact.

7.3.1 The impact of tick seasonal dynamics

The time-course of parasite transmission

Time delays in the life cycle of parasites are ubiquitous but these can be considerable in the transmission of tick-borne infections, as pathogens acquired from one host cannot be transmitted until after the tick's total inter-stadial period. This includes the development period and also the time it takes for the newly moulted tick to find a host. The total period lasts for months rather than days, and may vary 3- to 4-fold for the same tick species, depending on whether environmental conditions permit direct development or induce a period of diapause. Ticks are renowned for their high daily survival rates, which are essential to sustain such interrupted seasonal feeding activity, and especially for the longevity of the starving stages. Although long survival ensures an enduring reservoir of infection, any prolongation of the inter-stadial period will slow down the pace of disease transmission, thereby reducing the risk of infection.

R_0 for tick-borne parasites

The seasonally variable development rates of the ticks also have a direct effect on the basic reproduction number (R_0) of the transmitted pathogen (Boxes 3.4, 4.2; §5.3). The equation for estimating R_0 for tick-borne parasites is summarized in Box 7.1 (Randolph 1998). An interesting and important term is p^n, the proportion of ticks that survive their inter-stadial development period of n days, at a daily survival rate p. This term is functionally equivalent to the proportion of insect vectors that survive the extrinsic incubation period of insect-borne parasites, because the tick is not infective until it is ready to feed again. Even with the high daily survival of ticks, the natural range in the value of n has the single greatest effect on R_0 values; since it enters the equation as a power function it has a disproportionate impact on the value of R_0. This term, specific to the vector rather than the transmitted parasite, may theoretically introduce huge variations in transmission potential by ticks. For example, R_0 values may vary 50-fold for the many parasites of veterinary and medical significance transmitted by the pan-European tick *Ixodes ricinus*, and up to 80-fold for the *Theileria parva* group of blood protozoans that causes East Coast Fever and other cattle diseases throughout eastern and southern Africa, transmitted by the tick *Rhipicephalus appendiculatus* (see Randolph 1998). These figures point to the far greater potential for disease transmission in places where the climate permits continuous tick development.

On these grounds, it is clear that the spatially variable seasonal dynamics of the ticks exert a major quantitative influence on the transmission dynamics of tick-borne parasites. In addition, tick seasonality has qualitative effects, constraining the use of alternative transmission routes by parasites (§7.3.2) and thereby limiting the distribution of certain tick-borne infections. From these perspectives we can see that having a tick vector does make a difference to the transmission dynamics of the pathogen, and we have the first evidence (Watt and Walker 2000) of the crucial non-linearities referred to by Dye and Williams (1995).

Boxf 7.1 Equation for R_0 values for tick-borne parasites

Estimation of R_0 is an important step towards understanding the dynamics of any disease. For vector-borne diseases this is more complicated than for directly transmitted diseases, since we must consider transmission from host to vector and then back to the host, and the special components of the tick life-cycle. As such, this definition of the basic reproduction number includes details of the tick-population biology (especially survival and reproduction). If these are omitted, this definition can be summarized in the form given in Box 7.4.

$$R_0 = \frac{Nf\beta_{V-T}\beta_{T-T}\beta_{T-V}p^nF}{H(r+h)}, \qquad (7.1)$$

- N/H ratio of vectors to hosts (often referred to as m);
- f probability of a tick feeding on an individual of a particular host species;
- β_{V-T} transmission coefficient from vertebrate host to tick;
- β_{T-V} transmission coefficient from tick to vertebrate;
- β_{T-T} transmission coefficient from tick to tick trans-stadially or trans-ovarially;
- p tick's daily survival probability;
- n tick's inter-stadial development period in days;
- r daily rate of loss of infectivity in the host;
- h host's daily mortality rate;
- F tick's reproduction rate, equal to 1 when N refers to larvae or nymphs.

7.3.2 Co-feeding transmission: Lyme disease and tick-borne encephalitis

For a long time the transmission of tick-borne pathogens was thought to depend on the development of a systemic infection of the vertebrate host that reached a threshold level necessary for transmission back to a feeding tick, in other words a detectable viraemia, bacteraemia, or parasitaemia. Ticks could become infected when they were feeding anywhere on the host's body within the duration of the systemic infection. However, studies over the past decade have demonstrated that pathogen transmission can occur from infected to uninfected ticks feeding together on the same host ('co-feeding') in the absence of a systemic infection in the host (Fig. 7.2; Jones *et al.* 1987). Infection is limited to those parts of the body where ticks feed in close proximity and transmission depends on specific cellular events, such as the invasion of cells of the skin's immune system by the virus, facilitated by components of the tick's saliva (Labuda *et al.* 1996). This phenomenon was first demonstrated for several tick-borne viruses, notably for Thogoto virus (Orthomyxoviridae) (Jones *et al.* 1987) and the flavivirus that causes tick-borne encephalitis (TBE) in large parts of Eurasia (Labuda *et al.* 1993a) and more recently for other flaviviruses such as Louping ill (Jones *et al.* 1997).

Non-viraemic transmission between co-feeding ticks has also been found for the motile bacterium *Borrelia burgdorferi*, causative agent of Lyme disease (Box 7.2; Gern and Rais 1996). This transmission route is sufficient on its own to allow the maintenance of endemic cycles of *B. burgdorferi s.l* in sheep on upland moorlands (Ogden *et al.* 1997). In this case a new host species, sheep, that does not develop systemic infections, was added to the range of vertebrate hosts that contribute significantly to the transmission of this pathogen (Randolph *et al.* 1996). This is distinct from most of the vertebrate host species known to be competent to transmit *B. burgdorferi s.l.* (notably small rodents, squirrels, passerine birds, pheasants; Gern *et al.* 1998) that go on to develop systemic infections that persist for many months, giving a very high transmission potential (Randolph *et al.* 1996). As a result, Lyme borreliosis occurs extensively throughout the northern hemisphere, more or less wherever competent vector ticks occur (e.g. Korenberg 1994; O'Connell *et al.* 1998). The bacteria can exploit a wide range of ecologically diverse tick–host systems to maintain a high prevalence, typically 2–20% in questing nymphal ticks (Gray *et al.* 1998).

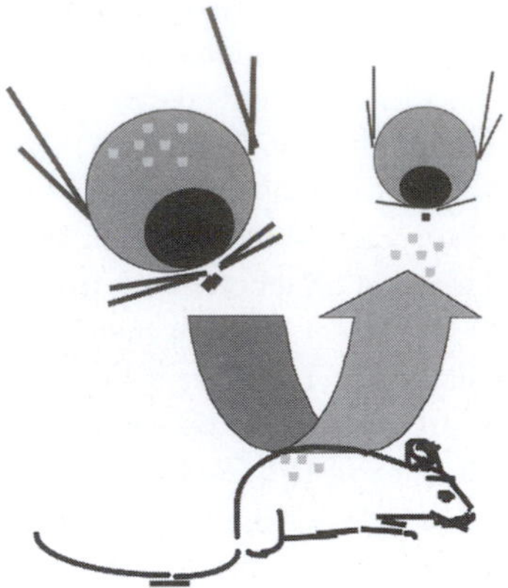

Figure 7.2 Tick-borne pathogens may develop a non-systemic infection whose infectivity to uninfected ticks is limited in space to the infected tick's feeding site, or in time to the infected tick's feeding period. (Re-drawn from an original by Lise Gern in Randolph *et al*. 1996.)

In contrast, TBE virus (Box 7.3) utilizes the same species of tick vectors (*I. ricinus* and *I. persulcatus*) and ubiquitous small rodent species (*Clethrionomys* and *Apodemus* spp.) but is far more focal in its distribution (Korenberg 1994; Immuno 1997) with very low prevalence in ticks, typically 0.1–5% (Kunz 1992). Within northern Italy, for example, Lyme disease is widely reported (Genchi *et al.* 1997), but recorded clinical cases of TBE are concentrated in a few restricted foci in the provinces of Trento and Belluno (Bassetti *et al.* 1993; Caruso *et al.* 1997). In neighbouring areas in each province, no cases of TBE have been recorded. The TBE virus isolated from ticks in positive areas of Belluno was identical to the Western TBE serotype, with the exception of a single amino acid (Hudson *et al.* 2001). This suggests that the conditions for its maintenance are more narrowly defined. The principal biological constraint is the very much shorter duration of infectivity within the vertebrate host, only 2–3 days (Kozuch *et al.* 1981). Consequently the transmission potential is low, some 60-fold lower than for Lyme bacteria (Randolph *et al.* 1996). The contribution by the non-systemic route is highly

Box 7.2 Lyme disease

Lyme disease is named after the small town of Lyme in Conneticut where in 1975 more than 12 children developed a rash, headaches, and arthritis. Initially the disease was ignored but a vigilant mother brought the world's attention to the New England 'arthritis epidemic' and this eventually led Dr Willy Burgdorfer to describe the bacterial spirochaete *Borrelia burgdorferi*. Far from being restricted to the gardens of middle-class Americans, the spirochaete has been found extensively across the United States, Europe, Russia, China, Japan, and Australia, transmitted by infected ticks from the *Ixodes ricinus* complex.

Such a wide distribution arises from the very high transmission potential of the spirochaetes under a variety of ecological conditions, due to the long period of infectivity in a wide range of the vertebrate species upon which the competent *Ixodes* tick vectors feed. These include shrews, mice, voles, hedgehogs, squirrels, chipmunks, blackbirds, robins, pheasants, and sheep. So far, only deer, important for adult ticks and therefore for the survival of tick populations, appear to be incompetent as hosts for *Borrelia*, although in Japan nymphal ticks feeding in groups on Sika deer had a higher infection prevalence than unfed nymphs on the vegetation (Kimura *et al*. 1995). The possibility of the transmission of non-systemic infections between co-feeding nymphs on deer, as was shown for sheep (Ogden *et al*. 1997), cannot yet be completely discounted. Control strategies have focused on reducing tick populations either directly or reducing wildlife hosts involved in the transmission cycle. More interesting approaches have been the release of sterile male ticks and the use of pheromone-attractants to disrupt mating behaviour. In an ingenious and effective system developed in the USA, deer may treat themselves with acaricide by rubbing their necks on four vertical acaricide-impregnated rollers as they come to feed on bait corn from bins (Pound *et al*. 2000). A vaccine against the strain of *Borrelia* common in America is also available. Further details in §7.3.5.

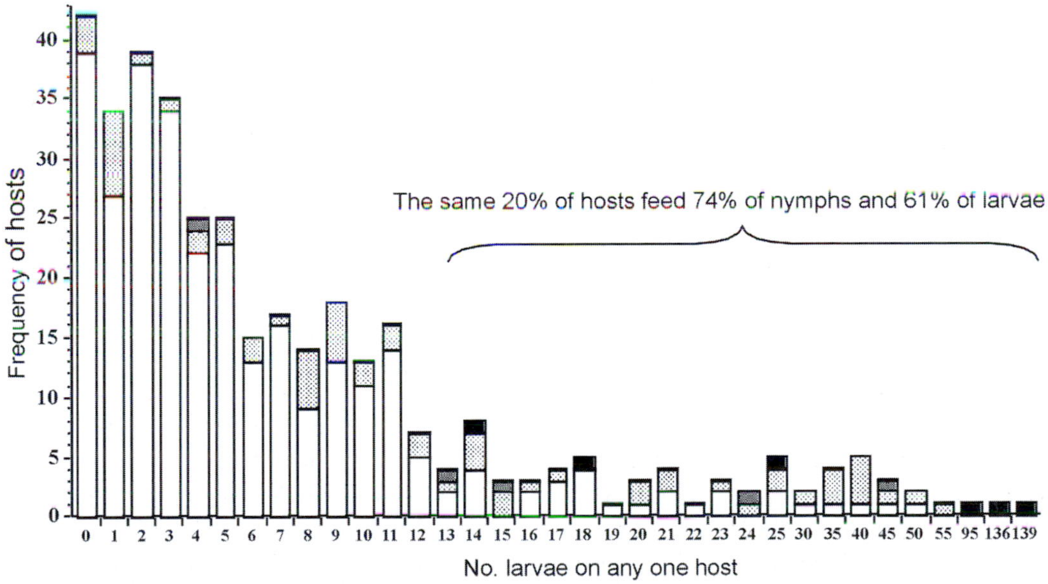

Figure 7.3 The coincident aggregated frequency distributions of larval and nymphal *Ixodes ricinus* ticks on rodents (*Clethrionomys glareolus* and *Apodemus flavicollis*) in the Borská nízina lowland, Slovakia. At each intensity of larval infestation (*x*-axis), the numbers of hosts coincidentally feeding zero (*open*), 1–2 (*stippled*), 3–4 (*shaded*) or 5–23 (*black*) nymphs are shown (after Randolph *et al*. 1999).

Table 7.1 Comparison of estimates of absolute R_0 values for TBE virus in four regions of Slovakia depending on the route of transmission (systemic or co-feeding non-systemic) and the pattern of tick infestations (independent or coincident aggregated distributions of larvae and nymphs) (after Randolph *et al.* 1999)

	Systemic infection[a]; independent distributions of larvae and nymphs (null hypothesis)		Non-systemic infection[b]; coincident distributions of larvae and nymphs (observed)	
	R_0 index $= 0.98N/H^c\mu^d$		R_0 index $= 1.65N/H\mu$	
	$\mu = 0.1^e$		$\mu = 0.1$	
	N/H	R_0	N/H	R_0
Danube lowland	1.7	0.17	4.5	0.74
Small Carpathians	6.7	0.66	13.4	2.21
Borská nízina	8.7	0.85	19.8	3.27
Danube Steppe	36.5	3.58	64.9	10.71

[a] Transmission of systemic infections.
[b] Transmission of non-systemic infections via co-feeding ticks.
[c] Mean vector-host ratio, where N is the total number of larval ticks feeding on H hosts.
[d] As defined and estimated in Randolph *et al.* (1996).
[e] Equivalent to an approximately 10% inter-stadial survival from larvae to nymphs.

significant; it allows >50% greater amplification of TBE virus in the tick population than the systemic route, largely because a non-viraemic infection avoids killing small mammal hosts before the ticks' blood meal is completed (Labuda *et al.* 1993b).

In certain parts of the vector's range, e.g. within foci of TBE in Slovakia, patterns of tick infestations on rodents are such that they facilitate co-feeding transmission (Randolph *et al.* 1999). Specifically, both immature tick stages show highly aggregated distributions on their host populations and these aggregated distributions are coincident rather than independent; those hosts feeding large numbers of larvae were simultaneously feeding the greatest numbers of nymphs. As a result, about 20% of hosts feed about three-quarters of both larvae and nymphs (Fig. 7.3; Randolph *et al.* 1999), and the number of infectible larvae feeding alongside potentially infected nymphs is double what it would be if the distributions were independent. Overall, co-feeding transmission under these circumstances increases the chance of TBE virus survival 3- to 4-fold, bringing the R_0 values to between 0.7 and 10.7 (Table 7.1). Even when the number of infected nymphs feeding on rodents is extremely low, a high degree of coincidence appears to permit endemic cycles, as seen in the weak, but stable, most westerly focus of TBE in Alsace (Perez-Eid 1990; Perez-Eid *et al.* 1992).

These first crude estimates of R_0 values for TBE virus suggest that the maintenance of this virus in defined foci in Europe depends on a high degree of coincident feeding by larvae and nymphs on the transmission-competent rodent hosts. Synchronous seasonal activity of nymphs and larvae, essential for such coincidence, only occurs in certain parts of the geographical range of *I. ricinus* where the appropriate seasonal temperature profile induces this pattern. It is indeed consistently seen within TBE foci in central Europe and Scandinavia, but not outside them (Randolph *et al.* 2000). Here we have an example of a clear link between climate and a specific transmission mechanism that determines the epidemiology of a tick-borne pathogen.

7.3.3 From tick seasonal dynamics to risk maps via population models

One versatile way to predict current and future risk of infection with tick-borne pathogens is to develop process-based tick population models as a foundation for disease-transmission models. Population models that predict patterns of tick seasonal dynamics driven by geographically variable cli-

Box 7.3 Tick-borne encephalitis

Tick-borne encephalitis is caused by viruses of a distinct antigenic complex of the Flaviviridae that includes Louping ill, Kysanur forest disease, Omsk haemorrhagic disease, and Powassan virus. Western tick-borne encephalitis (TBE) is a zoonosis of significance in many parts of Europe, from Switzerland to Russia and from Slovenia to southern Scandinavia. In many regions, particularly in Poland and the Baltic States, the incidence of infection increased by an order of magnitude during the early 1990s. TBE virus infections are neurotropic in both the vertebrate hosts, where they cause encephalitis, and in the tick vectors, although invasion of the arthropod nervous system is thought to have no consequences for transmission.

Although several tick species are competent vectors, natural ecological constraints make *I. ricinus* the only significant vector in the wild (Labuda and Randolph 1999). There are a number of transmission routes. Humans may become infected by drinking raw milk from infected goats, but neither humans nor goats can pass the virus back to ticks. Most commonly the virus is transmitted to vertebrates by nymphal ticks. Feeding larvae can also infect a host if they themselves have been infected by vertical transmission of the virus from the female ticks via their ovaries and eggs. Although this route is relatively rare, when it does occur multiplication of the number of infected ticks could be significant as the infected larvae pass the virus to other larvae that typically feed together in large groups. While ticks are feeding, their salivary glands develop and virus particles multiply and enter the host with the saliva. The salivary gland secretions include substances that change the nature of the skin site, suppress immunity, increase blood flow, and generally assist the ticks feeding but also facilitate virus transmission. The most efficient route is this saliva-activated non-viraemic transmission between co-feeding ticks. While ticks bite many species of hosts, rodent species such as *Apodemus flavicollis* have undetectable viraemias, no clinical signs of infection, and yet give rise to substantial numbers of infected ticks. These rodents are thus the principal amplifying host of this disease. Other hosts, such as roe deer, are important in maintaining sufficient tick populations for the disease to persist. A number of rodents species exhibit natural fluctuations in their relative abundance that may influence infection prevalence in ticks and therefore possibly also the number of human cases.

matic conditions address both the spatial and temporal aspects of infection risk. Given that we need to predict the numbers of three different life stages—larvae, nymphs, and adults—arising simultaneously from overlapping generations, simulation models seem the most reasonable approach. These include matrix models that become complex when extended to multiple matrices, a different one for each tick stage (Awerbuch *et al.* 1992; Sandberg *et al.* 1992). Of the tick population models based on the construction of age-specific life-tables, those developed by Haile and Mount (1987) for North American tick species use site-specific habitat structure and host assemblages, and estimates of development and mortality rates derived from a large variety of field and laboratory studies. Of more interest to general epidemiologists are models that are kept as simple as possible, designed to capture the essence rather than every detail of the complex interactions of ticks with their hosts and their abiotic environment.

Progress in this direction has been made in the simpler tropical systems, where tick generations are continuous and overlapping (Randolph 1994, 1997). All stages of the brown ear tick, *R. appendiculatus*, feed on a wide range of large hosts, both wild and domestic, but long-term population counts are only available for ticks on cattle. Detailed analyses of such data revealed that natural tick populations are regulated by very intense density-dependent processes, resulting in a degree of population stability remarkable for an organism with such a large reproductive output (Randolph 1997). This is most likely to be effected by an intimate biological process, such as the host's immune response, that is known to result in acquired resistance to ticks (reviewed by Rechav 1992). Mortality over one stage of the life cycle, from adults to larvae, however, is dominated by density-independent responses to climatic factors, principally moisture conditions.

A generic population model for *R. appendiculatus*, based on these analyses, gives predictions that match the observations very closely throughout the north–south range of the tick, from Uganda to the Cape (Fig. 7.4; Randolph and Rogers 1997). The

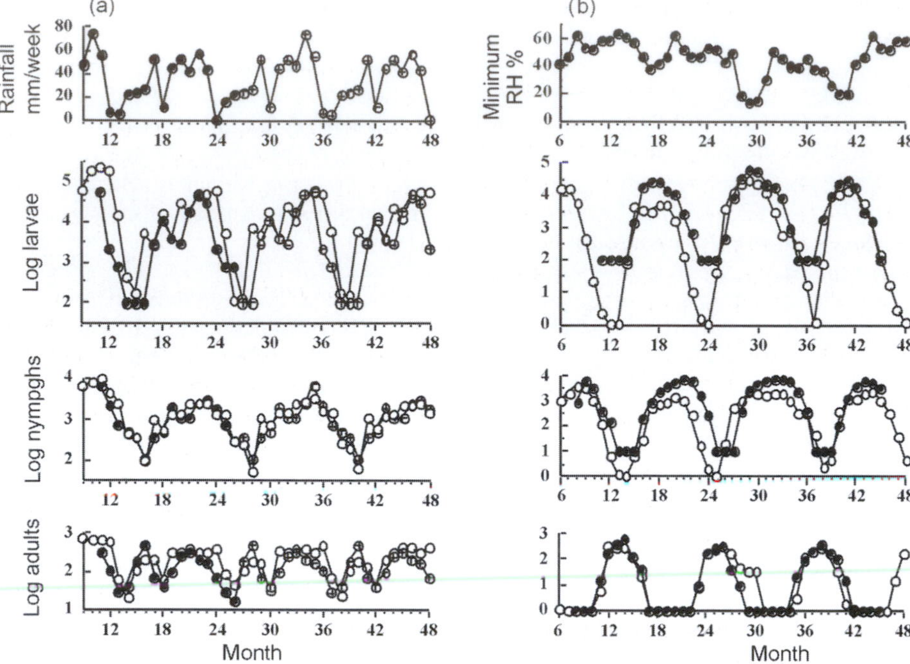

Figure 7.4 (a) Baker's Fort, Uganda. Seasonal variation in rainfall and the mean number of *Rhipicephalus appendiculatus* larvae, nymphs, and adults per host observed (*closed circles*) and predicted by the model (*open circles*). For the observations, month 11 is November 1977 (from Kaiser *et al*. 1991), and the observed pattern is repeated (⊕) for comparison with predictions up to month 48. (b) Gulu, South Africa. Seasonal variation in minimum relative humidity and the mean number of *R. appendiculatus* larvae, nymphs and adults per host observed (*closed circles*) and predicted by the model (*open circles*). For the observations, month 6 is June 1975 (from Rechav 1981). Observed larvae × 100, observed nymphs × 10 (after Randolph and Rogers 1997).

conclusion is that geographically variable tick population dynamics are determined principally by climatic factors, temperature that drives development, and moisture availability that determines mortality. Superimposed on this, host availability influences tick abundance. This factor is included very simply in this model as the tick's daily feeding probability; the scarcer the hosts, the longer the ticks must quest, thus prolonging the inter-stadial period and reducing proportional survival. The persistence or extinction of local tick populations can therefore be predicted by using local climate data to run the model. In other words, the distributional limits of the tick can be defined biologically in terms of the rates of its population processes. This is a major advance on some of the statistical methods (see Chapter 6). Maps that incorporate all three factors fundamental to risk of infection—the vector's distribution, abundance, and seasonality—can be generated across extensive areas. For example, using an interpolated climate database (Hutchinson *et al*. 1995), the predicted spatial variation in relative mean numbers of adult ticks in Kenya, Uganda, Rwanda, and Burundi shows vanishingly low tick abundance coinciding with the north-eastern edge of the recorded distribution (Randolph and Rogers unpublished). At any one locality, the map can be interrogated for the tick's seasonal population patterns.

The structure and components of the African tick model are equally applicable to any temperate tick species, including *I. ricinus*. The integration of a population model for *I. ricinus* with synoptic (1982–2000) processed satellite images for the full Eurasian range of Lyme disease and TBE (Green and Randolph 1998a, b), will provide a powerful tool for a process-based spatial analysis of these tick-borne diseases, and a template for other such systems (see Chapter 6 for further details).

7.3.4 Transmission heterogeneities: the 20–80 rule

One important factor that these population models so far fail to capture is the heterogeneity between hosts in the rate at which they encounter the vectors. We have seen (§7.3.2) that overdispersed distributions of ticks on their hosts are an important factor in transmission. In fact transmission heterogeneities are relevant to all infectious diseases (Chapters 2 and 5), but they are more apparent and more easily quantified in the field when they involve feeding by vectors on hosts. A study by Woolhouse *et al.* (1997) showed for three examples of vector-borne diseases (two malaria and one leishmania) that rates of contact between vectors and hosts are highly variable and are markedly overdispersed, with a variance much greater than the mean. Similar patterns were found for distributions of water contact rates (for transmission of schistosomiasis) and for rates of partner change (for transmission of sexually transmitted diseases). These heterogeneities have a substantial impact on the basic reproductive number which, for vector-borne infections, can be quantified using the general relationship:

$$R_0 \propto \sum_{i=1}^{m} v_i^2 = 1 + \mathrm{SV}(v), \tag{7.2}$$

where v_i is the proportion of bites on host i, m is the number of hosts and SV is the standardized variance (variance/mean^2). Clearly, R_0 is at a minimum when there is no variance in biting rates. In practice, using surrogate measures of insect biting rates for insect-borne pathogens the observed variance increases R_0 substantially, by a factor between three and four, remarkably similar to that observed for tick-borne encephalitis virus (§7.3.2). This illustrates the general result that transmission heterogeneities increase R_0. With respect to the effect of heterogeneities in biting rates, ticks *are* equivalent to insect vectors (*pace* Randolph 1998) simply because, in both cases, R_0 is related to the vector to host ratio.

There is a second consequence of these transmission heterogeneities, which is that individual hosts make different contributions to R_0 and, therefore, the removal of these hosts from the transmission cycle (e.g. by vaccination) has a disproportionate effect in reducing R_0. This effect has been quantified as the 20–80 rule, which states that removing the contribution of 20% of hosts typically reduces R_0 by at least 80%—the rule holds for all 10 datasets analysed by Woolhouse *et al.* (1997). Again, this rule also appears to apply to TBE virus, where 20% of the rodents feed three-quarters of the immature ticks. The 20–80 rule has two implications for control. The first is that control measures may be efficiently targeted at the 'core' 20%. The problem with this approach lies in identifying the core 20%. The second implication is that if untargeted control measures are implemented, then it is extremely important that coverage is very high within the core. If not, the impact of control is likely to be much less than would be anticipated from analyses that ignores heterogeneities in transmission.

7.3.5 Biotic factors: host relationships in Lyme disease

When considering the epidemiology of a tick-transmitted pathogen, it is important to appreciate the different roles played by the tick and the vertebrate hosts. The tick is both a vector and a reservoir for the pathogen. The vertebrate host amplifies the infection by acting as a source of infection for feeding ticks. Relationships between vertebrates and both ticks and pathogens play a significant role in determining the transmission dynamics of tick-borne infections. Nowhere is this better illustrated than in the Lyme borreliosis system.

The considerable slack in the Lyme borreliosis system (Box 7.2), with R_0 values well in excess of 1, permits *B. burgdorferi s.l.* to be maintained under a wide variety of conditions. Consequently, spirochaete-infected ticks are widespread and common throughout the Northern Hemisphere, but there is still considerable spatial variation in the level of infection risk as measured by the density of infected ticks. Not surprisingly, the density of ticks varies with habitat type (Gray *et al.* 1998). There is also a tendency for infection prevalence to be higher in Eastern Europe than in the West, although longitude explains only 11% of the variance in infection in nymphs and 8% in adults. In any one geographical zone, the recorded infection prevalence varies from 0 (at 7 of 94 sites) to 50% in nymphs, and from

0 (at 5 sites) to 40% in adult ticks (Gray *et al.* 1998). The predominant impression is one of extreme variability. This arises from ecological diversity within the web of transmission, introduced by spatially variable biotic factors that modify the risk of infection as determined fundamentally by abiotic factors.

Many of the major host species for *I. ricinus* contribute to the transmission of *B. burgdorferi s.l.*, but they do so in different, complementary ways because they feed different fractions of the tick population. For example, within woodland habitats, small rodents (mice and voles) and passerine birds feed mainly larvae but extremely few nymphs (Humair *et al.* 1993a, b; Kurtenbach *et al.* 1995; Randolph and Craine 1995). In contrast pheasants show the reverse pattern, feeding large numbers of nymphs but many fewer larvae (Kurtenbach *et al.* 1998a). Alongside these abundant rodent and bird hosts in woodlands are squirrels, *Sciurus carolinensis* or *S. vulgaris*, less abundant but feeding large numbers of both larvae and nymphs, and known to be *B. burgdorferi*-competent hosts (Craine *et al.* 1995, 1997; Humair and Gern 1998). In addition, roe deer, *Capreolus capreolus*, play an important role in supporting the tick population by feeding large numbers of all three life stages, although they are thought to be incompetent as *B. burgdorferi*-transmission hosts (Jaensen and Talleklint 1992).

A further element in this ecological diversity is the genetic diversity of *B. burgdorferi s.l.*, manifested phenotypically as strain-specific interactions with each host species. Five different genospecies of *B. burgdorgferi s.l.* circulate in Europe (Postic *et al.* 1994), whose distribution appears to be associated with their differential transmissibility from vertebrate to tick. This is due to a species-specific borreliacidal activity of the host's serum, mediated via the alternative complement system (Kurtenbach *et al.* 1998c). Rodent serum, for example, is lethal for Western European strains of *B. garinii* and *B. valaisiana*, but permissive for *B. afzelii* and *B. burgdorferi s.s.*, pheasant serum shows the reverse pattern, deer serum is lethal for all *Borrelia* strains, and sheep serum is partially permissive for *B. afzelii* and *B. burgdorferi s.s.* (Table 7.2) (the fifth genospecies, *B. lusitaniae* from southern Europe, was not tested). This matches the known reservoir status of many vertebrate species and has a marked impact on the infection pattern in tick populations and therefore the risk of infection to humans (Genchi *et al.* 1994; Humair *et al.* 1995, 1998; Hu *et al.* 1997; Craine *et al.* 1997; Ogden *et al.* 1997; Kurtenbach *et al.* 1998b; Humair and Gern 1998).

In European woodlands, *B. afzelii* circulating via mammals and *B. garinii* circulating via birds are typically both common in tick populations (Genchi *et al.* 1994; Humair *et al.* 1995, 1998; Hu *et al.* 1997; Humair and Gern 1998), yielding *c.*15% mixed infection prevalence in questing nymphs (Zhioua *et al.* 1994; Humair *et al.* 1995). Hedgehogs, for example, feed all stages of *I. ricinus* (and *I. hexagonus*), and amplify *B. burgdorferi s.l.* infections approximately 8-fold in larvae and 3-fold in nymphs and adults (Fig. 7.5(a); Gern *et al.* 1997). In contrast, in woodlands in southern Britain, where released farm-reared pheasants are abundant they feed such a large proportion of the tick population, especially nymphs, that they appear to have driven the R_0 value for *B. afzelii* below 1 (zero infection in

Table 7.2 Bacteriolysis of different genospecies of *Borrelia burgdorferi s.l.* when incubated for 22 h with untreated sera from a variety of vertebrate host species (after Kurtenbach *et al.* 1998c)

	% Mortality of spirochaetes when exposed to host serum			
	Borrelia burgdorferi s.s.	***Borrelia afzelii***	***Borrelia garinii***	***Borrelia valaisiana***
Mice	25	< 5	> 95	> 95
Voles	< 5	< 5	> 95	> 95
Squirrels	< 5	< 5	> 95	> 95
Pheasants	40	> 95	< 5	< 5
Deer	> 95	> 95	> 95	> 95
Sheep	44	28	> 95	> 95

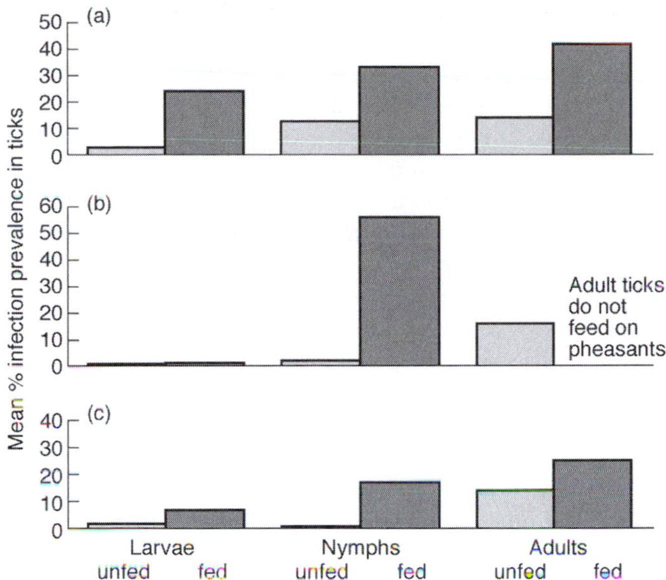

Figure 7.5 The effect of host–tick–parasite interactions on the risk of Lyme disease as measured by the mean prevalence of infection of *Borrelia burgdorferi s.l.* in questing unfed *Ixodes ricinus* ticks collected from the vegetation and engorged ticks taken from hosts. (a) In suburban/woodland habitats in Switzerland, engorged ticks taken from hedgehogs infected with *B. afzelii* and *B. garinii* showed amplification of infection at each life stage. (Data from Zhioua *et al.* 1994; Gern *et al.* 1997.) (b) In woodlands in Dorset, engorged larvae taken from rodents were infected with *B. burgdorferi s.s.*, while engorged nymphs from pheasants were infected with *B. garinii* and *B. valaisiana*. The reduction in infection prevalence in unfed adults was probably caused by nymphs also feeding on other hosts not competent to transmit these genospecies. (Data from Kurtenbach *et al.* 1998b.) (c) On moorlands in Cumbria, engorged ticks all taken from sheep were infected with *B. burgdorferi s.s.* (after Ogden *et al.* 1997).

questing ticks), leaving a high prevalence of *B. garinii* (10%) and *B. valaisiana* (6%). As most amplification of infection occurs in nymphs feeding on pheasants, however, the result is a high-infection prevalence (16%) only in adult ticks (Fig. 7.5(b); data from Kurtenbach *et al.* 1998b), which are less abundant and more noticeable than nymphs and therefore may pose less of a risk to humans. A similar outcome, but for quite different reasons, is seen on the high moorland areas where sheep graze in northern England. Here, the vertebrate fauna is very impoverished and sheep are now the only significant host to ticks. Sheep support Lyme borreliosis cycles by permitting the transmission of non-systemic localized infections from infected adult to spatially co-feeding uninfected nymphal ticks during the months of peak tick questing activity (Odgen *et al.* 1997). The degree of amplification of infection is 5-fold for larvae, 25-fold for nymphs, and nearly 2-fold for adult ticks. Once again, the precise pattern of tick feeding combined with a (hitherto undefined) host–parasite interaction that inhibits systemic infections in sheep, results in a high-infection prevalence (*c.*15%) only in adult ticks (Fig. 7.5(c); Ogden *et al.* 1997).

Alongside the principal European vector, *I. ricinus*, other tick species are also competent to transmit *B. burgdorferi s.l.*, including *I. uriae*. Each tick species has a characteristic host range; while *I. ricinus* feeds on a wide range of mammals and birds, *I. uriae* is confined to seabirds. It was only relatively recently that *B. burgdorferi s.l* infections were identified in *I. uriae* within a seabird colony (Olsen *et al.* 1993), prior to which seabirds and seabird ticks had not been implicated in the epizootiology of Lyme disease. However, the mere presence of the spirochaete does not necessarily mean that seabirds can maintain a transmission cycle. The question remains as to whether the seabird–tick association is sufficient for Lyme disease to persist, and to what extent any such cycle is isolated from those of *I. ricinus*.

Using the framework of Randolph and Craine (1995), a current study is investigating whether seabirds can maintain Lyme disease and the relative contribution of different species to R_0. Guillemots (*Uria aalge*) are numerically dominant and form extremely dense colonies on the Isle of May and have the highest infection prevalence of *B. burgdorferi s.l* in their feeding ticks. Initial results suggest that the amplification of *B. burgdorferi s.l* infections occurs at the larval feeding stage; unengorged larvae are completely uninfected, but about 7% of engorged larvae harbour *B. burgdorferi s.l.* This provides the first clear evidence that seabirds on this island are involved in the epizootiology of Lyme disease, though in exactly what capacity has yet to be determined.

7.3.6 Mixed infections within hosts: white-footed mice

In North America, the interest in tick-borne zoonoses has resulted in intensive studies of the white-footed mouse (*Peromyscus leucopus*). This mouse is susceptible to a range of pathogens including the spirochaete *B. burgdorferi s.s.*, the protozoan *Babesia microti* that causes human babesisosis and a protobacterium, probably *Ehrlichia equi*, which is a cause of human granulocytic ehrlichiosis (Walker and Dumler 1996; Spielman 1988; Magnarelli *et al.* 1997). The white-footed mouse appears to be an important source of infection for *Ixodes scapularis*, the tick that transmits all three organisms to people. Infections of humans with *B. burgdorferi s.s*, *B. microti*, and *E. equi* are found most frequently in the north-eastern USA (CDC 1995, 1998; Walker and Dumler 1996), implying that this region is also one of the endemic foci for these infections in the white-footed mouse. Serological surveys indicate that mice in this region may be exposed to one, two, or all three of these infections during their relatively short lifetime. (Table 7.3; Magnarelli *et al.* 1997).

We know very little about the consequences of any of these infections in the white-footed mouse, which is unfortunate because survival of infective hosts is a significant factor in transmission dynamics. It is commonly claimed that mice experimentally infected with *B. burdorferi* are clinically unaffected by the spirochaete, although clinical signs have been observed, particularly in subadult mice, and post-mortem examination fairly consistently reveals abnormal lesions (Burgess *et al.* 1990; Czub *et al.* 1992; Moody *et al.* 1994). Unfortunately, it is not at all clear what effect the experimental conditions have on the outcome; infection methods and routes, the source of the spirochaete, the infecting doses, and the age and sex of the recipient, all vary. There is certainly no information on whether *B. burgdorferi* in particular, or any of the other infections alone or in concert, affect the fitness of the white footed mouse in the wild and therefore its overall potential to transmit natural infections.

Table 7.3 Seroprevalence of mice with antibodies to *Borrelia burgdorferi*, *B. microti* and probably *Ehrlichia equi*

	B. burgdorferi	*B. microti*	*E. equi*
B. burgdorferi	20%	–	–
B. microti	12%	4%	–
E. equi (?)	1%	0.7%	0.7%

In the laboratory, mice infected with *B. burdorferi* become infective about one week after infection and appear to remain capable of infecting feeding ticks for most of their lifetime (Donahue *et al.* 1987; Lindsay *et al.* 1997), although the infection success changes from day to day. This has led some to suggest that spirochaetemia in white-footed mice may exhibit cyclical characteristics (Burgdorfer and Schwann 1998). The tick-mediated transfer of infection from mouse to mouse is not in dispute, whereas other modes of transmission (e.g. direct horizontal transmission via urine, sexual transmission, or true vertical transmission) have not been definitively demonstrated (Mather *et al.* 1991; Moody *et al.* 1994).

Although the white-tailed deer (*Odocoileus virginianus*) is the most significant host for *I. scapularis*, feeding all developmental stages, they are thought to be non-competent to transmit infections of *B. burgdorferi s.s.*, *B. microti*, or the probable *E. equi* to ticks (Piesman *et al.* 1979; Telford *et al.* 1996). In addition, there is little trans-ovarial transmission of these parasites (Piesman 1989). Accordingly, transmission in the north-eastern USA is most com-

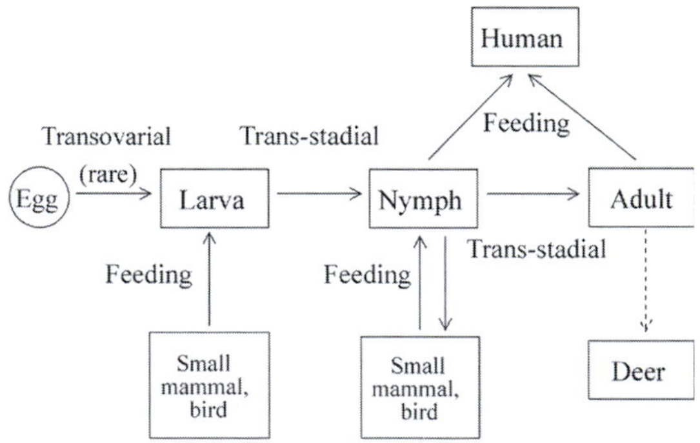

Figure 7.6 Opportunities for the transmission of *Borrelia burgdorferi* in the North American wildlife system. Not shown, but possible, is transmission by cofeeding larvae and nymphs (the stages overlap to some extent) and adults. Also not shown, is the range of large mammal hosts that adult *Ixodes scapularis* ticks are known to feed upon (cats, dogs, bears, racoons, etc.). Whether or not these hosts could play any direct role in spirochete transmission is unknown.

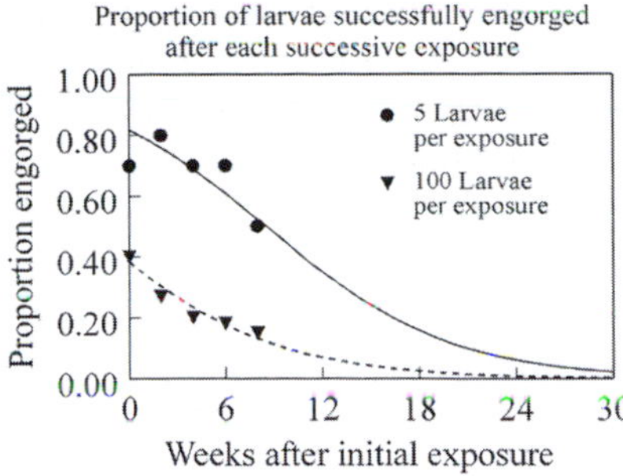

Figure 7.7 Feeding success in *Borrelia* spp.-free larval *Ixodes scapularis* declines when the white-footed mouse host (*Peromyscus leucopus*) has been previously exposed to larval ticks. In this experiment, two groups of mice were anaesthetized and exposed, at 2-week intervals, to *I. scapularis* larvae. One group was repeatedly exposed to 5 larvae, the other to 100 larvae. There were five mice in each group.

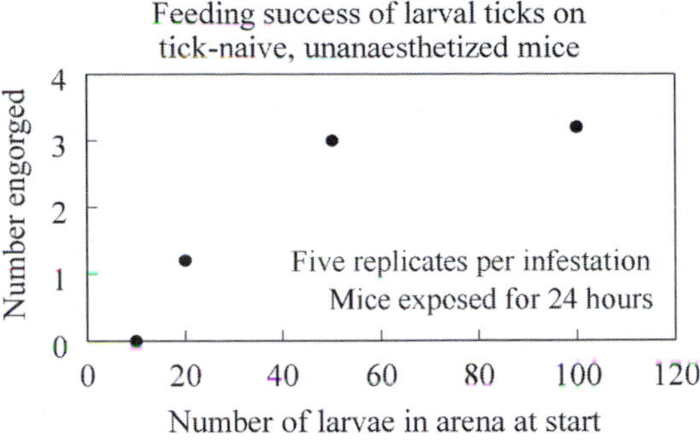

Figure 7.8 When unanaesthetized mice were exposed once only for 24 h to batches of larval *Ixodes scapularis* ticks in experimental arenas, the proportion of ticks that fed successfully declined with the number of ticks in the arena at the start (after Vail *et al*. 2000).

monly nymph to mouse, and mouse to larva or nymph (Fig. 7.6). The cycle of *B. burgdorferi* here is facilitated by questing nymphs appearing earlier in the year than questing larvae, due to the particular developmental cycle that involves an overwintering nymphal diapause. The spirochaete survives the winter months in both infected mice and infected fed larvae, and when the nymphs resume questing behaviour in the spring they infect susceptible mice, which serve as a source of infection for the larvae feeding in the summer (Fig. 7.6; Ostfeld *et al*. 1996; Vail *et al*. 2000). In endemic foci, the prevalence of ticks with evidence of exposure to *B. burgdorferi s.s*, *B. microti*, and *E. equi* may be as high as 50% and, as might be expected, the prevalence of exposure increases with the stage of tick development (Anderson *et al*. 1987; Piesman *et al*. 1987; Walker and Dumler 1996).

The white-footed mouse has previously been regarded as extremely tolerant of *I. scapularis* infestations. Recent experimental evidence, however, suggests that larvae are less likely to complete feeding successfully on mice repeatedly exposed to the tick (Fig. 7.7; Richter *et al*. 1998; Vail *et al*. 2000). One component of this process may be that grooming activity is proportional to the number of ticks attempting to attach, so that the mortality of these ticks increases in a density-dependent manner (Fig. 7.8; Levin and Fish 1998; Vail *et al*. 2000). Not only will this process tend to reduce the likelihood that individual mice will be infected with tick-borne pathogens, it also provides a possible mechanism by which tick density may be regulated (Vail *et al*. 2000). *I. scapularis* also feeds readily on a great many mammals, birds, and reptiles, and other undiscovered opportunities for regulatory processes must surely exist (Fish and Dowler 1989).

7.4 Transmission models of tick-borne infections and their uses

7.4.1 Basic tick model: heartwater model

Model design depends very much on the questions being addressed. Simplified models that can be moved between systems and that do not incorporate such details as tick dynamics may allow us to understand the relative roles of host species in year-to-year changes in disease prevalence, whereas incorporating tick dynamics may identify the relative importance of seasonality. Making models complex does not necessarily improve understanding and, as a general rule, starting simple and adding the important heterogeneities provides the most meaningful approach.

Models of biting arthropod vectors (principally insects) are often derived from the MacDonald (1957) model. They are built around the density of vectors and contain parameters such as the probability of infection to and from the vector, vector life expectancy, duration of infectiousness of infected vector, etc. (see: Dye and Williams 1995). In the simplest case, this allows an individual vector any number of bites, limited only by its lifespan. There is no attempt to impose any structure on the vector population because it is impossible to tell, without dissection, how many times an insect has fed previously. With ticks, however, the restriction of a single feed to each instar, and the obvious difference between instars, allows us to impose some restrictions on the vector population. In the model in Box 7.4, therefore, a three-stage tick (larva, nymph, and adult) has the opportunity to be infected twice and to infect twice, assuming that ticks lose infectiousness when they infect (i.e. there is no continuing trans-stadial transmission).

The results obtained when the tick population is considered without specific tick-stage structure is identical to the results from susceptible-infected-resistant (SIR, §5.2) models of directly transmitted infections (Anderson and May 1991). This basic model structure has been used to describe the transmission dynamics of *Theileria parva* and *Cowdria ruminantium* in cattle (Medley *et al.* 1993; O'Callaghan *et al.* 1998, 1999). In this model, all stages are treated equally, although in reality the amount of host tissue consumed (note that ticks feed from a sub-cutaneous haemorrhage containing immune-cell enriched sera) varies between stages and if this is related to the probability of transferring infection, then the stages may be important.

One assumption within the model in Box 7.4, is that infected ticks become infectious instantaneously, whereas in reality there is considerable developmental delay between the feeding tick stages. This developmental delay can be effectively modelled by incorporating a time delay, which does not affect the equilibrium levels, but may have a significant impact on the dynamics: in particular, the progress from introduction to equilibrium, and the consequences of introducing control methods.

Other specific aspects of the tick life-cycle can be introduced, such as trans-ovarial transmission. In the absence of sufficient knowledge to model population dynamics of the vector tick (*Amblyomma* spp.) in any detail, we assume that the tick population is at equilibrium density. It is then possible to derive equations describing the rate of change of the proportions of infected ticks of each stage, the quantities of primary interest when modelling transmission. Here, the infection prevalence in the questing vector population will equilibrate at the same level as the infection probability of the vectors (as in the simple model above).

The rate of progress to that stable equilibrium, however, will be determined by vector survival such that the shorter the time the vector survives,

Box 7.4 Simplified model of tick-borne infections applied to heartwater

Let a be the attachment rate of nymphs/adults (number of ticks/day). If we do not distinguish between instars, then the probability that this tick is infected is cx, where x is the proportion of hosts infected and c is the probability of a tick becoming infected when it feeds on an infectious host. The latter parameter can also be regarded as a composite probability of feeding on appropriate hosts and probability of infection if the host is infected. Assuming a constant loss of infectiousness of hosts, γ, a probability of host infection when bitten by an infectious tick, b, and the instantaneous infectiousness of infected ticks (i.e. no developmental delay), the differential equation describing the rate of change of the proportion of infectious hosts is:

$$\dot{x} = abcx(1-x) - \gamma x. \tag{7.3}$$

This is the logistic equation with the following solution:

$$x(t) = \frac{B}{1 + \frac{B - x_0}{x_0} e^{-ABt}},$$

$$\text{where } A = \gamma R_0,\ B = 1 - \frac{1}{R_0} \text{ and } R_0 = \frac{abc}{\gamma}. \tag{7.4}$$

The equilibrium proportion of infected vectors is *cxdot*, guaranteeing that the prevalence of infection in vectors is less than the prevalence in hosts. The eventual equilibrium is given by B, and the initial growth rate of the proportion infected is given as $(R_0 - 1)$.

If ticks carry over infection through two moults, or if infecting a host results in re-infection of the tick, then this framework requires modification. Essentially the term cx becomes:

$$c_1 x + c_s x = x(c_1 + c_s), \tag{7.5}$$

to account for possibilities of tick infection two feeds previously. It does not change the form of the equations.

the faster the vector infection-prevalence changes and the closer it tracks the host population infection-prevalence. In the case of a tick population with high rates of turnover, it would be possible to assume that the rate of change of infection in the vector population was at constant equilibrium. In order to make this assumption, however, infectiousness within the host population must be considerably longer than the questing tick life-expectancy, which is not true for all tick-borne pathogens (see above). Only if all these simplifying assumptions are tenable does the vector population again become redundant and the simple model is recovered.

In the model of heartwater (*Cowdria*) transmission (O'Callaghan *et al.* 1998), the principal input parameter was the adult tick attachment rate (TAR). This was based on the tick attachment rates observed in field studies in heartwater endemic areas of Zimbabwe. The nymphal attachment rate is taken as a simple multiple of the tick attachment rate. The rate at which cattle are infected is the sum of infection proportions from the different infectious tick stages, nymphs, and adults, weighted by the tick attachment rate, the relative nymphal attachment rate and the probabilities of infection by each. This model was used to identify the importance of the host-related factors in the persistence of this pathogen.

7.4.2 Natural history of infection and patterns of tick attachment

Tick-borne diseases of humans tend to be acute, with a relatively high proportion of case mortalities. This may be indicative of the fact that they are infections for which humans are not the 'usual' host. By contrast, the 'usual' host (or host complex) is typically thought to coexist with the infection without observable mortality or morbidity rates, although in reality the data we have on this is often weak. Indeed, many of these infections have only been discovered when exotic species (such as humans or domestic livestock) have come into contact with the tick population and the infection it carries.

The natural history of the infection within the usual host is a determinant factor in the epidemiology and transmission dynamics of the infectious agent. Typically, for any microparasite population, there is an initial parasitaemia (or viraemia or bacteraemia) following infection. In those individuals that survive, this tends to be controlled by the immune response (§5.2), although individuals may remain chronically infected if the infection is not eliminated. For most infections, it is implicitly assumed that the level of parasitaemia is positively correlated with infectiousness. Thus, in the case of tick-borne pathogens, infection in ticks that feed on hosts during the initial, acute period of their infection will be more prevalent and of greater intensity. This is no longer accepted for many tick-borne viruses transmitted non-systemically, but still appears to be true of protozoan infections (§7.3.2; Randolph 1995; Young *et al.* 1996).

Under many circumstances—when the average time between tick attachments is much shorter than host life-expectancy—hosts will be re-exposed to infection. Disease may occur on first infection, but subsequent infections are less likely to induce disease, unless the challenge is heterologous in some sense. For cattle infections, hosts in areas with high tick-attachment rates (e.g. one adult tick per 2 days) tend to be continuously infected (the so-called carrier state). Although this state has been shown to persist for years in the absence of re-infection in the case of *Theileria parva*, it remains unknown what effect continuous re-challenge has on infectiousness (i.e. parasitaemia).

Tick-borne infections are therefore characterized by two contrasting epidemiologies. In some circumstances, infections are capable of inducing acute, morbid infections, and consequently large outbreaks and epidemics (e.g. heartwater and theileriosis in domestic cattle; Norval *et al.* 1992). In other situations, infection may not induce disease, which has been termed the 'endemically stable state'. The relationship between morbidity, mortality, and tick-attachment rate was discussed by O'Callaghan *et al.* (1998). In essence, the mortality rates increase with the tick-attachment rate and then fall again at higher rates of attachment. The reason for the fall in morbidity and mortality is 2-fold. First, there appears to be an age-related resistance to disease (although not necessarily infection), which may derive from maternal factors in calves whose dams were infected. The higher the tick-attachment rate, the earlier the average age at infection and the less disease. Second, there appears to be a dose–response relationship, such that the higher the dose of infection, the more likely the infection will be clinical or fatal. There is evidence that the infection level within the tick depends on the stage of the tick (adults carry a higher dose than nymphs), as well as the parasitaemia of the infecting host. One complication is the apparent overdispersed distribution of parasites within vectors that has been widely observed (Medley *et al.* 1993; Ochanda *et al.* 1996). In areas of high tick-attachment rate, it is possible that several infected ticks may be feeding simultaneously, resulting in an increased dose of infection. The affect of this on dynamics has not been considered.

7.4.3 Mechanisms of Louping ill transmission

Another system in which the dynamics of tick–host interactions have been modelled specifically in order to consider their implications for persistence and control, is the case of Louping ill infection in red grouse and other hosts (Hudson *et al.* 1995) (Box 7.5). The results could be generalized to describe any system in which such mechanisms occur.

Within the first stage model (Box 7.6) there are two important equilibria: first, the equilibrium of the ticks where the value of the basic reproductive number is $R_{0,\,\mathrm{ticks}} > 1$; and, second, an equilibrium where grouse, ticks, and disease all persist and $R_{0,\,\mathrm{disease}} > 1$.

These R_0 terms can also be interpreted as joint threshold curves, since they depend on both the carrying capacity of the grouse and the density of the hares. If we plot the two curves for $R_0 = 1$ in the hare and grouse abundance parameter space (Fig. 7.9), then at low grouse and hare densities it is not possible for the virus to persist, since hare density is too low to maintain sufficient ticks. At higher grouse densities there are three possibilities. If there are too few hares, then there are insufficient ticks to maintain the tick population and so the virus will die out. When there are moderate numbers of hares, then the disease can persist; but if there are many hares, then the loss of virus to the hares outweighs the amplification by that grouse and the virus dies out. This is a phenomenon sometimes referred to as wasted bites, since the hares are effectively acting as a virus sink.

Box 7.5 Louping ill

Louping ill is a disease of sheep and red grouse caused by a flavivirus within the tick-borne encephalitis group and transmitted between hosts by the tick, *I. ricinus* (Box 7.3). The disease is controlled in sheep through vaccination, and exposure is prevented through the direct application of acaricides. In general the role of the sheep in the amplification of the virus can be discounted, since most shepherds actively attempt to control the disease, and it is assumed that wildlife maintain the disease. The only other animal known to produce a strong viraemia regularly is the red grouse. However, mortality of infected grouse is 80% higher than in uninfected birds (Reid *et al.* 1978; Hudson and Dobson 1991; Hudson 1992). This high level of mortality, coupled with the observation that adult ticks rarely feed on grouse, makes it unlikely that the virus will persist unless there is some other reservoir or mechanism aiding persistence. On moorland areas of England, the removal of sheep from the system through vaccination and dipping has been enough to reduce the levels of disease, but in Scotland this has not been the case.

Our understanding of the dynamics of Louping ill virus is based on the life cycle of the tick. There are three key features. First, despite investigation there is no evidence of trans-ovarial transmission of the pathogen from the female tick through the egg stage, so we assume no unfed larvae are infected. Second, as adult ticks do not feed on birds, a large mammalian host must be present to complete the tick life-cycle. Third, infection for the grouse is only derived from the nymph stage, so levels of seroprevalence in young grouse will be a function of the number of nymphs per host. As expected, seroprevalence in grouse is positively related to the abundance of tick nymphs on the grouse, but seroprevalence in grouse was consistently higher when mountain hares (*Lepus timidus*) were also present (Hudson *et al.* 1995). At first this may appear surprising since hares do not produce a significant viraemic response to Louping ill. There are two, not mutually exclusive, hypotheses to explain this: first, the hares may be sustaining and increasing the tick population while the grouse amplify the virus; second, the virus may be amplified through non-viraemic transmission on the hares (Jones *et al.* 1997).

Box 7.6 Basic model of Louping ill

In the simplest model, we assume that in the absence of disease both the ticks and the grouse will reach their carrying capacities. The approach we undertook is based on the SIR model (§5.2.), where the grouse hosts are defined as either susceptible (G_s) infected (G_I), or immune (G_z), to Louping ill virus, and the ticks are either susceptible (T_s) or infected (T_I). The different life stages are not specifically considered. Sheep and other hosts are assumed absent or effectively absent. The equations that describe the system can be written as:

$$\frac{dG_s}{dt} = (a_g - s_g G)G - b_g G_s - \beta T_i G_s \tag{7.6}$$

$$\frac{dG_i}{dt} = \beta T_i G_s - \Gamma G_i \tag{7.7}$$

$$\frac{dG_z}{dt} = \gamma G_i - b_g G \tag{7.8}$$

$$\frac{dT_s}{dt} = (a_T - s_T T)T\beta_4 H - b_T T_s - \beta_2 T_s G_i - \beta_4 H T_s \tag{7.9}$$

$$\frac{dT_i}{dt} = \beta_2 T_s G_i - b_T T_i - \beta_4 H T_i. \tag{7.10}$$

Where $G = G_s + G_I + G_z$ is the total number of grouse; $T = T_s + T_i$ is the total number of ticks; a_g is the per capita birth rate of grouse; s_g is a measure of density dependence in grouse; b_g is the per capita natural death rate of grouse; β is the probability per unit time of an infected (nymph) tick biting a grouse and infecting it; $\Gamma = \alpha + \gamma + b_g$, where α is the death rate of the grouse due to the disease and γ is the per capita rate at which grouse recover from the disease and become immune. With respect to the ticks: a_T is the per capita birth rate of the ticks; s_T is a measure of the density dependence in ticks; β_4 is the probability of an adult tick biting a host and reproducing; b_T is the per capita natural death rate of the ticks; β_2 is the probability per unit time of a tick (larvae or nymph) biting an infected grouse and becoming infected and H is the density of hosts (hares) on which ticks can feed. The dynamics of this model are explored in the text.

Box 7.7 Deriving R_0 in Louping ill-type systems

The two basic reproductive numbers can be derived directly from the equations in Box 7.6. If we add the two tick equations we get:

$$\frac{dT}{dt} = (a_T - s_T T)T\beta_4 H - b_T T - \beta_4 T H \tag{7.11}$$

$R_{0,\text{ticks}}$: When grouse are at their carrying capacity and ticks are introduced then that tick will live for:

$$\frac{1}{b_T + \beta_4 H} \tag{7.12}$$

units of time and produce:

$$\frac{a_T \beta_4 H}{b_T + \beta_4 H} \tag{7.13}$$

new ticks, hence the basic reproductive number of ticks:

$$R_{0,\text{ticks}} = \frac{a_T \beta_4 H}{b_T + \beta_4 H}. \tag{7.14}$$

$R_{0,\text{disease}}$: To consider the basic reproductive number of the disease, then the grouse lives for $1/\Gamma$ units of time and produces:

$$\frac{\beta_2 K_T}{\Gamma} \tag{7.15}$$

infected ticks. They each live for:

$$\frac{1}{b_T + \beta_4 H} \tag{7.16}$$

units of time and produce:

$$\frac{\beta K_g}{b_T + \beta_4 H} \tag{7.17}$$

infected grouse. Therefore the basic reproductive number of the disease is:

$$R_{0,\text{disease}} = \frac{\beta_2 K_T \beta K_g}{\Gamma(b_T + \beta_4 H)}. \tag{7.18}$$

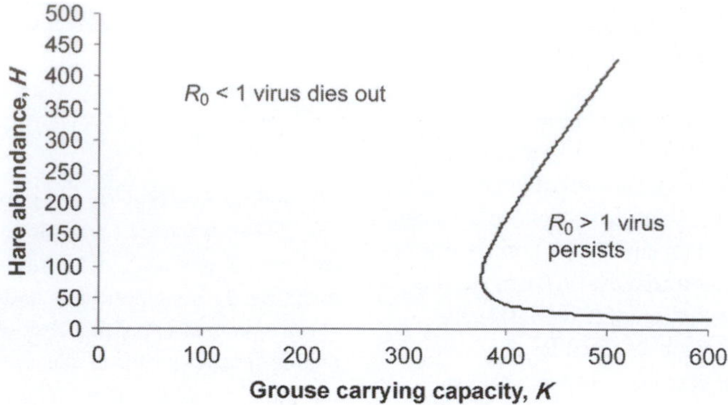

Figure 7.9 Curve which represents $R_{0,\text{disease}} = 1$ in hare-grouse space with no non-viraemic transmission. The parameters used are estimated from knowledge of the system where possible and are as follows: $a_g = 0.153$, $b_g = 0.087$, $b = 0.000112$, $a = 7.5$, $g = 1.875$, $a_T = 83.33$, $s_T = 0.001$, $b_2 = 0.0002828$, $b_4 = 0.00012656$, and $b_T = 0.0833$ (see Boxes 7.6 and 7.7).

To add more reality to the system we can incorporate the non-viraemic transmission route to the model by allowing susceptible ticks to become infected when they feed on hares, if there are infected ticks present (Box 7.8). The R_0 for the ticks remains the same but that for the disease changes. If we now re-draw the joint threshold curve (Fig. 7.10) when there are no longer wasted bites, so now the curve collapses against the vertical axis. In fact, if the rate of non-viraemic transmission is high, then it is possible for the hares to maintain the disease in the absence of grouse. That is, the virus can invade a hare population and persist even if no viraemic host could be sustained by the habitat. Whether this occurs for realistic values of q, however, is not known since we do not yet have proper estimates of the amount of non-viraemic transmission that occurs in the field.

7.5 Synthesis

Does having a tick vector make a difference to pathogen transmission? The answer is clearly 'Yes', if we want to predict spatial and temporal variation in the risk of zoonotic infection to humans or their livestock. In this chapter, we have shown how the pattern of the distribution and abundance of infection is often closely linked to constraints imposed on the ticks, just as it is in insect vector systems. To answer the question as Dye and

Box 7.8 Incorporating non-viraemic transmission in the model

Non viraemic transmission can be incorporated into the model by simply allowing susceptible ticks to become infected when they are co-feeding with infected ticks. Now the model equations become:

$$\frac{dG_s}{dt} = (a_g - s_g G)G - b_g G_s - \beta T_i G_s \quad (7.19)$$

$$\frac{dG_i}{dt} = \beta T_i G_s - \Gamma G_i \quad (7.20)$$

$$\frac{dG_z}{dt} = \gamma G_i - b_g G \quad (7.21)$$

$$\frac{dT_s}{dt} = (a_T - s_T T)T\beta_4 H - b_T T_s - \beta_2 T_s G_i - \beta_4 H T_s - qHT_s T_i \quad (7.22)$$

$$\frac{dT_i}{dt} = \beta_2 T_s G_i + qHT_s T_i - b_T T_i - \beta_4 H T_i. \quad (7.23)$$

Where q estimates the rate of non-viraemic transmission that is assumed to increase with the densities of hares, infected ticks, and susceptible ticks. In this case $R_{0,\text{ticks}}$ is the same but now:

$$R_{0,\text{disease}} = \frac{\beta_2 K_T \beta K_g}{\Gamma(b_T + \beta_4 H)} + \frac{qHT_s}{b_T + \beta_4 H}.$$

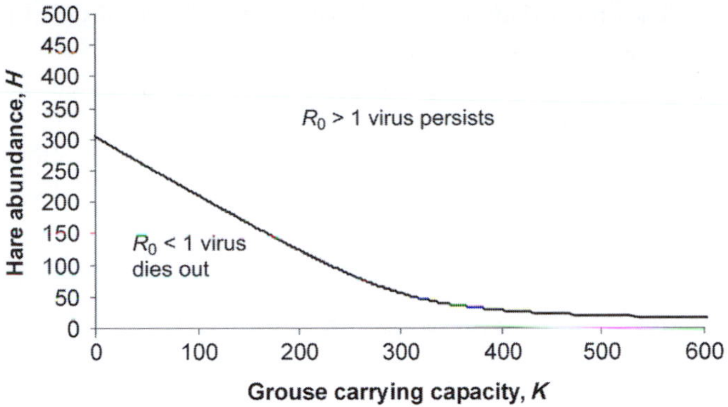

Figure 7.10 Curve which represents $R_{0,\text{disease}} = 1$ in hare-grouse space with non-viraemic transmission. The parameter values are as in Fig. 7.9 with $q = 0.000000005$ (see Box 7.8).

Williams meant it, the search for non-linearities in the transmission process has scarcely begun, but there are reasons to believe that the potential for their existence is greater amongst ticks than amongst some of the insect vectors. It is not that pathogens can afford to cause any loss in ticks' limited mobility and high survival rates, any more than they can afford to do so in insect vectors. Any diminution of either of these will reduce both the tick's fitness and the pathogen's reproduction number, and so will be selected against in both parties. The tick's biology, however, opens different doors, possibly even leading to positive non-linear effects of infection.

In comparison with other routes of transmission, the feeding habits and life cycle of ticks introduces an appreciable time delay in infection. This means that the rate of infection by ticks is not dependent on the *current* infectiousness of the hosts, but subject more to the environmental conditions experienced during the period of delay. Vector performance over this period, both in terms of their survival and development of infectivity, may vary with infection status (Watt and Walker 2000). There is also scope offered by the trans-ovarial transmission route. Given the potential amplification afforded by this route, it is not surprising that such vertical transmission is common amongst vector-borne parasites, but more surprising that filial infection rates are so low—typically <20% of any one egg batch is infected (Turell 1988; Nuttall *et al.* 1994). To infect the larvae, ingested parasites must escape from the gut and cross several membrane barriers to reach the ovaries and invade the developing oocytes before the eggs are protected by the outer chorion. This may be why trans-ovarial transmission by insects is limited to viruses. Perhaps only viruses multiply and migrate sufficiently rapidly to negotiate this tortuous route in time to catch the second, or even later, ovarian cycle after the infected blood meal, by which time the insect's lifespan may be almost played out. The longer interval between feeding and oogenesis in ticks offers greater trans-ovarial potential to all types of transmitted microparasites. At the same time, trans-ovarial transmission reduces reproduction rates in both insects (Turell 1988) and ticks (Davey 1981). This would constitute a powerful selection force:

- against increased filial infection rates by vector-borne parasites, given that they can alternatively rely on horizontal amplification; and
- for vector defence mechanisms against oocyte invasion.

At such low filial infection rates, vertical transmission cannot maintain infections over more than a few generations without horizontal amplification via vertebrates. High vector-fecundity can, however, make trans-ovarial transmission a quantitatively significant route. For ticks, which each lay thousands of eggs, yielding hundreds of larvae but only tens of nymphs (Fig. 7.1), the absolute number of trans-ovarially infected vectors can equal the number infected trans-stadially. In addition, co-feeding transmission of non-systemic infections may be particularly significant in amplifying a low infection-prevalence in larvae, as few infected larvae feed alongside much larger numbers of uninfected larvae picked up when an individual host contacts larvae from a single egg batch. Such dividends are clearly enough to maintain this vertical transmission route alongside the horizontal route common to all vectors.

Who knows what other non-linear phenomena may lurk within these systems? That is the challenge.

CHAPTER 8

The role of pathogens in biological conservation

S. Cleaveland, G. R. Hess, A. P. Dobson, M. K. Laurenson, H. I. McCallum, M. G. Roberts, and R. Woodroffe

Pathogens cause mortality and can be transmitted rapidly between hosts leading to short-term epidemics that can threaten small vulnerable populations. Pathogens are also believed to be a major driving force in generating biodiversity. This chapter focuses on the conservation aspects of disease control and the special problems with reservoirs and small populations. How large a threat are pathogens? How should conservationists manage pathogens, given that data are scarce?

8.1 Background

This chapter examines the complex role of pathogens in biological conservation, focusing on both the threat that infectious diseases pose to endangered populations, and the role of infectious agents as a force generating and maintaining biodiversity. In comparison with habitat loss, over-harvesting, and pollution, the threat posed by disease may appear to represent a minor problem for conservation. Habitat loss and fragmentation are the driving forces most frequently indicted in the loss of biodiversity (e.g. Saunders and Hobbs 1991). However, disease risks for animals living in the remaining habitat fragments are likely to increase as contact with human and domestic animal populations becomes more frequent, and with alterations in microclimate and landscape ecology (Scott 1988; Saunders and Hobbs 1991; Forman 1995). Here we outline the circumstances under which infectious diseases represent a conservation problem, and draw attention to the insights that are needed to permit informed decisions about future management strategies.

During the past 10 years, it has become clear that infectious diseases can represent a serious threat to endangered species, occasionally causing sudden and unexpected local declines in abundance. Pathogens have been implicated in the extinction or decline of species as ecologically and taxonomically diverse as rainforest frogs (Box 8.1), desert tortoises (Jacobson 1994), island land birds (Van Riper III *et al.* 1986), and marine turtles (Herbst *et al.* 1995), and grassland carnivores (Thorne and Williams 1988; Woodroffe 1997). There is growing concern that disease may be impeding efforts to conserve threatened and endangered species (Dobson and May 1986; May 1988; Scott 1988; Cohn 1991; Wolff and Seal 1993; Holt 1993; Viggers *et al.* 1993; Laurenson *et al.* 1998). Yet, conservationists are ill-equipped to manage disease risks for several reasons. First, it is difficult to identify or to predict the circumstances under which disease might pose a risk. Second, management strategies to deal with these risks are not well established. Third, techniques and tools for preventing or responding to disease threats are not well developed. Finally, because disease is often overlooked as a conservation problem until a major problem occurs, there is little funding or political will for putting ideas into practice.

Box 8.1 Disease and the decline of Australian rainforest frogs

Over the last 20 years, there have been precipitous declines and extinctions in at least 14 species of Australian rainforest frogs (Laurance *et al.* 1996; Campbell 1999). The declines have occurred largely among stream-dwelling montane species, living in rainforest habitats. In a given area, declines and disappearances have usually occurred in several species simultaneously. Disappearances have progressed from south to north, commencing just north of Brisbane (28°S) in 1985, with the most recent declines north of Cairns (15°S) in 1994.

Laurance *et al.* (1996) suggested that the declines might have been caused by disease. They used several lines of argument to support this hypothesis: most 'disappearing' species had similar ecology, the 'extinction wave' moved northward, declines of affected species were rapid and simultaneous, and pathological evidence of disease was found among declining populations. However, both the factual basis of some of these arguments and the inference of disease impact has been strongly challenged (Alford and Stephens 1997; Hero and Gillespie 1997; Laurance *et al.* 1997). Berger *et al.* (1998) isolated and identified a novel fungal pathogen from morbid frogs from both Australia and Panama. The fungus was formally named as *Batrachochytrium dendrobates* by Longcore *et al.* (1999). Berger *et al.* (1998) and Daszak *et al.* (1999) have suggested that *B. dendrobates* is the causal agent of frog declines in both Australia and Panama. The fungus has now been identified in morbid individuals of both declining frog species and stable populations of anurans, including the cane toad (*Bufo marinus*: an introduced species in Australia). Berger *et al.* (1998) also showed that frogs exposed to skin scrapings from infected animals died, whereas control animals survived.

What does epidemiological theory predict about this pathogen and what does it suggest are the critical tests? McCallum and Dobson (1995) reviewed the predictions that theory would make about the properties of a pathogen capable of causing population declines. The high prevalence of the pathogen amongst morbid frogs is not, in itself, evidence that the fungus is responsible for the declines, but disease impact is suggested if prevalence is higher in morbid individuals than in the general population. A pathogen affecting only an endangered species would not be expected to drive it to extinction, because the disease would be expected to die out once host density became sufficiently low. If a pathogen has a reservoir host within which it is relatively benign, then extinction of an endangered host is possible when the disease is transmitted from the reservoir host to the endangered host. One would therefore predict that, as is observed, the fungus should be present both in frogs that have declined and also in species that have not declined. One would further predict that the prevalence of the fungus should be higher in species with stable populations than in species that have declined. At present, there are no reliable data on the prevalence of the fungus in frog populations (as distinct from amongst morbid individuals). Collecting such data is clearly a high priority.

Laurance *et al.* (1996) cite the 'extinction wave', or northern movement of extinctions, as a major line of evidence supporting their disease theory. One needs to explain the northward movement between rainforest patches isolated by 500 km of more from each other, at a rate of about 100 km/year. The patches are sufficiently isolated that much of the terrestrial vertebrate fauna has speciated. Cane toads moved through this habitat 30 or more years previously, at a rate of about 25 km/year. Movement of a pathogen by direct contact between anurans at four times this rate is thus not entirely implausible (Van den Bosch *et al.* 1992), but it would require that the pathogen should have a very high rate of increase (and therefore high prevalence and low pathogenicity) among lowland anurans. Another possibility would be a flying vector, either birds or insects. Alternatively, it may be that the fungus has always been present, but a change in some environmental factor is now causing it to become pathogenic to some frog populations. As the immune systems of amphibia are greatly affected by temperature and other environmental factors (Carey *et al.* 1999), it is important that observations and experiments are done under a variety of environmental conditions.

Finally, a crucial test that a pathogen is responsible for population decline, rather than merely being associated with it, is whether declines in experimental populations can be prevented by treatment with an appropriate fungicide (McCallum and Dobson 1995).

8.2 When is disease a conservation problem?

In the past, control of wildlife diseases has focused on infections that affect people (e.g. rabies) and livestock (e.g. bovine tuberculosis). From the conservation perspective, disease-control measures have generally been implemented only in the wake of epidemics that have caused sudden declines in endangered populations (Table 8.1).

However, to prevent future population declines, we need to identify and quantify risks before the disease causes a significant problem. Risk assessments must address not only the impact of pathogens that cause high mortality, but those that are more chronic and persistent in nature, and may cause population declines through effects on fecundity and recruitment.

Pathogens make their host populations vulnerable to extinction in two major ways. They may be a direct cause threatening population extinction. This has occurred in populations of black-footed ferrets (*Mustela nigripes*; Thorne and Williams 1988), bighorn sheep (*Ovis canadensis*; Berger 1990; Jessup *et al.* 1995), and African wild dogs (*Lycaon pictus*; Gascoyne *et al.* 1993; Kat *et al.* 1995; Woodroffe 1997) among others. Alternatively, pathogens may suppress the size or resilience of their host populations, increasing the probability of extinction due to other factors. This has been suggested for various endemic land birds in Hawaii (Van Riper III *et al.* 1986), as well as grey wolves (*Canis lupus*; Mech and Goyal 1995) and Mednyi arctic foxes (*Alopex lagopus semenovi*; Goltsman *et al.* 1996).

Direct extinction of a host population caused by a pathogen has attracted little attention from epidemiologists in the past. This is probably because, at first sight, understanding such systems appear to offer few intellectual challenges. Pathogens with short infectious periods are predicted to persist in large, spatially structured host populations with high turnover (Dye *et al.* 1995). It is not surprising that such pathogens cannot persist in the small, isolated populations typical of many threatened species (Lyles and Dobson 1993). Yet, this phenomenon has enormous interest for conservationists because 'failure to persist' may involve extinction of the host population as well as the pathogen.

Since pathogens that cause high mortality cannot persist in small populations, it is the generalist pathogens as opposed to the host-specific pathogens that present the most serious extinction risks. Pathogens maintained in relatively common species might cause population extinction in threatened species when they 'spill over' into threatened hosts (Chapter 3; Begon and Bowers 1995; Hudson and Greenman 1998). Most local extinctions and population crashes in threatened species follow this pattern (Table 8.1).

Disease is likely to become increasingly important in wildlife management for several reasons, although most are ultimately related to human demographic patterns. Rapid population growth and increased mobilization of humans and domestic animals not only threatens habitats and vulnerable species but also favours the spread

Table 8.1 Infections transmitted from more common sympatric populations that have caused population declines in threatened mammals

Threatened host	Pathogen	Reservoir host	References
Ethiopian wolf	Rabies, canine distemper virus	Domestic dogs	Sillero-Zubiri *et al.* (1996); Laurenson *et al.* (1998)
African wild dog	Rabies, canine distemper virus	Domestic dogs	Alexander and Appel (1994); Alexander *et al.* (1993); Alexander *et al.* (1996); Gascoyne *et al.* (1993); Kat *et al.* (1995)
Baikal seal	Canine distemper virus	Domestic dogs	Mamaev *et al.* (1995)
Caspian Seal	Canine distemper virus	Terrestrial carnivores (?)	Forsyth *et al.* (1998)
Black-footed ferret	Canine distemper virus	Various wild carnivores	Williams *et al.* (1988)
Arctic foxes	Otodectic mange	Domestic dogs	Goltsman *et al.* (1996)
Chimpanzee	Polio	People	Van Lawick-Goodall (1971)
Bighorn sheep	Pasteurella	Domestic sheep	Foreyt and Jessup (1982); Foreyt (1989)
Chamois	Infectious keratoconjunctivitis	Domestic sheep	Degiorgis *et al.* (2000)
Monk seal	Morbillivirus	Dolphins (?)	Osterhaus *et al.* (1998)
Mountain gorilla	Measles	People	Hastings *et al.* (1991)
Rainforest toads	Chytridiomycosis	Cane toads	Berger *et al.* (1998)

and persistence of infectious diseases. In many instances domestic animals are brought into increasingly closer contact with wildlife (Box 8.2). At the same time there has been a general deterioration in the infrastructure for domestic animal disease control in many developing countries. This

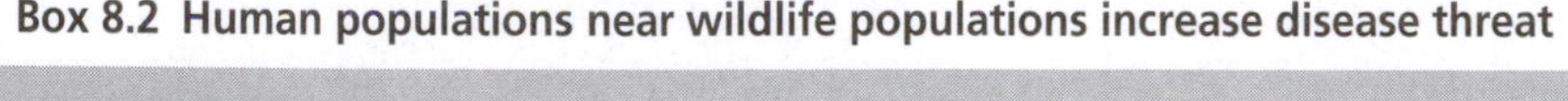

Box 8.2 Human populations near wildlife populations increase disease threat

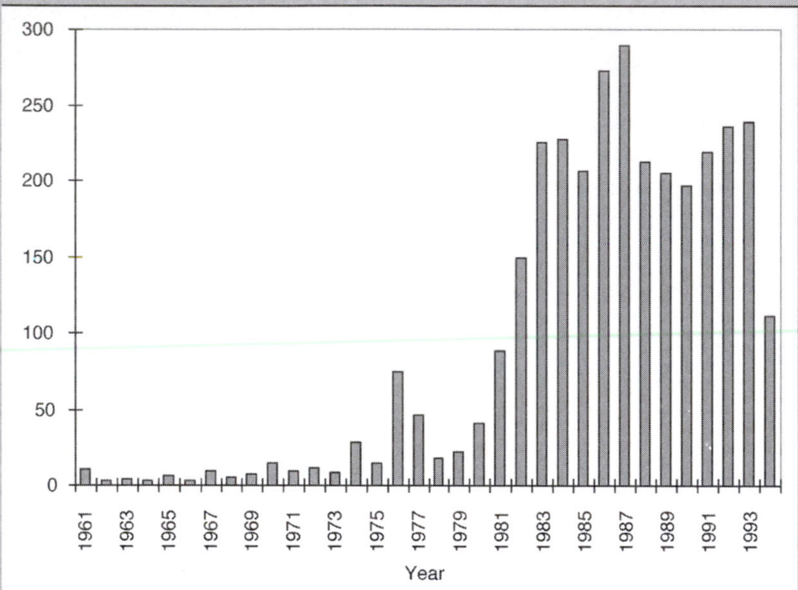

Figure 8.1 Confirmed cases of rabies in Kenya, 1978–94 (after Cleaveland 1998).

The expansion of human populations surrounding protected areas has led to fragmentation and isolation of wildlife populations, increasing the vulnerability of endangered populations to disease threats. Such threats are illustrated by outbreaks of rabies in endangered populations of the Ethiopian wolf (*Canis simensis*) and the African wild dog (*Lycaon pictus*), which significantly reduced survival in critical populations (Gascoyne *et al*. 1993; Kat *et al*. 1995; Sillero-Zubiri *et al*. 1996; Laurenson *et al*. 1998).

Within Africa, the growth of the human population has also led to a substantial increase in the abundance of domestic dogs. The growth in domestic dog populations has altered the dynamics of infectious diseases in carnivores and heightened the risk of transmission between domestic and wild carnivores. This is most evident where human population expansion occurs close to protected areas, as in the Serengeti. During the past two decades, the annual rate of human population growth in villages close to the Serengeti's protected area boundary (<10 km) has been substantially higher than the national and regional average (Campbell and Hofer 1995). Similarly, in the Bale Mountains National Park, Ethiopia, the human population density inside the park has approximately doubled in the last ten years (Stephens 1997).

In Africa, rural domestic dog populations are characterized by rapid growth and high turn-over rates, conditions that are ideal for maintenance of viral infections. Brooks (1990) calculated an annual increase of 4.7% between 1954 and 1986 in Zimbabwe, a rate exceeding that of the human population. Similar estimates have been obtained from other African populations, with an annual increase of 5–10% reported in Tanzania (Cleaveland 1996) and 9% in Kenya (Kitala and McDermott 1995). Consequently, populations that could previously sustain only sporadic, short-lived epidemics may now be large enough to maintain infection and provide a persistent source of infection for wildlife. Evidence for a threshold density for maintenance of dog rabies was found in studies conducted in Zimbabwe (Foggin 1988; Brooks 1990), Natal (Bishop 1995), and Serengeti (Cleaveland and Dye 1995), where rabies appears to persist only in populations exceeding five dogs/km^2.

Box 8.3 Canine distemper virus: a new disease threat for big cats?

Canine distemper virus (CDV) is well recognized as a disease threat to wild canids and mustelids. Indeed, the CDV epidemic in black-footed ferret populations of north America remains a classic example of disease threatening an endangered species. Historically, CDV has not been considered pathogenic in cats, but recent epidemics among large cats in captivity and in the wild have raised concerns about the emergence of a new disease threat for wild felids. In 1994, a CDV epidemic killed 30% of lions (approximately 1000 individuals) in the Serengeti National Park and Masai Mara National Reserve (Roelke-Parker *et al*. 1996; Kock *et al*. 1998). The presence of CDV in domestic dogs bordering the Serengeti National Park prior to the epidemic suggested that this population was a possible source of infection for wild carnivores (Roelke-Parker *et al*. 1996; Cleaveland *et al*. 2000). CDV isolates from four species in the Serengeti (lion, domestic dog, spotted hyena, and bat-eared fox) were closely related to each other but genetically distinct from CDV isolates from other locations (Carpenter *et al*. 1998). This suggests that a new, more virulent strain of CDV may have emerged for large felids within recent years. The threat of transmission of pathogenic strains of CDV from sympatric domestic dog reservoirs raises concerns for a number of endangered populations, such as the Asiatic lions of Gir Forest.

arises not only from a lack of resources available to government veterinary services, but also because of changing socio-political conditions. These trends are well illustrated by rabies, which has been increasing throughout much of sub-Saharan Africa, both as a result of the demographics of the domestic dog populations and because of a lack of effective disease control measures (Fig. 8.1).

A further impact of human activity occurs through environmental degradation, which may alter the dynamics of disease in natural populations through direct effects on host immunity (Lloyd 1995) or through destabilization of social systems. Human persecution of coyote and jackal populations, for example, has led to increased levels of aggression in disrupted populations (Springer 1982; McKenzie 1993), conditions that favour the spread of diseases such as rabies. In addition, reduced species diversity in degraded habitats means that opportunistic species, such as jackals, may attain densities far higher than those in conservation areas with greater species diversity (Cleaveland and Dye 1995).

The appearance of new pathogens with increased virulence for wildlife may also be a factor in the emergence of new disease threats for endangered populations. For example, the high mortality among lions of the Serengeti during a recent epidemic of canine distemper has been attributed to infection with a new strain of virus that is more pathogenic to lions (Box 8.3). However, in many wildlife epidemics, it is likely that the changing ecology and dynamics of the host populations, including the reservoir host species and vectors, play the central role in disease emergence.

8.2.1 What strategies can be used to minimize disease risks?

For endangered species management, the key issues are to identify the risk of infection entering the host population and the probability that the infection will result in host extinction. Population viability analyses have demonstrated that infections, such as rabies and canine distemper, significantly increase the risk of extinction of the Ethiopian wolf (Mace and Sillero-Zubiri 1997). Although they may be equally important in conservation efforts, much less is known about the effect of chronic and endemic infections in endangered populations or the impact of infections that reduce host fecundity.

Faced with the possibility that infectious disease might contribute to the extinction of a threatened population, conservation managers have several options. Using the example where infection is maintained in a domestic dog reservoir, we summarize the advantages and disadvantages of each strategy in Table 8.2.

8.2.2 Managing reservoir hosts

When disease in a reservoir population threatens an endangered species, it may be possible to devise

Table 8.2 Management strategies for the control of disease in endangered populations, using the example where disease is maintained in a domestic dog reservoir (after Laurenson *et al*. 1997b)

Option	Advantages	Disadvantages	Chance of Success
1. Do nothing.	• Cheap, easy, avoids identified risks of intervention (e.g. handling). • Avoids risks of intervention yet to be determined (e.g. alteration of selection pressure).	• Continued risk of disease outbreaks. • No benefits or involvement of local people.	• Disease risk unchanged.
2. Reduce density of domestic dogs (culling or sterilization).	• Reduces pathogen transmission by decreasing contact likelihood and lowering number/density of susceptibles. • Also reduces chance of hybridisation (Ethiopian wolf).	• Potentially severe cultural and logistic problems. • Expensive. • No direct protection. • Need for continuing and sustainable program. • May exacerbate disease risks if policy results in increased immigration of new susceptibles.	• Dubious or limited in short-term, perhaps better in the long-term if coupled with other strategies (3 or 4).
3. Education programme and/or restrictions placed on dog movements.	• No direct intervention. • Greater responsibility of dog ownership may reduce uncontrolled breeding and lower birth rates.	• Long-term input needed to develop good relations between local people and wildlife authorities. • Continuing programme needed.	• As above.
4. Vaccination of domestic dog reservoir.	• Reasonable feasibility in most communities. • Benefits for local people through reduced risk to themselves or their dogs (e.g. rabies). • Provides protection for other wildlife populations. Increased life expectancy of dogs may reduce demand for new pups, reducing the rate of generation of susceptibles through birth or immigration.	• Expensive. • Need for continued and sustainable programme. • Dog population may increase. • Lower risk for wildlife but does not guarantee protection. • Efficacy greatest if infection maintained only in domestic dog reservoir.	• Good.
5. Vaccinate wildlife population directly.	• Direct protection of individuals.	• Expensive. • Logistic difficulties in some populations. • Need for intensive monitoring. • Need for sustained programme. • Most available vaccines require injection which carries some handling risk. • Modified live vaccines require extensive safety trials. Oral vaccines (e.g. rabies) would require safety, efficacy and delivery trials for population at risk.	• Good, if safe vaccines available (e.g. rabies).

management strategies that minimize infectious contacts between host populations. Such management has taken a number of forms in the past. For example, the Parc National des Volcans, in Rwanda, does not permit sick park rangers, guides, or tourists to visit habituated groups of mountain gorillas (Hastings *et al.* 1991). Management plans for bighorn sheep in California recommend a 15-km buffer zone to separate bighorn populations from domestic sheep carrying potentially threatening infections (Jessup *et al.* 1995).

If it is impossible to prevent contact between threatened hosts and reservoir hosts, reducing the population size of the reservoir host might reduce the risk of disease transmission. Alternatively, management could focus on the eradication of the infection by vaccination. The fact that eradication of rinderpest from cattle also (and unintentionally) achieved eradication from several wild ungulate species suggests that this is possible (Plowright 1982). Similar, but smaller scale, vaccination programmes have been started among domestic dogs on the borders of Serengeti National Park, Tanzania, and around the Bale Mountains National Park, Ethiopia, to protect wild carnivores from rabies and canine distemper (Cleaveland 1997; Laurenson *et al.* 1998).

Strategies for managing reservoir host populations present a number of challenges. Since little is known about the epidemiology of natural infections in animal populations, the level and spatial extent of vaccination coverage required to achieve local eradication are usually unknown. For example, high birth and death rates, and rapid turnover characterize domestic dog populations in rural Africa and this requires a high frequency of coverage to maintain population immunity (Cleaveland 1996). The high costs of implementing and maintaining a sufficient level of coverage (Box 8.4) may be unsustainable in developing countries. In some situations, however, vaccination programmes may be judged cost-effective when they create significant gains for local human communities. For example, rabies vaccination programmes, even if initiated because of conservation concerns, would reduce the loss of human and animal life and alleviate the economic burden of rabies, which arises principally from the need for highly expensive post-exposure treatment in humans.

A number of concerns about domestic dog vaccination have been raised. For example, eliminating a source of mortality through vaccination in dogs might result in an increase in the size of the reservoir host populations (Moutou 1997). The consequences of such an increase could be serious if vaccination were halted, as the large number of susceptible individuals could lead to more severe epidemics. Increasing population size would also favour persistence of other infectious pathogens, with the possible emergence of new disease threats. However, preliminary data from the Serengeti indicate that, although dog survival has increased with vaccination, populations are probably limited by dog owners and are now more stable (Cleaveland *et al.* 2001). In some cases, the species of conservation concern are also suspected to be reservoir hosts for pathogens that threaten species of economic importance to people. This can be a flash point for controversy between conservationists and farmers calling for the removal of wildlife that might infect their livestock. For example, *Brucella abortus* has been present in the Yellowstone bison (*Bison bison*) herd since its introduction by infected cattle in the early twentieth century (Dobson and Meagher 1996; Meyer and Meagher 1997). Ironically, it is now

Box 8.4 Fraction to vaccinate

Suppose a perfect vaccine is available and suppose we keep a fraction v of the population vaccinated. The minimum fraction to achieve elimination of the disease agent from the population can be calculated as follows. Let R_v be the reproductive number in a partially vaccinated population. If a fraction v remains vaccinated, only a fraction $1 - v$ of the susceptibles remains accessible for the agent. Therefore:

$$R_v = (1 - v)R_0. \quad (8.1)$$

The agent cannot sustain itself in this population if $R_v < 1$, which, after rewriting, translates into the well known expression:

$$v > 1 - \frac{1}{R_0}. \quad (8.2)$$

perceived to be a threat to the local cattle industry (Baskin 1998).

Determining the risk of transmission of *B. abortus* from wildlife to cattle requires an understanding of the transmission dynamics of the pathogens (§5.4.6). Three routes of transmission between bison have been described:

1 from mother to calf;
2 as a sexually transmitted disease;
3 through environmental contamination when an infected bison cow aborts (Williams *et al.* 1997).

Brucella abortus can only be transmitted indirectly to cattle from wildlife through environmental contamination. However, in contrast to the situation with cattle, brucellosis-induced abortion is rare in bison (Meyer and Meagher 1997). Further, because vertical and horizontal transmission are a function of the fraction of individuals infected (frequency-dependent), systematic culling of the Yellowstone bison herd will not lead to the eradication of the disease (Getz and Pickering 1983).

Bovine tuberculosis, caused by *Mycobacterium bovis*, is another disease of cattle where wildlife reservoirs may play an important role (Box 5.8). The principal species implicated in the British Isles is the badger (*Meles meles*; Clifton-Hadley *et al.* 1995; Mairtin *et al.* 1998) and a heated debate between the farming community and conservationists about the problem and its solution continues (Krebs *et al.* 1998). Direct contact between badgers and cattle appears to be unusual, and pasture contamination with faeces and urine has been suggested as the principal route of infection (Brown *et al.* 1992). Although evidence that badgers transmit tuberculosis to cattle is equivocal (the transmission might be from cattle to badgers), there are indications that culling of badgers is effective in reducing the incidence of tuberculosis in cattle (Krebs *et al.* 1998).

8.2.3 Vaccination of threatened hosts

Vaccinating threatened hosts is the most certain and direct method of protecting individuals from disease and has been attempted for a variety of host-pathogen systems. These include rabies in Florida panthers (*Felis concolor coryi*; Roelke and Glass 1992) and African wild dogs (Gascoyne *et al.* 1993; Kat *et al.* 1995), canine distemper in black-footed ferrets (Williams *et al.* 1992), measles in mountain gorillas (*Gorilla gorilla beringei*; Hastings *et al.* 1991), polio in chimpanzees (Van Lawick-Goodall 1971), anthrax (*Bacillus anthracis*) in roan antelope (*Hippotragus equinus*; De Vos and Scheepers 1996), and para-influenzavirus III in bighorn sheep (Jessup *et al.* 1991). Although a clear benefit of vaccination has not been demonstrated in any of these cases, it is questionable whether such a study could be justified, given the need to leave control groups untreated in the face of a disease threat.

Direct vaccination may be impractical for several reasons. First, no safe, effective vaccine may be available. For example, inactivated canine distemper vaccines fail to provide protection (Williams *et al.* 1988) and modified live vaccines can carry a risk of inducing clinical distemper and death (Carpenter *et al.* 1976). Indeed, administration of a live distemper vaccine to black-footed ferrets in captivity induced canine distemper in one of the last remaining populations (Thorne and Williams 1988). However, new generation vaccines appear to be much safer and less likely to revert to pathogenicity (Williams *et al.* 1996). In recent years, new canine distemper vaccines have been administered to a wide range of wildlife species including African wild dogs (Spencer and Burroughs 1992) and lions (Kock *et al.* 1998), and appear to be both safe and effective. Nonetheless species' susceptibility to disease induced by modified live canine distemper vaccines varies widely, with species such as the Gray fox (*Urocyon cineroargenteus*) appearing to be highly susceptible (Henke 1997).

Administering vaccines to endangered populations may also lead to logistical difficulties. Most available vaccines must be delivered by injection, which may not be feasible for low-density populations. Handling or darting animals for administration of injectable vaccines invariably carries some degree of risk, which must be carefully evaluated against potential benefits. The development of oral vaccines against rabies (Box 8.5) and anthrax (Rengel and Bohnel 1994) may overcome some of these difficulties. However, obstacles still remain with respect to efficacy, safety in both target and non-target species and the design of appropriate strategies for bait delivery.

Vaccination protocols that are safe and effective for use in captivity may be inappropriate for field use. Available evidence suggests that effective vaccination of harbour seals (*Phoca vitulina*) against phocine distemper virus (Harwood and Hall 1990) and wild dogs against rabies (Woodroffe 1997) may require multiple doses of vaccine administered over several weeks or months. This greatly increases the financial costs associated with a vaccination programme, and multiplies any risk associated with vaccine administration.

Although complete vaccination coverage of endangered populations is probably impracticable, a low coverage may be sufficient to improve the population viability of endangered populations. In a population viability analysis of Ethiopian wolf populations, direct vaccination of as few as 20–40% of wolves against rabies was sufficient to eliminate the largest epidemics, protect populations from falling to very low densities and almost always ensured population viability (Haydon *et al.*, unpublished). Similar improvements in population viability required reduction of rabies in the reservoir population of 80–100%. Thus, despite the undoubted logistic difficulties of vaccinating wildlife, this approach to disease control may still be worthwhile.

Direct vaccination may also be cheaper than controlling disease in a reservoir population. In a preliminary cost-effectiveness analysis of protecting Ethiopian wolf populations from rabies outbreaks, the cost of protecting each wolf was 1.5–4 times more expensive by vaccinating dogs than by directly vaccinating wolves, depending on the size and geography of the wolf population (Laurenson, unpublished).

In summary, although there are important logistic, efficiency and safety issues associated with

Box 8.5 Oral vaccination: the future of wildlife disease control?

The success of wildlife rabies vaccination campaigns in Europe and North America demonstrates the potential of oral vaccines for the control of wildlife diseases. Oral rabies vaccines and bait delivery systems have now been developed for several populations, including red foxes in Europe, raccoons, skunks, coyotes in North America, jackals in Zimbabwe, and domestic dogs.

The pioneering field trials were carried out in Switzerland between 1978 and 1982 using an attenuated SAD strain of rabies virus in chicken-head baits (Wandeler 1994). In the first trial, baits were strategically laid to create an immune barrier to the spread of advancing fox rabies. Later baits were distributed in already infected alpine valleys in an attempt to eliminate infection. Both trials proved successful. Vaccination of 50–80% of foxes halted the spread of rabies and led to the elimination of disease in infected areas. In areas freed of fox rabies, the disease also disappeared from all other species, clearly demonstrating the role of the red fox as the sole maintenance host for rabies in these areas.

Widespread oral vaccination campaigns were subsequently adopted in large areas of Western Europe, using both modified live and recombinant vaccines in manufactured baits. Millions of doses of oral vaccines have now been distributed leading to a sharp decline in the incidence of rabies in red foxes and other species (Pastoret and Brochier 1998). Similar results have been reported in Texas, with a decline in canine rabies following oral vaccination campaigns of coyotes (Krebs *et al.* 1999). It is clear that oral immunization of wildlife can eliminate rabies from endemic areas with wildlife reservoirs and can prevent the spread of disease to new areas.

Oral rabies vaccines also provide a potential solution to some of the problems of domestic dog rabies control, particularly in areas where a large proportion of dogs cannot be handled. However, oral vaccines are not a panacea to the dog rabies problem. To be effective, oral vaccination programmes require a well-developed infrastructure, substantial economic resources, and a sound understanding of dog biology—the very problems that beset traditional vaccination campaigns (Cleaveland 1998).

A novel application of oral vaccines is the development of vaccines to control fertility. From the perspective of wildlife disease control, immunocontraception offers a potential solution to several emerging disease problems. First, controlling the size of reservoir populations is likely to reduce the risk of transmission from reservoir hosts (such as domestic dogs) to endangered wildlife populations. Second, reducing fertility in rapidly growing populations may prevent the emergence and establishment of new infections, such as sarcoptic mange in red foxes.

wildlife vaccination that still need to be addressed, in some circumstances, the approach offers significant advantages for conservation management and is likely to become an increasingly feasible option in the future.

8.3 The effect of population size and structure

Disease is one of several stochastic or catastrophic factors that can drive small populations to extinction. Large populations that are better able to withstand the effects of environmental stochasticity are also better able to recover from epidemics. For example, distemper epidemics affecting a population of approximately 3000 Serengeti lions (*Panthera leo*; Roelke-Parker *et al.* 1996) and a population of more than 18 000 North Sea harbour seals (Harwood and Hall 1990) were followed by population recovery. In contrast, a distemper epidemic affecting a population of 58 black-footed ferrets caused the species to become extinct in the wild (Thorne and Williams 1988). Larger populations of bighorn sheep have persisted for longer than smaller populations in the face of disease risk (Berger 1990).

Population viability analyses often predict that extinction is less likely if corridors that permit movement of animals among populations join small, isolated populations. However, several researchers have noted that the establishment of corridors might allow epidemics to spread between sub-populations, potentially causing metapopulation extinction (Chapter 6; Simberloff and Cox 1987; Hess 1996a). Epidemics caused by wildlife translocations have decimated both free-ranging and captive wildlife populations in the past (Table 8.3). In managed populations, individuals may also be moved among captive breeding and wild populations by people (e.g. Gipps 1991; Viggers *et al.* 1993). Among the potential problems is the introduction of a new disease into a population, the presence of diseases in existing populations that may affect introduced animals, exposure to new diseases, and diseases acquired during the translocation process (Ballou 1993; Meltzer 1993; Wilson *et al.* 1994; Cunnigham 1996). The importance of such a phenomenon depends upon the extent to which populations are *epidemiologically* (rather than demographically) isolated. A generalist infection may spread between apparently isolated populations through their mutual contact with a more continuously distributed reservoir host. Under such circumstances, the negative impact of corridors is likely be minimal.

8.4 Evolutionary considerations of disease control in wildlife

Concerns have been raised that, by protecting threatened hosts from infection, vaccination may halt natural selection for heritable resistance to disease. This might ultimately make the population more susceptible to infection, if vaccination were halted at some point in the future. The same logic applies to any form of disease control measure that decreases exposure to infection and hence reduces selection pressure, including vaccination of reservoir populations and reducing contact between reservoir and endangered populations (see Table 8.2). A key factor is the frequency of exposure to the pathogen, which must be great enough for alleles conferring resistance to gain a competitive advantage and become more frequent in the population (Laurenson *et al.* 1997b). Loss of heritable resistance is most likely to be a problem for less virulent diseases, and of comparatively little concern for highly virulent diseases for which little natural immunity may exist (e.g. rabies).

This concern is minimized to some extent by the imperfections of vaccines. An individual that would mount a poor immune response to infection is also likely to mount a poor response to vaccination. It is therefore likely that selection would act against individuals with little natural immunity, if a vaccinated population were exposed to infection. From the perspective of evolutionary considerations, the benefits of vaccination are likely to outweigh the costs when there is a high risk of contact with a highly virulent pathogen.

A further factor concerns the relative time-scales for evolution of disease resistance. For endangered species, we are often dealing with small populations, vulnerable from a number of potential threats and facing an immediate extinction threat. There is therefore little opportunity for evolution of disease resistance. Furthermore, rapid evolution of patho-

Table 8.3 Movement of species among free-ranging and captive wildlife populations that has exacerbated the spread of disease

Disease	Species of concern	Movement	References
Canine distemper	Black-footed ferret	Wild to captive	Thorne and Williams (1988)
Haematozoa (*Plasmodium kempi)*	Wild turkeys	Wild to wild	Castle and Christensen (1990)
Lungworm (*Bronchonema magna*)	Bontebok, springbok	Wild to wild	Verster *et al*. (1975)
Mycoplasma spp.	Gopher and desert tortoises	Captive to wild; wild to wild	Jacobson (1994); Jacobson *et al*. (1995)
Parvovirus	Racoons	Wild to wild	Allen (1986) cited in Woodford (1993)
Rabies	Racoons, skunks	Wild to wild	Jenkins *et al*. (1988)
Rinderpest	Cloven-hoofed artiodactyls (e.g. wildebeest, eland, buffalo, bushbuck)	Domestic to wild; wild to wild	Lowe (1942); Thomas and Reid (1944); Sinclair and Norton-Griffiths (1979); Scott (1981)
Tuberculosis	Arabian oryx	Captive to wild	Kock and Woodford (1988)
Warble flies, nostril flies	Caribou	Captive to wild	Rosen (1958) cited in Woodford (1993)
'Whirling disease'	Rainbow trout	Captive to wild	Trust (1986)

gens could result in pathogens overcoming any host resistance at the next encounter.

There is a growing awareness of the important role that diseases have played, and continue to play, in shaping biodiversity. However, the sudden outbreaks of disease that threaten endangered populations probably play only a minor role in the evolution of this diversity. The growth and expansion of human populations has inexorably led to degradation of ecosystem diversity. The result is that pathogens are now more likely to spread and infect vulnerable wildlife populations, forcing a number towards extinction.

8.5 Synthesis: what do we need to know?

Conservation managers are poorly equipped to manage disease risks to threatened species. Some of the information needed is technical. For example, there is an urgent need for testing of vaccines to establish safe and effective vaccination protocols for field use. Disease monitoring of threatened populations is also needed.

Perhaps the most urgent need is for a better understanding of pathogens in multi-host systems. This is crucial for determining the management strategies that will be most appropriate for field use (Table 8.2). When should management focus upon the threatened species itself, and when is management through the reservoir host(s) more appropriate? How does the existence of multiple hosts influence the vaccination cover needed to achieve disease eradication? How common are interspecies transmission events, and how important are rare events?

In addition to these general questions, there is a need for epidemiological studies of animal disease. In the past, veterinary and wildlife epidemiologists have focused only on pathogens that are economically important, either because they can be transmitted to people or because they affect livestock production. For example almost nothing is know about the epidemiology of canine distemper in domestic dog populations, presumably because it cannot infect people, and vaccination is left to the individual dog-owner. However, canine distemper virus is a serious disease that has had a profound effect upon wildlife in recent years (Thorne and Williams 1988; Grachev *et al*. 1989; Alexander, K. *et al*. 1996; Roelke-Parker *et al*. 1996; Kennedy *et al*. 2000). Epidemiological investigations of such infections will be extremely valuable for conservation management of disease.

Another area for future research is the application of risk assessment to conservation issues.

While there have been considerable advances in development of risk assessment models for control of domestic animal diseases, few quantitative analyses have been conducted on the risk of disease introduction in threatened populations. A valuable component of this approach would be the development of guidelines for identifying and monitoring potential disease threats. Application of these techniques would not only provide a more robust basis for cost–benefit evaluation of disease control programmes, but would also guide the collection of appropriate field data. Wildlife managers are increasingly aware of potential disease threats for wildlife populations, but have little quantitative data on which to base practical decisions about disease management.

CHAPTER 9

Visions for future research in wildlife epidemiology

B. T. Grenfell, W. Amos, P. Arneberg, O. N. Bjørnstad, J. V. Greenman, J. Harwood, P. Lanfranchi, A. R. McLean, R. A. Norman, A. F. Read, and A. Skorping

'There are epidemics of opinion as well as of disease' Sir B. Brodie (1856).

In recent years modern parasite ecology has experienced an epidemic of ideas. We now examine where we think the focus of work will concentrate over the next five years.

9.1 Introduction

Prediction is a notoriously difficult task in any field of biology. The editors of this volume decided wisely to avoid this task, instead appointing as seers the doughty individuals who write below. Their prognostications are diverse, but share two common characteristics.

The first is a call for more *integration,* both conceptually (§9.4) and across disciplines (§9.2). In particular, there is a clear message that the field should seek cross-discipline collaboration between field, theoretical, and laboratory-based epidemiologists and specialists in other fields (mathematicians, human geneticists, immunologists, etc.). This is true for both fundamental research problems (e.g. the measurement of immunocompetence; §9.3) and for applied questions (§9.6). In particular, in the modelling section the authors stress the need for a 'meeting of minds' between empiricists and theoreticians (§9.2; see also §9.6.2). The former often needs a better appreciation of the power and—especially—the limitations of theoretical approaches. The latter frequently need to understand more completely the importance of empirical data and expertise, the difficulties of collecting data, and the key skill of balancing mathematical elegance and tractability with biological realism and relevance.

The second—though not always explicit—characteristic of the following sections is how much progress there *has* been in many areas since the 1993 Isaac Newton meeting and the book that arose from that meeting (Grenfell and Dobson 1995). Inevitably, given the uncertainties of research, this is truer for some areas than others. For example, there is currently much theoretical interest in modelling complex pathogen strain dynamics, as well as fitting epidemiological models to data, and in understanding the stochastic persistence of pathogen metapopulations. As emphasized in the modelling section, the most fruitful ground for theoretical developments is frequently provided by the coevolution of models and data, as witnessed by the theoretical developments that have built on the current explosion of immuno-epidemiological data (§9.3 and §9.5). A particular focus for future immuno-epidemiological work should be the collection of detailed spatio-temporal data for infections more immunologically complex than measles. This is currently in progress for many human and wildlife infections (Mills *et al.* 1999). However, such studies need to be part of a long-term research investment, since many important epidemiological questions can only be addressed with data over several decades (Grenfell and Dobson 1995).

In a particularly clear vision of evolutionary aspects of the subject (§9.4), the authors begin in a more pessimistic vein, proposing that none of the questions that Read *et al.* (1995) raised in Grenfell and Dobson (1995) have been answered. However, as the section proceeds, their enthusiasm gets the better of them and they anticipate a time when parasite ecology might have as complete a degree of conceptual unification as behavioural ecology. This is arguably too negative a perspective. As the authors note, many problems in host–parasite evolutionary biology arise at the interface of evolution, population genetics, and population dynamics, and a relatively simple evolutionary 'toolkit' is not available to solve most of these problems.

However, the corollary here is that host–parasite interactions provide unique perspectives on a central question in biology: how do the ecological and evolutionary time-scales interact? Again, collaboration across disciplines is of great importance in tackling this major problem. The Trento meeting provided the ideal setting to initiate, and in some cases rekindle, such collaborations.

9.2 Perspectives on modelling epidemiological systems

Since the seminal work of Anderson and May (1991) there has been a significant body of work applying mathematical models to study wildlife epidemiological systems. Progress has been achieved both in the areas of model construction and model validation. The theoretical interest in wildlife epidemiology is reflected by the fact that over 300 papers have been published in these areas over the last five years. There are two broad areas where further advances are both urgently required and likely to be biologically exciting: allowing for stochastic variation (§5.3.3) and incorporating biological complexity (§5.3.4).

9.2.1 Statistical developments

There are two main areas of challenge for numerical epidemiology and ecology over the next five years and these are united by their common requirement of ecologically realistic stochastic models. The first challenge arises from the realization that real parasite populations are governed by a combination of deterministic and stochastic forces. The key regulatory, density-dependent, deterministic, non-linear processes include such processes as mass-action transmission (§5.3; Grenfell and Dobson 1995), competition/interference (Rohani *et al.* 1998), and parasite-induced host mortality. At the same time, important stochastic forces arise from birth and death processes (Bartlett 1956, 1960), environmental stochasticity (Rand and Wilson 1991), and spatial or temporal heterogeneities. This duality is now widely recognized across ecology (Turchin 1995), yet elucidating the relative influence of deterministic versus stochastic forces has become a major challenge. Finkenstdt and Grenfell's (2000) study of pre-vaccination dynamics of measles in England and Wales suggest that the large amplitude fluctuations in numbers of infected people in big cities may be astonishingly *non-stochastic*—many apparent irregularities are due to birth-rate variations. The challenge of understanding the noise-to-signal issue is heightened by the intricate interaction between non-linear regulatory processes and demographic (e.g. Wilson and Hassell 1997) or environmental stochasticity (Rand and Wilson 1991). Rand and Wilson (1991), for instance, demonstrated that the deterministically stable annual seasonal dynamics from an SEIR model (§5.2) could interact with stochastic forcing to produce chaotic stochasticity (also called chaotic transience). Drepper *et al.* (1994) suggested that such complex transience might also govern the dynamics of measles in New York. These and related phenomena, that arise from the interaction between regulatory and stochastic forces, are becoming the focus of a range of ecological investigations (McCann and Yodiz 1994; Higgins *et al.* 1997; Cushing *et al.* 1998). Understanding the dynamics of real populations, which are invariably affected by process-inherent stochasticity, will provide an exciting and rewarding playground for ecological theoreticians. As an example, recent studies in the statistical and physical literature indicates that *even* the classical (in a non-linear dynamics sense) notions of chaos and sensitivity-to-initial-conditions may require separate interpretation and methodology once dynamics are stochastic (Paladin *et al.* 1995; Loreto *et al.* 1996, Tong 1996; Bjørnstad and Grenfell 2001).

A second area that is currently exciting both for statisticians and empiricists is the tightening link between models and data. With recent developments in statistics, and in particular computer-intensive re-sampling procedures (Efron and Tibshirani 1993; Gilks *et al.* 1996), the tools are now available to fit complex mechanistic models to data. The last few years have seen some exciting efforts to bridge the gap between theory and data. Begon *et al.* (1998) have used time-series data to shed light on basic assumptions of mixing and mass-action (§5.3). Similarly, Ellner *et al.* (1998) and Finkenstdt and Grenfell (2000) have developed statistical methods to directly link models for microparasite transmission dynamics to mechanistic (or semi-mechanistic) models. These efforts parallel recent work in other areas of population ecology (for example see: Dennis *et al.* 1995; Costantino *et al.* 1997; Kendall *et al.* 1999). However, the quality of data, and of the biological insight, which underlies its collection, is still the key to successful developments in this area. Carefully designed experimental epidemiology, and the need to grasp the nettle and undertake brave experiments at the right scale, is a particular priority here (e.g. §3.5; Hudson *et al.* 1998).

A particularly interesting avenue of research has been the development of discrete state-space models (i.e. modelling population size rather than population density) that now start to address some of the challenges laid down by Bartlett (1956). The advent of Markov Chain Monte Carlo methods naturally lend themselves to statistical estimation. They have the dual role that they can be seen as theoretical models for real populations (having a finite number of individuals, affected by demographic, and environmental stochasticity) and at the same time as statistical models for abundance data. Abundance models (as a complement to density models) bear a promise of aiding our understanding of the dynamics of small populations with respect to extinction, critical community sizes and recurrent epidemics (§5.2, §5.4, §8.3).

A key decision arising in the model-construction process itself is whether the formulations have to be stochastic to capture the essential features of the ecological system or whether the statistical aspects can be approximated by deterministic (Holt and Pickering 1985) or a pseudo-stochastic formalism (Anderson and May 1978). Moment closure methods offer a promising way forward here.Nevertheless, even these become very complex when we start to consider the full heterogeneities of real parasite population dynamics, especially in the macroparasites (Chapter 2; Grenfell *et al.* 1995; Smith 1999). Keeping in mind that deterministic formulations can be obtained by taking the expectation (or the skeleton; *senu* Tong 1990) of stochastic models, we expect an exciting and pluralistic research programme over the next few years that will focus on the mechanistic link between classical (deterministic) theory, stochastic models, and sound data.

9.2.2 Complexity in multi-strain and multi-host systems

The dynamics of systems with many host species and/or multiple pathogen strains has been an exciting area of development over the last five years.

Multiple pathogen strain interactions is currently a major area of focus, both empirically (§9.3, §9.5) and theoretically. For microparasites, the main problem has been the curse of dimensionality—as we increase the number of strains considered (n), the maximum number of uninfected compartments we have to consider goes up by 2^n (Gupta *et al.* 1996; Andreasen *et al.* 1997). There is currently a major effort to address this problem—particularly for malaria and influenza, though the implications for wildlife infections are also profound. The main strategy involves efforts to decrease the dimensionality of the problem, by coevolving mathematical tractability and biological realism in terms of how we 'label' the immunological or susceptibility status of individual hosts (Gupta *et al.* 1996; Andreasen *et al.* 1997; Ferguson *et al.* 1999).

There are equally interesting problems for interaction of macroparasite strains and species (Chapter 4), with the added complexity of modelling the stochastic dynamics of parasite intensity (Grenfell and Dobson 1995; Smith 1999).

Multiple host species The other side of the problem of multiple strains is the structuring of the parasites and strains between host species. Much effort has been focused on models of apparent competition, mediated by parasites (Holt and Pickering 1985; Begon and Bowers 1995; Hudson and Greenman 1998). Again, dimensionality is a problem leading to analytical intractability of models. Since it is only in exceptional cases that globally valid analytic solutions are possible, most studies have been restricted to a local analysis about the point equilibria of the model. However, with higher dimensional models being developed, algebraic limits are being encountered even for local analysis (Holt and Pickering 1985; Begon and Bowers 1995). This has forced researchers to investigate model structure by exhaustive numeric search.

At the 1993 Isaac Newton Conference, Begon and Bowers (1995) set out the difficulties in analysing multi-host, multi-pathogen models, thereby providing a challenge to modellers to develop more powerful tools of analysis that could enable at least a local analysis about the system equilibria to be completed. There has been some response to this challenge with the further exploitation of bifurcation methods to address some of these difficult problems. Details are provided elsewhere (§3.8; Hudson and Greenman 1998; Greenman and Hudson 2000). These new techniques also have relevance to stochastic moment equation systems and the reaction–diffusion equation describing spatially heterogeneous systems. Their development raises the hope that there may be other areas of under-developed mathematical theory that could be of use in unravelling the complex behaviour of existing models.

While workers have tended to undertake stability analysis of system equilibria, some important insights can be obtained by examining transient behaviour in the system and, in particular, the speed of convergence onto the equilibrium state. This is certainly the case when one is looking, for example, at pest control or the effects of a vaccination programme. Ideally one would like to have explicit analytic expressions for the dominant eigenvalue at the equilibrium in question but even if this were available it would not provide information about the system trajectory far from equilibrium. The unavailability of tractable eigenvalue formulae and the difficulty of analysing non-local behaviour are further challenges to the mathematician to develop more sophisticated analytic tools.

Perhaps we are being too gloomy about the difficulties involved in looking at local and transient behaviour. Even for complex systems, one can still derive important results about its sub-systems, as evidenced by the work on R_0-threshold conditions (§3.2, §5.3; Heesterbeek and Roberts 1995b; Roberts and Heesterbeek 1995; Diekmann and Heesterbeek 2000). In some cases one can also make reasonable assumptions that simplify the structure sufficiently to allow extraction of analytic threshold conditions (Chapter 4; Dobson 1985). The work of Gupta *et al.* (1994) also shows that multipathogen models can be solved analytically when vector transmission is involved.

In summary, the next five years promise to be exciting ones for wildlife-disease modelling; in particular, mathematical and statistical analysis are essential tools for understanding the increasing flood of genetic and automatically recorded data on population and spatial heterogeneities in hosts and pathogens. There has been much progress in these areas; however, we all agree that we still seek a meeting of minds between empiricists and theoreticians.

9.3 Challenges for immuno-epidemiology

This book reflects a number of recurring themes, which fall under the broad heading of immuno-epidemiology. In this section we begin by considering the status and future prospects of this area and the section ends with an immunological perspective on the much-discussed issue of measuring immunocompetence.

There are historically two branches of parasite biology concerning the immune response: immuno-epidemiology and parasite immunology. The former was essentially about the role of diagnostic markers of infection and the information that can be gleaned from biological specimens about a host's infection status, i.e. immunology informing epidemiology. Although the term has been coined for macroparasites in particular, it is little different in aims and practice from, for example, using immunological markers to assess the progression of an individual between HIV infection and AIDS. Parasite immunology concerns the role and mechanisms of the immunological response to parasites within individual hosts. This approach has largely ignored epidemiology—experiments to understand the immunology can be made considerably simpler by removing the epidemiology. Consequently, a divide has grown up, with little epidemiological or ecological information being used to inform immunological approaches: immunology being 'proper' biology, you understand.

It is a major challenge to build and analyse a model starting from an immune response within a host through to an epidemiological description of spread between hosts. There are very few examples of such an analysis, but those that are available show that having both aspects in a model can lead to substantial changes in the dynamics compared to a model disregarding immune response (e.g. Roberts and Heesterbeek 1998). The situation allies most strongly to interactions in which immunity is not lifelong and absolute (as in measles) but requires 'boosting' by subsequent infection (as for many helminth parasites). The point is that the infectious output of an infected individual is a result of the battle with the immune system, but this output determines how much infection gets back into other hosts and with that also the level of challenge of the immune system. This presents not only a modelling challenge, but also a challenge in getting the right tools to analyse these models and, equally importantly, in being able to carry out the right sort of experiments. This is certainly an area where experimenters and theoreticians should collaborate on animal infections, since there, in contrast to human epidemiology, the necessary experiments could actually be carried out.

To some extent, both epidemiology and immunology are given impetus by having to explain infections that are not as simple as measles. This infection is characterized by a short infectious period during which an effective immunological response develops and clears the infection (if the host does not die). Antibody markers of current infection are well known and form the basis for diagnostics and those for past infection remain for life. In the case of infections that persist, e.g. hepatitis-B virus, or those in which re-infection is common, e.g. rotavirus and macroparasites, the basic questions related to the measles paradigm are what are the mechanisms permitting persistence or reinfection (parasite immunology) and what markers of infection can be used to determine the recent or current state of the host (immuno-epidemiology)?

A criticism of the parasite immunologist's approach is that more often than not, the immunological response is studied without the context within which infection normally occurs. For example, single-challenge infections are still commonly used rather than continual or trickle exposure. Hosts are raised and kept in environments of varying degrees of sterility, although usually hosts are subject to frequent challenges by a variety of parasites. Hosts are frequently super-optimally nourished, although all but a minority of human and domesticated hosts are sub-optimally nourished. Hosts are usually born to mothers that kept in the same clean, ideal environment, although hosts are usually born to mothers with the same conditions as above. Social conditions with experiments usually restrict interactions between individuals, despite the frequently observed relationships between stress, social status, and infection/disease. The list could continue.

This criticism of parasite immunology is built on the ecological assumption that the immune response has developed and performs within a context, and that this context is ignored more often than not. Ecology has to inform immunology. Thus, for example, parasite immunologists might regard pregnancy as an abnormal state, whereas, for the great majority of host species (all if we ignore humans in post-war established market economies), sexually reproductive females spend the majority of their lives pregnant. Any results obtained from never pregnant, sexually mature female hosts must be of questionable relevance.

9.3.1 Questions in immuno-epidemiology

A persisting theme of this book is the importance of multiple-host and multiple-parasite systems, both systems where one parasite can infect many different hosts and those where one host is infected with many different parasites. Within the broad field of immuno-epidemiology, the second set of host–parasite systems shines out as a beacon marking interesting questions where new theoretical models and new methodologies will allow significant progress. By investigating the rules determining the susceptibility and infectiousness of multiply-infected hosts, immuno-epidemiology may be able to create an important bridge between within-host and between-host parasite population dynamics (Gupta *et al.* 1996). Immuno-epidemiology should ideally encompass both immunology and epidemiology—what are the consequences of each to the other? What needs to be considered—immunological interactions?

What types of interactions take place between parasites sharing a single host?

Competition for resources, apparent competition to escape from immune responses (innate and specific), genetic changes through mutation, recombination, and sex. The list is long, but already a number of such interactions are well-documented—sometimes well-quantified. So far most of what has been done concerns monotypic infections or mixed infections with closely related parasites—for example, competition between different strains of HIV for CD4 cells (McLean and Nowak 1992), and between different strains of malaria for erythrocytes (Gravenor *et al.* 1995). But some work concerns interactions between unrelated pathogens—for example, the impact of live influenza vaccine upon the within-host dynamics of an established HIV virus population (Staprans *et al.* 1995). Generally, of course, hosts provide an environment for parasites within which the whole range of ecological interactions might be expected—from competition to symbiosis (White *et al.* 1998).

Buried within the question of apparent competition between parasites to avoid shared immune responses is the huge question of constraints upon the immune response. Immune responses are tightly regulated, but what is the impact of that regulation upon the outcome of the host–parasite interaction?

What should we consider—ecological interactions?

Once it is accepted that immunological responses have developed and evolved to increase the overall fitness of the host, a whole variety of questions follow. Perhaps the biggest consequence is that immunity becomes just another factor to be traded off against processes such as growth and reproduction (Box 2.8). It has been recognized for some time that the neuroendocrine system regulates immunity and that there are complex relationships with, for example, circadian rhythms.

There is increasing ecological evidence that immunocompetence is regulated by sex hormones, such as testosterone, and often clear gender differences in morbidity and mortality rates, with males usually doing worse than females (Boxes 2.7, 2.6, and 2.9). Is this a result of male-mate competition indicating that growth is a better fitness investment for males than females resulting in reduced investment in immunity? Is the apparent reduction in immunocompetence of pregnant females a consequence of investment of resources in offspring rather than immunity? Is there an ecological explanation here somewhere for the greater incidence of auto-immune disease in women than men? Do individual hosts vary the proportion of investment in immunity by age (and reproductive status)? Is there a trade-off between reproductive potential and immunity? Many of these questions might be addressed by careful comparative studies and development of life-history models that include infectious disease explicitly as a morbidity/mortality cause.

In multiple parasite infections of a single host (the norm), do hosts differentiate between parasites in determining the level of immune response to each? For example, it would make ecological sense to react quickly to a multiplying virus that would kill if uncontrolled, which in the context of constraints may imply reduction of immunity directed against parasites that are less dangerous. This would imply mechanisms (within the immune system) that are capable of detecting changes in abundance and spatial distribution of parasites within the host, and determining morbid consequences of each parasite population.

One of the continually re-occurring questions is how long-lived hosts can survive in a world of rapidly evolving pathogens. Part of the explanation must lie in the ability of immune system to rapidly respond to new antigens, providing a mechanism that changes on a time-scale equivalent to that of the microparasites. However, there should also be a role for the dissemination of information from mother to offspring. It has been recognized for a long time that offspring derive protective immunoglobulins (either trans-placental or trans-mammary in all mammal species) against viral infections. Could not other mechanisms exist to pass information about the pathogenicity of other infections? The neonate that receives 'immunological wisdom' from its mother is in a much better position to juggle resource allocation between growth and immunity than one that is ignorant.

9.3.2 Estimating an immuno-epidemiological response

Molecular biology provides us with tools to count numbers of parasites and numbers of responding immune cells within infected hosts. In a number of cases such tools have been used to measure not only number of parasites but also their rate of change under specific interventions. In this way it is becoming possible to define *in vivo* demographic rates for parasites and responding cells (Haase 1999). The subject is in its infancy and so far most progress has been made in the field of understanding the pathogenesis of HIV. However the methods used are of wide application and presage the long-overdue return of an ecological approach to immunology and microbiology (Anderson 1998).

Can we measure immunocompetence?

As part of an ongoing effort to explain the benefits of brightly coloured plumage in some birds, behavioural ecologists have hypothesized that such ornamentation may play a role in signalling immunocompetence to potential mates (Saino *et al.* 1997; Sorci *et al.* 1997). A spin-off of these intriguing studies has been that we are forced to address the general question: what is immunocompetence? It is clear that the ability to withstand parasitic insults is an important component of fitness. Even in the apparently clean and ordered world inhabited by humans of the industrialized West, people with severely impaired immunity have greatly shortened expectation of life imposed by their burden of opportunistic infections (Grant and Feinberg 1996). Indeed, the rise of the epidemic of HIV infection has opened up a window into the array of infections that healthy people routinely withstand or suppress. What, then, is immunocompetence? A sensible definition is that it is the component of fitness that can be attributed to having an immune response (Box 2.8). This in turn is the difference between benefits in terms of the ability to withstand parasitic insult and costs in terms of energy expenditure and immune induced pathology.

Instantly, under this definition, one must ask 'but what parasitic insult?' Existing studies have recognized the difficulty of this question and have neatly circumvented it by making very broad measures of the magnitude of the immune response: the size of a lesion induced by injecting a non-specific T cell growth stimulator or the protein content of a plasma sample as reflected in its colour (Sorci *et al.* 1997; Norris and Evans 2000). However, none of the measurements used to date have been shown to correlate with the ability to withstand a relevant parasitic insult. Existing studies may not be measuring relevant quantities. There are many technologies available for counting the magnitude of specific responses, some old (e.g. Elisas for measuring circulating specific antibody concentrations), some very new (e.g. tetramer assays for counting specific T cells; Altman *et al.* 1996). How numbers of specific cells relates to the net within host growth rate of parasites (and hence to immune protection) is an important outstanding question in the immunology of infectious disease. The recognition by ecologists of the role of immunocompetence in an individual's fitness once again highlights the need for a greater understanding of this relationship.

If specific rather than non-specific responses are to be used as surrogate for immunocompetence the question 'specific for what?' must be addressed. Here there might be an interesting lesson to learn from the experience of the AIDS epidemic. The presence of severely immunosuppressed people in a community acts as an indicator of the infections circulating in that community and controlled by healthy, immunocompetent people. Opportunistic infections are the parasitic insult that a healthy person's immunocompetence allows them to withstand. Might it not be possible to use immunosuppressed individuals re-introduced into the relevant sites as indicators of the relevant parasitic insult in a given environment? To be more specific, newly hatched birds would be taken and the relevant organs removed to suppress their immune response. When returned to the nest, any infection developing in the immune suppressed chicks would be very strong candidates to use as a challenge to test the immunocompetence of their nest-mates.

9.4 Evolutionary epidemiology—visions of an exciting future

In 1993, the evolution and genetics working group at the Newton meeting was asked to produce a list of promising avenues for research over the subsequent five years (Read *et al.* 1995). Five years later, that list remains an exciting vision of the future (see also Lively and Apanius' chapter in the same volume); so far as we can see, none of the issues have yet been resolved, and there has only been progress on the minority of them. Rather than rehash the issues here, and rather than simply extending that already long list, we want to discuss a more general vision. We urge readers interested in the case for studying issues such as coevolutionary dynamics, genetics in epidemiology, intra-generation (somatic, immune) versus intergeneration evolution, virulence, resistance, and local coadaptation to look at the report from the Newton Institute and to section §9.5 in this chapter.

A striking feature of host–parasite interactions is overwhelming diversity. Consider, for example, the enormous variation in life cycles of parasites, of invasion routes, and strategies for survival with hosts, and the staggering variation in key life-history variables, such as age to maturity and daily fecundity, even in groups as morphologically similar as nematodes (Skorping *et al.* 1991). Variation in host responses to infection is equally striking. The worm burden that triggers inflammatory responses, for instance, depends on worm species, host species, the organ system involved, and host factors such as age, sex, condition, and reproductive status. Much of this sort of diversity is catalogued in often excruciating detail in many textbooks, and further description of this variation is proceeding apace. To us, a critical challenge is to make sense of it all.

Consider how textbook writers guide students through this enormous diversity of natural-history facts. Perhaps the most common way to do it is to arrange things taxonomically—indeed phylogeny remains our best way of cataloguing this diversity (Smith 1994). Others, such as immunology textbooks, are lists of mechanisms, again sometimes arranged by parasite taxa (Wakelin 1996a, b). Some arrange things by the problems parasites (or hosts) have to solve, but do little more than list the range of solutions they use (e.g. Matthews *et al.* 1999). Clearly, these authors are not interested in a general synthesis, and for their purposes the details are sufficient in themselves. This does not, however, obviate the general issue: there is no synthesis of this diversity, which would allow predictions about which of these solutions, arises and when. Parenthetically, our best way to predict what might be happening for any given host or parasite is from knowledge of what its phylogenetic relatives do—a phenomenological approach that hardly explains why they do it.

If an aim is to generate principles general enough to be applied to a range of different systems, but also of sufficient realism to provide workable predictions about the real world, classic parasitology has not been very successful. An alternative approach, which seems to us to offer the most exciting possibility, is a functional one aimed at determining how natural selection acts on host and parasite traits, and the constraints on the outcome of selection. This approach has allowed the Neo-Darwinian synthesis to successfully make sense of the hugely disparate facts collected by traditional natural historians. The adaptationist programme, particularly in the formal guise of optimality models, has been particularly successful in explaining variation—both qualitative and quantitative—in previously inexplicable phenomena, such as the diversity of foraging behaviour and breeding systems of free-living organisms. These sort of cost–benefit analyses ought to be able to explain, for example, differences in virulence or why some parasites exploit two hosts, not one. At least one recent textbook summarizes attempts to date to employ a functional approach (Poulin 1998). Notably, the textbook is thin but it does point to an exciting future.

Understanding how different selective pressures will affect the evolution of important parasite traits is not only of academic interest, but is urgently needed from a practical point of view. Efforts to control parasites, for example by drugs, inevitably impose new selection pressures. Our current ignorance of the outcome of these large-scaled evolutionary experiments is alarming. Recent papers have suggested that under given conditions, drug-use could lead to the evolution of more virulent strains within nematodes (Skorping and Read 1998) and bacteria (Wilkinson 1998). Other outcomes, depending on how the drugs are applied, could be more chronic diseases, changes in parasite habitats or adaptations to new hosts.

Given the success of evolutionary theory in explaining variation in free-living organisms, how would we apply the same line of reasoning to the often considerably more complicated world of host–parasite interactions? Take host specificity as an example. Why can a species like the digenean *Cryptocotyle lingua* use dozens of different host species within widely different orders and even classes, while others, like *Schistosoma haematobium*, are restricted to just one species? Generally, as complexity increases, so realism of epidemiological models declines, the more host species a parasite can exploit. An extreme generalist that can successfully use a diverse range of hosts must be a true nightmare to the practical epidemiologist. This suggests an obvious advantage of being a generalist: the risk of extinction should decline with the number of hosts. Fortunately, most parasites tend to have a restricted number of potential host species, suggesting that there are costs of exploiting several hosts. But we have few guiding principles, so determining host specificity, requires substantial, and often undirected, field work. Clearly, if we could make realistic assumptions on host specificity based on some general ecological information about the parasite, epidemiology would be easier. It might also be possible to predict when parasites might jump to new host species.

Comparative analysis may be one way to proceed. Using parasites where host specificity has been well studied, and assuming that problems with confounders such as sample size and study effort can be overcome, this method could, for example, tell us if specificity is related to location within the host, to geographical gradients, to species diversity within host taxa, or to life-cycle complexity. Such correlative studies could help us discover the broad patterns, but if we want to understand how different selective forces affect the number of hosts a parasite can use, we need to identify any tradeoffs between the costs and benefits of specificity.

In epidemiological terms, the good thing (from the parasite's point of view) of using several hosts is that the effective host population density is increased. This means that the contact rate between transmission stages and hosts should be higher for a generalist than for a specialist, all other things being equal. The cost of using many hosts could be several: locating, infecting, growing, and reproducing within one particular host, may require different adaptations than dealing with another host. If the main problem is to grow and reproduce, we may assume that the average fecundity of a generalist is lower than that of a specialist. We should therefore expect a trade-off between the parasites' contact rate with hosts and fecundity. This will act as a constraint on the number of hosts a parasite should use. Although the shape of the trade-off curve is likely to vary between species, this framework suggests that the degree of specificity of a parasite depends on the relative importance of fecundity versus contact rates. What kinds of parasites would we expect to sacrifice high fecundity in order to increase contact rates? Monoxenous parasites in hosts of low densities, or parasites in highly fluctuating host populations are possible candidates, and we encourage the reader to think of others.

Parasites also show considerable variations in how many sites within the hosts' body they can exploit. For example, *Ascaris*-species can be found both in the liver, in the lungs, and in the intestinal system during their development, while *Trichuris spp*. are restricted to the large intestine. Comparative studies may be useful to test hypothesis for such variation (Read and Skorping 1995). An interesting question related to both site and host specificity is the following: what is the easiest for parasites to adapt to—new sites within a host species, or the same kind of sites within new hosts? Phylogenetic analysis, studying radiations of taxa in relations to habitats and hosts may solve this issue.

The study of the evolutionary ecology of free-living organisms has progressed by a combination of theoretical modelling, field observations, elegant experiments that manipulate causal variables, and comparative (cross-species) studies. It seems likely that the same approach will bear substantial fruit in the study of host–parasite interactions. In this context we see the principle problems as follows. Many traits of interest lend themselves to relatively straightforward optimality models. For example, the optimal level of host responsiveness to a given challenge is likely to be a simple consequence of the microeconomics of the fitness consequences of responding or not. But frequently the value of these traits in a population will have population dynamic consequences, which will in turn affect the optimal value of the traits. Continuing the example, the optimal level of responsiveness is likely to depend on the probability of re-infection; which will be in part determined by levels of host responsiveness in the population. Optimality modelling for traits with population dynamic consequences is not insurmountable (in the context of virulence see: Van Baalen and Sabelis 1995), but how often general answers will emerge is unclear.

Cross-species associations between trait values and ecology have played a key part in evolutionary ecology (Harvey and Pagel 1991); arguably the majority of our knowledge of adaptation is based on comparative studies. Such studies rely on the existence of substantial cross-species variation, which is uncorrelated with phylogeny. This is frequently not the case for major parasite taxa: variation in life cycles, migration routes, intra-host habitats, etc., is great within mammalian nematodes, for example, but the variation often lies between higher taxa rather than within them, substantially reducing the ability to statistically separate different factors. Whether this is an artefactual consequence of convergence in life histories and the morphological characters on which the phylogenies are based is unclear; modern molecular techniques may reveal more frequent transitions between character states than are revealed by morphological phylogenies (e.g. Blaxter *et al*. 1998).

The primary aim of both comparative studies and optimality models is to generate hypotheses that must, when possible, be tested experimentally. Experimental studies in evolutionary ecology have frequently involved selection experiments or experimental evolution (e.g. Verhulst *et al*. 1999). These are not always a panacea—they assume the genetic variation found in the laboratory is typical of that available to natural selection. In any case, such experiments will be technically more difficult where at least two species are involved. Another experimental route, successfully employed in studies of free-living organisms, is to look at trait polymorphism within species. This can involve comparisons of genetically distinct morphs, though arguably more profitable have been studies of adaptive phenotypic responses to environmental variation. For example, some of the best evidence for adaptive sex-ratio theory is that parasitoid wasps adjust brood sex ratio under conditions predicted by theory (Herre 1985). In some special circumstances, this may be possible with parasites (eg. Read *et al*. 1995; Gemmill *et al*. 1997), but these may prove to be exceptional. But for many traits of interest—like number of hosts per life cycle—there is frequently no polymorphism within species. Quite why that is so is an extremely interesting question in its own right, but if it is generally true closes off an approach that has proved highly successful elsewhere.

Surmounting (or side-stepping) these difficulties should allow us to explain the diversity of traits such a life cycle, age to maturity, immunoresponsiveness, mechanisms of resistance, virulence, migration routes, intra-host habitats, mechanisms of immune evasion, and so on. It would be very exciting if in ten years time it was possible to teach the natural history of host–parasites in the way in which the huge number of previously disparate facts concerning animal behaviour are synthesized by behavioural ecologists. Indeed, the benefits of general principles can be seen in ecological epidemiology (see the rest of this volume), where the same framework can be successfully employed in many different contexts, allowing working scientists (and students) to recognize generalities in the usually overwhelming sea of biological details.

9.5 Challenges from a genetic perspective

There are a number of different approaches that could be taken to consider the genetic inter-relationships between a pathogen and its host. Several of these have already been examined in detail in the previous two sections. In this section we examine a series of thoughts and questions that focus on the context of genetic aspects of wildlife diseases, specifically the evolution of mutation rate, the role of the multiple histocompatibility complex (MHC), and genetic polymorpohisms. Many of these arose from reading the inspiring chapters by Lively and Apanius (1995) and by Read *et al*. (1995) from the Newton Institute meeting (Grenfell and Dobson 1995).

9.5.1 Evolution of mutation rate

Every organism faces a barrage of ever-changing challenges over evolutionary time. On the one hand, pathogens and predators tend to evolve to overcome methods of defence, whilst on the other, hosts and prey evolve to resist attack. Consequently, it is almost a truism to state that any organism that is unable to adapt will sooner or later go extinct. Equally, there can be too much novelty. Any organism that mutates too fast runs the risk of causing enough disruption of function to outweigh the benefits brought by adaptability. Somewhere in between these limits exists an optimal rate of change, sufficient to allow adaptation but not enough to cause disruption.

Different organisms are likely to vary greatly in their optimal mutation rates. High mutation rates will be favoured in species that produce large numbers of energetically cheap offspring, each of which has a small probability of surviving, and species where rapid adaptation is at a premium. Low mutation rates will be associated with species that invest heavily in small offspring and which live in relatively stable environments. Parasites will tend to fall within the former class, with their constant need to overcome host defences.

Although it is clear that species do vary in their underlying mutation rates, intriguing questions remain to be answered such as: how fast do mutation rates evolve? An optimal system might involve mechanisms for producing mixed offspring, some of which are like their parents whilst others show great diversity. Have such mechanisms evolved? Equally, an optimal genome might be one that is partitioned into regions with different intrinsic mutation rates. To some extent, this has already happened, in the form of, for example, the mammalian MHC genes. However, what about subtler effects—perhaps a general tendency for genes associated with infection and resisting the immune response to lie in regions with higher mutation rates? To study these important effects, there is a need to develop improved technologies for mutation detection. PCR and point-mutation detection systems offer some hope. An exciting possibility should effective technology emerge would be, at the risk of some circularity, to turn the question on its head. If critical regions associated with pathogenicity are indeed more variable, methods capable of identifying regions with enhanced mutation pressure will identify candidate genes that are likely to be important in pathogen—host interactions.

9.5.2 How important is the MHC?

It is a widespread belief that high MHC diversity is useful for resisting disease, and hence that species that have lost diversity through, for example, a genetic bottleneck, will be at risk over evolutionary time. However, this argument is called into question by the observation that many diseases are highly species-specific. A pattern composed primarily of one-to-one interactions appears more consistent with hosts being generally well-protected against most infections. Extant diseases in the main probably represent continuations from rare, ancient events, in which a pathogen found and exploited a weakness in new host. Persistence then continued through round after round of an evolutionary arms race in which the host strove to eliminate its weakness and the parasite strove to counter each innovation and so maintain its access. The MHC seems unlikely to play a prominent role in such arms races because most MHC diversity is not inherited but is generated anew each generation. Instead, the MHC may play very much a second line of defence, against pathogens and parasites that have already gained access to the body. Further analysis of long-term studies that show MHC-parasite resistance correlations are essential to teasing out the role of the MHC (Paterson *et al*. 1998).

In this light, an interesting future line of research would be to put more effort into genetic control of parasites. What are the key phenotypic and genetic differences that cause changes in virulence when parasites and pathogens are introduced to nave populations? What is the relative importance of each component of the infection pathway, from behaviours that influence host–pathogen encounter rates, through nasal biochemistry and architecture, feeding behaviour and preferences, entry into to establishment and persistence within the gut, blood, or other tissue? Recent studies on St Kilda sheep indicate the powerful potential links between detection of parasite genes in natural populations and modern, genomically-based animal breeding (Smith *et al*. 1999).

9.5.3 Using genetic polymorphisms as tools

The more we understand about how different pathogen sequences evolve, the more we can reconstruct historical and current patterns of transmission. For example, modelling transmission rates between alternative hosts in multi-host systems, inferring strain-specific virulences (should they exist), and mapping evolutionary timescales onto the patterns of spread within epidemics (e.g. comparing a phylogenetic tree of morbillivirus isolates from the harbour seal epizootic with spatio-temporal patterns of disease incidence). However, this requires knowing how each marker evolves, and generalisations from mammalian studies may not hold. Examples of actual and potential problems include:

1 Whole classes of vertebrate marker systems, which are either absent or not useful in micro-organisms (microsatellites and minisatellites are the most important genetic markers in vertebrate studies but these are rare and often not polymorphic in micro-organisms).
2 Unusual modes of inheritance, such as the biparental inheritance of mitochondrial DNA in molluscs (may also be true for some nematodes), micro and macro nuclei in Protozoa and alternation between sexual and asexual forms.
3 Base-composition biases. Species vary greatly in their patterns of nucleotide base utilization and some of these difference may well be functional; for example, involved with infecting hosts with different body temperatures or living in guts regions with different pHs. Any shift in ecology could cause a radical change in the rate and nature of base substitutions that might be informative (if recognized) or confusing (if ignored).
4 It is easy to forget that the very large effective population sizes of some pathogens can almost negate the power of neutral genetic drift to cause changes in allele frequencies over the sorts of timescale we can study. Conversely, while the use of neutral markers may be of limited use in some systems, most differences can be attributed to natural selection.

9.6 The applied and veterinary perspective

Applied problems in wildlife epidemiology generally involve issues that can be broadly categorized as 'quality of life' (usually relating to conservation or animal welfare) or 'wealth creation' (perhaps more accurately 'wealth threat'). Both sets of issues may be involved in a particular problem.

Quality of life issues will arise if a wildlife disease can cause a dramatic decrease in its host population. Examples of such diseases are the phocine distemper virus, which caused up to 60% mortality in some European seal populations (Boxes 3.3 and 5.2; Harwood 1998; Swinton *et al*. 1998), and the calicivirus responsible for Rabbit haemorrhagic disease (Box 5.7). Often the management options, which are available to combat these diseases, have financial implications beyond those of the control operation itself, as may well be the case with Rabbit haemorrhagic disease. If they do, then wealth creation issues will also arise.

Wealth creation issues are created by diseases whose hosts include both wildlife species and domesticated animals. In extreme cases, such as swine fever (Box 5.6), the potential effect of the disease on domestic animals is so great that there is a national or international programme to eradicate the disease. If the proposed management solution to such a problem involves large-scale culling of the wildlife host, this will be unacceptable to some sections of society and immediately raises quality of life issues.

We believe that problems that combine quality of life and wealth-creation issues will become more and more important in wildlife epidemiology over the next decade. The problems faced by scientists who have to provide advice to managers and politicians in these complex situations constitute a major topic in themselves. Here, we consider only the scientific issues involved in investigating them.

9.6.1 Identifying objectives and priorities for management

It is important that all of the stakeholders in a particular problem are brought together at an early stage so that their expectations of what management intervention might achieve can be identified. Appropriate stakeholders include, not only those with a technical interest in the problem (wildlife epidemiologists, veterinarians, and managers) and those who will be effected economically, but also those who are concerned about the ethical implications of different management options, and those who are involved in the legal and political aspects of the problem.

An essential first step in such discussions is a critical review of the available information on the problem. An agreed and authenticated dataset should be established, which can then be used as the basis for further calculations and mathematical modelling. Ideally, a set of objectives for management, which is likely to be achievable, should be agreed and priorities assigned to each objective. However, since the different stakeholders will probably have rather different priorities and objectives, this may not be possible. Nevertheless, it should be possible to agree on a core set of objectives and to identify the constraints that the interests of the different parties are likely to impose on other objectives.

9.6.2 Mathematical models and their uses

We expect that most practical problems in the management of wildlife diseases will involve more than a single host and one of its pathogens. Begon and Bowers (1995) recognized that this was likely to be an area of growth in the mathematical modelling of wildlife diseases in the report from the Newton Institute meeting. Their view has been confirmed, not only in wildlife epidemiology but in ecology as a whole. However, they also recognized that extending simple host–pathogen models to more than two species raises formidable problems (Hudson and Greenman 1998). The number of parameters to be estimated grows very rapidly as more species are included in the model, and data to estimate these parameters is usually hard to obtain (see §9.2.2). In addition, it is often difficult to decide how many species must be included to provide a realistic representation of the situation. The latter difficulty is particularly problematic in other ecological contexts, such as food web analysis (e.g. Yodzis 1998), but it may be less so in epidemiology, because many systems are relatively well-defined.

One solution to this set of problems, which has been used successfully in other contexts such as the management of fisheries under uncertainty, is to concentrate on the analysis of a relatively simple mathematical model of the system but also to examine whether the conclusions drawn from this model are robust if additional complexity is added. This is analogous to the use of spatially implicit models to analyse the effects of spatial heterogeneity in epidemiology. Although such models can be mathematically tractable, their realism needs to be tested using more complex models, in systems where this is possible (Bolker and Grenfell 1996).

It should be appreciated that even the scientific stakeholders may have to make trade-offs in the modelling process. Theoretical ecologists prefer relatively simple models that are mathematically tractable and that can be generalized to apply to a wide range of problems. By contrast, wildlife managers will be most comfortable with models that bear a close resemblance to the specific problem that is being addressed. Clearly a spirit of compromise is needed in these circumstances. However, if mathematical models are to be used for management purposes, then the individual parameters of the model have to be interpretable and estimable. In general, the individual variables need to be measurable, although new statistical tools (§9.2.1), permit the estimation of 'hidden' (i.e. unobservable) variables.

9.6.3 Implementing models and management

The appropriate trade-off between complexity and simplicity is intimately linked to the appropriate use of the models. Developing an agreed mathematical model can help to improve understanding of a problem, because it forces stakeholders to be specific about the mechanisms that they believe are responsible for the problem. When these proposed mechanisms are recast into mathematical form it often becomes obvious that some do not, and perhaps cannot, provide a satisfactory explanation of the available observations and these can then be discarded.

However, the main use of models in applied situations is to make predictions and to provide advice for decision-making. Although more complex models usually provide a better explanation of the available data than simpler models, they often have only limited value for prediction and advice, because individual parameters will be rather poorly estimated (Hilborn and Mangel 1997). As a result, predictions made from these models will have greater uncertainty associated with them than the predictions of relatively simple models. Nevertheless, it is important to test the robustness of the predictions of these simpler models to violations of their basic assumptions. If the addition of complexity to the basic model results in very different predictions, then this should immediately raise doubts about the suitability of the simple model for management purposes. This kind of sensitivity analysis can also be used to assign priorities for future research. If predictions are particularly sensitive to changes in certain parameters in the model, then additional research is clearly required to improve the estimates of these parameters. This research may involve laboratory and field experiments, as well as additional data collection.

It is unlikely that all stakeholders will be able to agree on one basic model. However, developing more complex models, which allow the implications of different scenarios to be investigated, can accommodate the views of different parties. It will rarely be possible to estimate all, or even many, of the parameters of these complex models. But feasible ranges of values can usually be agreed, and it is then possible to see what outcomes these might lead to. Again, if certain scenarios result in predicted outcomes, which are very different from those obtained with the simple models, then further research is clearly required.

The modelling process can be considered a success if it helps to ensure that resources are used efficiently both in management and research relating to the problem, and if it help to identify the most effective control or preventative measures. For the latter, some kind of risk evaluation may be most appropriate, so that probabilities can be assigned to the potential outcomes of different management strategies.

References

Adler, F. R. and Kretzschmar, M. (1992). Aggregation and stability in host-parasite models. *Parasitology* **104:** 199–205.

Aitkin, M., Anderson, D., Francis, B. and Hinde, J. (1989). *Statistical modelling in GLIM.* Oxford, Oxford University Press.

Alexander, H. M., Thrall, P. H., Antonovics, J., Jarosz, J. M. and Oudemans, P. V. (1996). Population dynamics and genetics of plant disease: a case study of anther-smut disease. *Ecology* **77:** 990–996.

Alexander, J. and Stimson, W. H. (1988). Sex hormones and the course of parasitic infection. *Parasitology Today* **4:** 189–193.

Alexander, K. and Appel, M. (1994). African wild dogs (*Lycaon pictus*) endangered by a canine distemper epizootic among domestic dogs near the Masai Mara National Reserve, Kenya. *Journal of Wildlife Diseases* **30:** 481–485.

Alexander, K. A., Kat, P. W., Munson, M. A., Kalake, A. and Appel, M. J. G. (1996). Canine distemper-related mortality among wild dogs (*Lycaon pictus*) in Chobe National Park, Botswana. *Journal of Zoo and Wildlife Medicine* **27:** 426–427.

Alexander, K. A., Smith, J. S., Macharia, M. J. and King, A. A. (1993). Rabies in the Masai Mara, Kenya: a preliminary report. *Onderstepoort Journal of Veterinary Research* **60:** 411–414.

Alford, R. A. and Stephens, S. J. (1997). Lack of evidence for epidemic disease as an agent in the catastrophic decline of Australian rain forest frogs. *Conservation Biology* **11:** 1026–1029.

Allen, T. J. (1986). Evaluation of movements, harvest rate, vulnerability and survival of translocated raccoons in Southeastern Virgina. *Transactions of the Northeast Sector of Wildlife Society* **43:** 64.

Allen, J. C., Schaffer, W. M. and Rosko, D. (1993). Chaos reduces species extinction by amplifying local population noise. *Nature* **364:** 229–232.

Altizer, S. M. (1998). *Ecological and evolutionary interactions between Monarch butterflies and the protozoan parasite,* Ophryocystis elektroscirrha. Department of Ecology, Evolution and Behavior. Minnesota, University of Minnesota: 164.

Altizer, S. M., Oberhauser, K. S. and Brower, L. P. (2000). Associations between host migration and the prevalence of a protozoan parasite in natural populations of adult monarch butterflies. *Ecological Entomology* **25:** 125–139.

Altman, J. D., Moss, P. A. H., Goulder, P. J. R., Barouch, D. H., McHeyzer Williams, M. G., Bell, J. I. *et al.* (1996). Phenotypic analysis of antigen-specific T lymphocytes. *Science* **274:** 94–96.

Anderson, J. F., Johnson, R. C. and Magnarelli, L. A. (1987). Seasonal prevalence of *Borrelia burgdorferi* in natural populations of white footed mice, *Peromyscus leucopus*. *Journal of Clinical Microbiology* **25:** 1564–1566.

Anderson, R. M. (1995). Evolutionary pressures in the spread and persistence of infectious agents in vertebrate populations. *Parasitology* **111:** S15–S31.

Anderson, R. M. (1998). Complex dynamic behaviours in the interaction between parasite populations and the host's immune system. *International Journal for Parasitology* **28:** 551–566.

Anderson, R. C. (2000). *Nematode parasites of vertebrates. Their development and transmission.* (2nd edn). London, CABI Pubishing.

Anderson, R. M. and Crombie, J. A. (1984). Experimental studies of age-prevalence curves for *Schistosoma mansoni* infections in populations of *Biomphalaria galbrata. Parasitology* **89:** 79–105.

Anderson, R. M. and Gordon, D. M. (1982). Processes influencing the distribution of parasite numbers within host populations with special emphasis on parasite-induced host mortalities. *Parasitology* **85:** 373–398.

Anderson, R. M. and May, R. M. (1978). Regulation and stability of host-parasite population interactions. I. Regulatory processes. *Journal of Animal Ecology* **47:** 219–247.

Anderson, R. M. and May, R. M. (1982a). Coevolution of hosts and parasites. *Parasitology* **85:** 411–426.

Anderson, R. M. and May, R. M. (ed.) (1982b). *Population biology of infectious diseases.* Berlin, Springer Verlag.

Anderson, R. M. and May, R. M. (1985). Age-related changes in the rate of disease transmission: implications for the design of vaccination programmes. *Journal of Hygiene (Cambridge)* **94:** 365–436.

Anderson, R. M. and May, R. M. (1991). *Infectious disease of humans: dynamics and control*. Oxford, Oxford University Press.

Anderson, R. M., Jackson, H. C., May, R. M. and Smith, A. M. (1981). Population dynamics of fox rabies in Europe. *Nature* **289:** 765–777.

Anderson, T. J. C. (1995). Ascaris infections in humans from north-america—molecular evidence for cross-infection. *Parasitology* **110:** 215–219.

Anderson, T. J. C. and Jaenike, J. (1997). Host specificity, evolutionary relationships and macrogeographic differentiation among *Ascaris* populations from humans and pigs. *Parasitology* **115:** 325–342.

Anderson, T. J. C., Romeroabal, M. E. and Jaenike, J. (1993). Geneticstructure and epidemiology of *Ascaris* populations—patterns of host affiliation in Guatemala. *Parasitology* **107:** 319–334.

Anderson, T. J. C., Romeroabal, M. E. and Jaenike, J. (1995). Mitochondrial-DNA and Ascaris microepidemiology—the composition of parasite populations from individual hosts, families and villages. *Parasitology* **110:** 221–229.

Anderson, T. J. C., Blouin, M. S. and Beech, R. N. (1998). Population biology of parasitic nematodes: applications of genetic markers. *Advances in Parasitology* **41:** 219–283.

Andersson, M. (1994). *Sexual selection*. Princeton, New Jersey, Princeton University Press.

Andreasen, V., Lin, J. and Levin, S. A. (1997). The dynamics of circulating influenza strains conferring partial cross-immunity. *Journal of Mathematical Biology* **35:** 825–842.

Apanius, V., Penn, D., Slev, P. R., Ruft, L. R. and Potts, W. K. (1997). The nature of selection on the major histocompatibility complex. *Critical Reviews in Immunology* **17:** 179–224.

Arneberg, P., Skorping, A. and Read, A. F. (1997). Is population density a species character? Comparative analyses of the nematode parasites of mammals. *Oikos* **80:** 289–300.

Arneberg, P., Skorping, A. and Read, A. F. (1998a). Parasite abundance, body size, life histories and the energetic equivalence rule. *American Naturalist* **151:** 497–513.

Arneberg, P., Skorping, A., Grenfell, B. T. and Read, A. F. (1998b). Host densities as determinants of abundance in parasite communities. *Proceedings of the Royal Society of London Series B-Biological Sciences* **265:** 1283–1289.

Augustine, D. J. (1998). Modelling *Chlamydia*–koala interactions: coexistence, population dynamics and conservation implications. *Journal of Applied Ecology* **35:** 261–272.

Awerbuch, T., Sandberg, S. and Spielman, A. (1992). Critical abundance of hosts perpetuating the tick that transmits the agent of Lyme diseases. *American Journal of Tropical Medicine and Hygiene* **45:** S205.

Baker, W. L., Egbert, S. L. and Fiazier, G. F. (1992). A spatial model for studying the effects of climatic change on the structure of landscapes subject to large disturbances. *Ecological Modelling* **56:** 109–126.

Ballou, J. D. (1993). Assessing the risks of infectious diseases in captive breeding and reintroduction programs. *Journal of Zoo and Wildlife Medicine* **24:** 327–335.

Bancroft, D. R., Pemberton, J. M., Albon, S. D., Robertson, A., Maccoll, A. D. C., Smith, J. A. *et al.* (1995a). Molecular-genetic variation and individual survival during population crashes of an unmanaged ungulate population. *Philosophical Transactions of the Royal Society of London Series B-Biological Sciences* **347:** 263–273.

Bancroft, D. R., Pemberton, J. M. and King, P. (1995b). Extensive protein and microsatellite variability in an isolated, cyclic ungulate population. *Heredity* **74:** 326–336.

Barlow, N. D. (1993). A model for the spread of bovine tubercolosis in New Zealand possum populations. *Journal of Applied Ecology* **30:** 156–164.

Barlow, N. D. (1995). Critical evaluation of wildlife disease models. In: *Ecology of infectious diseases in natural populations*. (ed. B. T. Grenfell and A. P. Dobson). Cambridge, Cambridge University Press: 230–259.

Barlow, N. D. (1996). The ecology of wildlife disease control: simple disease models revisited. *Journal of Applied Ecology* **33:** 303–314.

Barlow, N. D. and Kean, J. M. (1998). Simple models for the impact of rabbit calicivirus disease (RCD) on australasian rabbits. *Ecological Modelling* **109:** 225–241.

Barnes, A. I. and Siva-Jothy, M. T. (2000). Density-dependent prophylaxis in the mealworm beetle *Tenebrio molitor* L-(Coleoptera : Tenebrionidae): cuticular melanization is an indicator of investment in immunity. *Proceedings of The Royal Society of London Series B-Biological Sciences* **267:** 177–182.

Barrett, T., Blixenkrone-Møller, M., Diguardo, G., Domingo, M., Duignan, P., Hall, A. *et al.* (1995). Morbilliviruses in aquatic mammals: report on round-table discussion. *Veterinary Microbiology* **44:** 261–265.

Bartlett, M. S. (1956). Deterministic and stochastic models for recurrent epidemics. *Proceeding of the third Berkely symposium on mathematical statistics and probability*. (ed. J. Neyman). Berkeley, University of California Press: 81–109.

Bartlett, M. S. (1957). Measles periodicity and community size. *Journal of the Royal Statistical Society (A)* **120:** 48–70.

Bartlett, M. S. (1960). *Stochastic population models in ecology and epidemiology*. London, Methuen.

Baskin, Y. (1998). Home on the range. *BioScience* **48:** 245–251.

Bassetti, D., Cristofolini, A., De Venuto, G., Costanzi, C., Chemini, C., Ciufolini, M. G. *et al.* (1993). Studi sulla diffusione del TBE virus in Trentino. *Rivista Medica Trentina* **31:** 57–63.

Baxby, D. and Bennett, M. (1997). Cowpox: a re-evaluation of the risks of human cowpox based on new epidemiological information. *Archives of Virology* **13:** S1–S12.

Beck, L. R., Rodriguez, M. H. and Dister, S. W. (1994). Remote sensing as a landscape epidemiologic tool to indentify villages at high risk for malaria transmission. *American Journal of Tropical Medicine and Hygiene* **51:** 271–280.

Beck, L. R., Wood, B. L. and Dister, S. W. (1995). Remote sensing and GIS—new tools for mapping human health. *Geographic Information Systems* **5:** 32–37.

Begon, M. and Bowers, R. G. (ed.) (1995). *Beyond host-pathogen dynamics. Ecology of infectious diseases in natural populations*. Cambridge, Cambridge University Press.

Begon, M., Harper, J. L. and Townsend, C. R. (1996a). *Ecology: individuals, populations and communities*. Oxford, Blackwell Scientific Publications.

Begon, M., Sait, S. M. and Thompson, D. J. (1996b). Predator-prey cycles with period shifts between two- and three-species systems. *Nature* **381:** 311–315.

Begon, M., Feore, S. M., Brown, K., Chantiey, J., Jones, T. and Bennett, M. (1998). Population and transmission dynamics of cowpox in bank voles: testing fundamental assumptions. *Ecology Letters* **1:** 82–86.

Begon, M., Hazel, S. M., Baxby, D., Brown, K., Cavanagh, R., Chantiey, J. *et al.* (1999). Transmission dynamics of a zoonotic pathogen within and between wildlife host species. *Proceedings of the Royal Society of London Series B-Biological Sciences* **266:** 1939–1945.

Bellaby, T., Robinson, K., Wakelin, D. and Behnke, J. M. (1995). Isolates of *Trichuris muris* vary in their ability to elicit protective immune-responses to infection in mice. *Parasitology* **111:** 353–357.

Bellaby, T., Robinson, K. and Wakelin, D. (1996). Induction of differential T-helper-cell responses in mice infected with variants of the parasitic nematode *Trichuris muris*. *Infection and Immunity* **64:** 791–795.

Bennett, M. and Baxby, D. (1996). Cowpox. *Journal Of Medical Microbiology* **45:** 157–158.

Berding, C., Keymer, A. E., Murray, J. D. and Slater, A. F. G. (1986). The population-dynamics of acquired-immunity to helminth infection. *Journal of Theoretical Biology* **122:** 459–471.

Berger, J. (1990). Persistence of different-sized populations: an empirical assessment of rapid extinctions in bighorn sheep. *Conservation Biology* **4:** 91–98.

Berger, L., Speare, R., Daszak, P., Green, D. E., Cunningham, A. A., Goggin, C. L. *et al.* (1998). Chytidiomycosis causes amphibian mortality associated with population eclines in the rainforests of Australia and Central America. *Proceedings of the National Academy of Science, USA.* **95:** 9031–9036.

Bishop, G. C. (1995). Canine rabies in south Africa. *Proceedings of the Third International Conference of the Southern and Eastern African Rabies Group*, March 1995, Harare.

Bjørnstad, O. and Grenfell, B. T. (2001). Noisy clockwork: time series analysis of population fluctuations in animals. *Science* (in press).

Blaxter, M. L., De Ley, P., Garey, J. R., Liu, L. X., Scheoleman, P., Vierstraete, A. *et al.* (1998). A molecular evolutionary framework for the phylum Nematoda. *Nature* **392:** 71–75.

Bliss, C. I. and Fisher, R. A. (1953). Fitting the negative binomial distribution to biological data. *Biometrics* **9:** 176–199.

Blouin, M. S., Dame, J. B., Tarcaut, C. A., and Courtney, C. H. (1992). Unusual population-genetics of a parasitic nematode—mtDNA variation within and among populations. *Evolution* **46:** 470–476.

Blouin, M. S., Yowell, C. A., Courtney, C. H. and Dame, J. B. (1995). Host movement and the genetic-structure of populations of parasitic nematodes. *Genetics* **141:** 1007–1014.

Blower, S. M. and Dowlatabadi, H. (1994). Sensitivity and uncertainty analysis of complex models of disease transmission. *International Statistical Review* **62:** 229–243.

Bolasfernandez, F. and Wakelin, D. (1992). Immunization against geographical isolates of *Trichinella spiralis* in mice. *International Journal for Parasitology* **22:** 773–781.

Bolker, B. and Grenfell, B. (1995). Space, persistence and dynamics of measles epidemics. *Philosophical Transactions of The Royal Society of London Series B-Biological Sciences* **348:** 309–320.

Bolker, B. M. and Grenfell, B. T. (1996). Impact of vaccination on the spatial correlation and persistence of measles dynamics. *Proceedings of The National Academy of Sciences of The United States of America* **93:** 12648–12653.

Bonsall, M. B. and Hassell, M. P. (1997). Apparent competition structures ecological assemblages. *Nature* **388:** 371–373.

Bonsall, M. B. and Hassell, M. P. (1998). Population dynamics of apparent competition in a host-parasitoid assemblage. *Journal of Animal Ecology* **67:** 918–929.

Boone, J. D., Otteson, E. W., McGwire, K. C., Villard, P., Rowe, J. E. and Stjeor, S. C. (1998). Ecology and demographics of hantavirus infections in rodent populations in the Walker River Basin of Nevada and California. *American Journal of Tropical Medicine and Hygiene* **59:** 445–451.

Booth, D. T., Clayton, D. H. and Block, B. A. (1993). Experimental demonstration of the energetic cost of parasitism in free-ranging hosts. *Proceedings of the Royal Society of London Series B-Biological Sciences* **253:** 125–129.

Boots, M. and Begon, M. (1993). Trade-offs with resistance to a granulosis-virus in the indian meal moth, examined by a laboratory evolution experiment. *Functional Ecology* **7:** 528–534.

Boulinier, T., Sorci, G., Monnat, J. Y. and Danclun, E. (1997a). Parent-offspring regression suggests heritable susceptibility to ectoparasites in a natural population of kittiwake *Rissa tridactyla*. *Journal of Evolutionary Biology* **10:** 77–85.

Boulinier, T., Sorci, G., Clobert, J. and Danchin, E. (1997b). An experimental study of the costs of reproduction in the kittiwake *Rissa tridactyla*: comment. *Ecology* **78:** 1284–1287.

Bouloux, C., Langlais, M. and Silan, P. (1998). A marine host-parasite model with direct biological cycle and age structure. *Ecological Modelling* **107:** 73–86.

Bouvier, G. and Burgisser, H. (1958). *La brucellose du chamois. Les maladies des ruminants sauvages de la Suisse*. Lausanne, Switzerland, Fondation B. Galli Valerio: 111–113.

Bowers, M. A. and Harris, L. C. (1994). A large-scale metapopulation model of interspecific competition and environmental change. *Ecological Modelling* **72:** 251–273.

Boxshall, G. A. (1974). The population dynamics of *Lepeophtheirus pectoralis* (Muller): dispersion pattern. *Parasitology* **69:** 373–390.

Boyd, H. E. G. (1999). *Early development of parasitism in Soay sheep*. Zoology Department, Cambridge, University of Cambridge.

Brabin, L. (1992). The epidemiological significance of chagas-disease in women. *Memorias Do Instituto Oswaldo Cruz* **87:** 73–79.

Bradley, D. J. (1982) Epidemiological Models-theory and reality. In: *Population dynamics of infectious diseases: Theory and applications* (ed. R. M. Anderson). Chapman and Hall, London. 320–333.

Braisher, T. L. (1999). *Genetic variation in trichostrongylid parasites of the Soay sheep of St. Kilda*. Genetics department, Cambridge, University of Cambridge: 190.

Breiman, L., Friedman, J. *et al.* (1984). *Classification and regression trees*. Pacifica Grove, CA, Wadsworth & Brookes/Cole.

Brooks, R. (1990). Survey of the dog population of Zimbabwe and its level of rabies vaccination. *Veterinary Record* **127:** 592–596.

Brower, L. P., Fink, L. S., Brower, A. V., Leong, K., Oberhauser, K., Altizer, S. *et al.* (1995). On the dangers of interpopulational transfers of monarch butterflies—discussion. *Bioscience* **45:** 540–544.

Brown, C. R., Brown, M. B. and Rannala, B. (1995). Ectoparasites reduce long-term survival of their avian host. *Proceedings of the Royal Society of London Series B-Biological Sciences* **262:** 313–319.

Brown, J. A., Cheeseman, C. L. and Harris, S. (1992). Studies on the spread of bovine tuberculosis from badgers to cattle. *Journal of Zoology* **227:** 694–696.

Brown, J. H. (1995). *Macroecology*. Chicago, University of Chicago Press.

Brown, J. H. and Kodrick-Brown, A. (1977). Turnover rates in insular biogeography: effect of immigration on extinction. *Ecology* **58:** 445–449.

Brown, S. P. (1999). Cooperation and conflict in host-manipulating parasites. *Proceedings of the Royal Society of London Series B-Biological Sciences* **266:** 1899–1904.

Buitkamp, J., Filmether, P., Stear, M. J. and Epplen, J. T. (1996). Class I and class II major histocompatibility complex alleles are associated with faecal egg counts following natural, predominantly *Ostertagia circumcincta* infection. *Parasitology Research* **82:** 693–696.

Bundy, D. A. P. (1988). Gender-dependant patterns of infection and disease. *Parasitology Today* **4:** 186–189.

Burdon, J. J. (1991). Fungal pathogens as selective forces in plant-populations and communities. *Australian Journal of Ecology* **16:** 423–432.

Burgdorfer, W. and. Schwan, T. G (1998). Lyme borreliosis: a relapsing fever-like disease. *Scandinavian Journal of Infectious Diseases* **77:** 17–22.

Burgess, E. C., French, J. B. and Fitzpatrick, A. G. (1990). Systemic disease in *Peromyscus leucpous* associated with *Borrelia burgdorferi* infection. *American Journal of Tropical Medicine and Hygiene* **42:** 254–259.

Bush, A. O. and Homes, J. C. (1986). Intestinal helminths of lesser scaup ducks: an interactive community. *Canadian Journal of Zoology* **64:** 142–152.

Bush, A. O. (1990). Helminth communities in avian hosts: determinants of pattern. In: *Parasite communities: patterns and processes*. (ed. G. W. Esch, A. O. Bush, and J. M. Aho). London, Chapmann and Hall: 197–232.

Bush, A. O., Lafferty, K. D., Lotz, J. M. and Shostak, A. W. (1997). Parasitology meets ecology in its own terms: Margolis *et al.* revisited. *Journal of Parasitology* **83:** 575–583.

Calder, W. A. (1987). Metabolic allometry—basic correlations are independent of units when properly converted. *Journal of Theoretical Biology* **128:** 523–524.

Campbell, A., (ed.) (1999). *Declines and disapppearances of Australian frogs*. Canberra, Environment Australia.

Campbell, K. and Hofer, H. (1995). People and wildlife: spatial dynamics and zones of interaction. *Serengeti II: Dyanmics, management and conservation of an ecosystem*. (ed. A. R. E. Sinclair and P. Arcese). Chicago, The University of Chicago Press: 534–570.

Carey, C., Cohen, N., Rollins-Smith, L. (1999). Amphibian declines: an immunological perspective. *Developmental and Comparative Immunology* **23:** 459–472.

Carpenter, J. W., Appel, M. J. G., Erickson, R. C. and Novilla, M. N. (1976). Fatal vaccine-induced canine distemper virus infection in black-footed ferrets. *Journal of the American Veterinary Medical Association* **169:** 961–964.

Carpenter, M. A., Appel, M. J. G., Roelke-Parker, M. E., Munson, L., Hofer, H., East, M. *et al.* (1998). Genetic characterization of canine distemper virus in Serengeti carnivores. *Veterinary Immunology and Immunopathology* **65:** 259–266.

Carter, K. C. and Wilson, P. A. G. (1989). The course of infection in rats given small primary doses of *Strongyloides ratti* and *Strongyloides venezuelensis*. *Journal of Helminthology* **63:** 107–114.

Caruso, G., Mondardini, V., Granata, C., Mei, R., Marcolina, D., Bentiveuna, S. *et al.* (1997). Risultati di un'indagine virologica e sieroepidemiologica sull'infezione da TBE virus (TBEV) in provinca di Belluno. *Giornale Italiano di Malattie Infettive* **3:** 50–52.

Castle, M. D. and Christensen, B. M. (1990). Hematazoa of wild turkeys from the midwestern United States: translocation of wild turkeys and its potential role in the introduction of *Plasmodium kempi*. *Journal of Wildlife Diseases* **26:** 180–185.

Cattadori, I. M., Hudson, P. J., Merler, S. and Rizzoli, A. P. (1999). Synchrony, scale and temporal dynamics of rock partridge (*Alectoris graeca saxatilis*) populations in the Dolomites. *Journal of Animal Ecology* **68:** 540–549.

Center of Disease Control (1995). Lyme Disease Surveillance Summary. *Surveillance Summary* **6:** 1–11.

Center of Disease Control (1998). *Statewide surveillance for Ehrlichiosis—Connecticut and New York, 1994*. Atlanta, CDC.

Charnov, E. L. (1993). Is maximum sustainable-yield independent of body size for mammals (and others). *Evolutionary Ecology* **7:** 309–311.

Chatfield, C. (1980). *The analysis of time series: an introduction*. London, Chapman and Hall.

Cheeseman, C. L., Wilesmith, J. W., Stuart, F. A. and Mallinson, P. J. (1988a). Dynamics of tuberculosis in a naturally infected badger population. *Mammal Review* **18:** 61–72.

Cheeseman, C. L., Cresswell, W. J., Harris, S. and Mallinsonson, P. J. (1988b). Comparison of dispersal and other movements in 2 badger (*Meles meles*) populations. *Mammal Review* **18:** 51–59.

Clayton, D. H. (1991). The influence of parasites on host sexual selection. *Parasitology Today* **7:** 329–334.

Clayton, D. H. and Tompkins, D. M. (1994). Ectoparasite virulence is linked to mode of transmission. *Proceedings of the Royal Society of London Series B-Biological Sciences* **256:** 211–217.

Clayton, D. H., Lee, P. L. M., Tompkins, D. M. and Blodie, E. D. (1999). Reciprocal natural selection on host-parasite phenotypes. *American Naturalist* **154:** 261–270.

Cleaveland, S. (1996). *The epidemiology of rabies and canine distemper in the Serengeti, Tanzania*. University of London, unpublished PhD thesis.

Cleaveland, S. (1997). Dog vaccination around the Serengeti. *Oryx* **31:** 13–14.

Cleaveland, S. (1998). The growing problem of rabies in Africa. *Transactions of the Royal Society of Tropical Medicine and Hygiene* **92:** 131–134.

Cleaveland, S. and Dye, C. (1995). Maintenance of a microparasite infecting several host species: rabies in the Serengeti. *Parasitology* **111:** S33–S47.

Cleaveland, S., Appel, M. G. A., Chalmers, W. S. K., Chillingworth, K., Kaare, M. and Dye, C. (2000). Serological and demographic evidence for domestic dogs as a source of canine distemper virus infection for Serengeti wildlife. *Veterinary Microbiology* **72:** 217–227.

Cleaveland, S., Fèvre, E., Coleman, P., Coen, P. and Mlengeya, T. (2001). Rabies in Tanzania: impact and control. *Proceedings of the 1st Annual Scientific Conference of the Tanzania Wildlife Research Institute*, December 2000, Arusha, Tanzania.

Clifton-Hadley, R. S. (1998). DNA fingerprinting of *Mycobacterium bovis* isolates using spoligotyping: epidemiological issues. *Proceedings of a Meeting held at the East County Hotel, Ennis Co. Clare* on the 25th, 26th and 27th March 1998, Society for Veterinary Epidemiology and Medicine.

Clifton-Hadley, R. S., Wilesmith, J. W., Richards, M. S., Upton, P. and Johnston, S. (1995). The occurrence of *Mycobacterium bovis* infection in cattle in and around an area subject to extensive badger (*Meles meles*) control. *Epidemiology and Infection* **114:** 179–193.

Clutton-Brock, T. H., Price, O. F., Albon, S. D. and Jewell, P. A. (1991). Persistent instability and population regulation in soay sheep. *Journal of Animal Ecology* **60:** 593–608.

Cohn, J. P. (1991). New focus on wildlife health: tracking and controlling animal disease in the wild may be important to conservation efforts. *BioScience* **41:** 448–450.

Coltman, D. W., Pilkington, J. G., Smith, J. A. and Pembertan, J. M. (1999). Parasite-mediated selection against inbred Soay sheep in a free-living island population. *Evolution* **53:** 1259–1267.

Colwell, R. R. (1998). Global climate and infectious disease: the cholera paradigm. *Science* **264:** 2025–2031.

Coman, B. (1997). *Rabbit control and rabbit calicivirus disease: a field handbook for land managers in Australia*. Gillingham, Adelaide.

Comins, H. N., Hassell, M. P. and May, R. M. (1992). The spatial dynamics of host-parasitoid systems. *Journal of Animal Ecology* **61:** 735–748.

Corner, A. H. and Connell, R. (1958). Brucellosis in bison, elk and moose in Elk Island National Park, Alberta, Canada. *Canadian Journal of Comparative Medicine* **22:** 9–21.

Corti, R., Gibert, P., Gindre, R., Landry, P. and Sarrazin, C. (1984). *Donnes sur la biometrie et l'état sanitaire du chamois (Rupicapra rupicapra) dans le nord-est du massif des Ecrins (Hautes Alpes)*. Revue O. N.C. Gibier et Faune Sauvage.

Costantino, R. F., Desharnais, R. A., Cushing, J. M. and Dennis, B. (1997). Chaotic dynamics in an insect population. *Science* **275:** 389–391.

Coulson, T., Milner-Gulland, E. J. and Clutton-Brock, T. (2000). The relative roles of density and climatic variation on population dynamics and fecundity rates in three contrasting ungulate species. *Proceedings of the Royal Society of London Series B-Biological Sciences* **267:** 1771–1779.

Courchamp, F., Pontier, D., Langlais, M. and Artois, M. (1995). Population dynamics of feline immunodeficieny virus within cat populations. *Journal of Theoretical Biology* **175:** 553–560.

Courchamp, F., Say, L. and Pointer, D. (2000). Detection, identification, and correction of a bias in an epidemiological study. *Journal of Wildlife Diseases* **36:** 71–78.

Cowan, P. E. (1990). Brushtail possum. In: *The handbook of New Zealand mammals*. (ed. C. M. King). Oxford, Oxford University Press: 68–98.

Cowan, P. E. (1996). Possum biocontrol: prospects for fertility regulation. *Reproduction fertility and Development* **8:** 655–660.

Coyne, M. J. and Smith, G. (1994). Trichostrongylid parasites of domestic ruminants. In: *Parasites and infectious diseases*. (ed. M. E. Scott and G. Smith). San Diego, Academic Press: 235–248.

Craine, N. G., Randolph, S. E. and Nuttall, P. A. (1995). Seasonal-variation in the role of grey squirrels as hosts of *Ixodes ricinus*, the tick vector of the Lyme-disease spirochete, in a british woodland. *Folia Parasitologica* **42:** 73–80.

Craine, N. G., Nuttall, P. A., Maniott, A. C. and Randolph, S. E. (1997). Role of grey squirrels and pheasants in the transmission of *Borrelia burgdorferi sensu lato*, the Lyme disease spirochaete, in the UK. *Folia Parasitologica* **44:** 155–160.

Crawley, M. J. (1993). *GLIM for ecologists*. London, Blackwell Scientific Publications.

Crnokrak, P. and Roff, (1999). Inbreeding depression in the wild. *Heredity* **83:** 260–270.

Crofton, H. D. (1971). A quantitative approach to parasitism. *Parasitology* **62:** 179–193.

Crombie, J. A. and Anderson, R. M. (1985). Population dynamics of *Schistosoma mansoni* in mice repeatedly exposed to infection. *Nature* **315:** 491–493.

Crouch, A. C., Baxby, D., McCracken, C. M., Gaskell, R. M. and Bennett, M. (1995). Serological evidence for the reservoir hosts of cowpox virus in british wildlife. *Epidemiology and Infection* **115:** 185–191.

Cunningham, A. A. (1996). Disease risks of wildlife translocations. *Conservation Biology* **10:** 349–353.

Cushing, J. M., Dennis, B., Desharnais, R. A. and Constantino, R. F. (1998). Moving towards an unstable equilibrium: saddle nodes in population systems. *Journal of Animal Ecology* **67:** 298–306.

Czub, S., Duray, P. H., Thomas, R. E. and Schwann, T. G. (1992). Cystitis induced by infection with the Lyme disease spirochete, *Borrelia burgdorferi*, in mice. *American Journal of Pathology* **141:** 1173–1179.

D'Amico, V., Elkinton, J. S., Dwyer, G., Burard, G. P. and Buonaccorsi, G. P. (1996). Virus transmission in gypsy moths is not a simple mass action process. *Ecology* **77:** 201–206.

Daszak, P., Berger, L., Cunningham, A. A., Hyatt, A. D., Green, D. E. and Speare, R. (1999). Emerging infectious diseases and amphibian population declines. *Emerging Infectious Diseases* **5:** 735–748.

Davey, R. B. (1981). Effects of *Babesia bovis* on the ovipositional success of the southern cattle tick, *Boophilus microplus. Annals of the Entomological Society of America* **74:** 331–333.

Davies, C. M., Webster, J. P., Kruger, O., Munatsi, A., Nocumba, J. and Woolhouse, M. E. J. (1999). Host-parasite population genetics: a cross-sectional comparison of *Bulinus globosus* and *Schistosoma haematobium. Parasitology* **119:** 295–302.

Degiorgis, M. P., Frey, J., Nicolet, J., Abdo, E. M., Fatzer, R., Schlatter, H. *et al.* (2000). An outbreak of infectious keratoconjunctivitis in Alpine chamois (*Rupicapra r. rupicapra*) in Simmental-Gruyeres, Switzerland. *Schweizer Archiv Fur Tierheilkunde* **142:** 520–527.

De Jong, M., Diekmann, O. and Heesterbeck, J. A. P. (1995). How does transmission of infection depend on population size? In: *Epidemic models, their structure and relation to data*. (In: D. Mollison). Cambridge, Cambridge University Press: 84–94.

DeLeo, G. A. and Dobson, A. P. (1996). Allometry and simple epidemic models for microparasites. *Nature* **379:** 720–722.

De Leo, G. A. and Gatto, M. (1996). *Model of optimal body size in parasites of wild mammals*, 10–15 Settembre, Società Italiana di Ecologia, Napoli.

De Vos, V. and Scheepers, G. J. (1996). Remote mass vaccination of large free-ranging wild animals for anthrax using Sterne spore vaccine. *Salisbury Medical Bulletin* **87:** 116–121.

Dennis, B. and Patil, G. P. (1984). The Gamma-distribution and weighted multimodal Gamma-distribution as model of population abundance. *Mathematical Biosciences* **68:** 187–212.

Dennis, B., Desharnais, R. A., Cosming, J. M. and Constantino, R. M. (1995). Nonlinear demographic dynamics: mathematical models, statistical methods, and biological experiments. *Ecological Monographs* **65:** 261–281.

Derothe, J. M., Loubes, C., Orth, A., Renauld, F. and Moulia, C. (1997). Comparison between patterns of pinworm infection (*Aspiculuris tetraptera*) in wild and laboratory strains of mice, *Mus musculus*. *International Journal for Parasitology* **27:** 645–651.

Diekmann, O. and Kretzschmar, M. (1991). Patterns in the effects of infectious diseases on population growth. *Journal of Mathematical Biology* **29:** 539–570.

Diekmann, O. and Heesterbeek, J. A. P. (2000). *Mathematical epidemiology of infectious diseases: model building, analysis and interpretation*. Chichester John Wiley and Sons.

Diekmann, O., Heesterbeek, J. A. P. and Metz, J. A. J. (1990). On the definition and the computation of the basic reproduction ratio R_0 in models for infectious diseases in heterogenous populations. *Journal of Mathematical Biology* **28:** 365–382.

Diekmann, O., de Jong, M. C. M., de Koeijer, A. A. and Reijnders, P. (1996). The force of infection in populations of varying size: a modelling problem. *Journal of Biological Systems* **3:** 519–529.

Dietz, K. (1982). Overall population patterns in the transmission cycle of infectious agents. In: *Population biology of infectious diseases*. (ed. R. M. Anderson and R. M. May). Berlin, Springer: 87–102.

Dietz, K. (1992). The estimation of the basic reproduction number for infectious diseases. *Statistical Methods in Medical Research* **2:** 23–41.

Dobson, A. P. (1985). The population dynamics of competition between parasites. *Parasitology* **91:** 317–347.

Dobson, A. P. (1986). Inequalities in the individual reproductive success of parasites. *Parasitology* **92:** 675–682.

Dobson, A. P. (1988a). The population biology of parasite-induced changes in host behavior. *Quarterly Review of Biology* **63:** 139–165.

Dobson, A. P. (1988b). Restoring island ecosystems: the potential of parasites to control introduced mammals. *Conservation Biology* **2:** 31–39.

Dobson, A. P. (1990). Models for multi-species parasite-host communities. In: *Parasite communities: patterns and processes*. (ed. G. Esch, A. O. Bush and J. M. Aho). London, Chapmann and Hall: 261–288.

Dobson, A. P. (1995). Infectious diseases and human population history.Throughout history the establishment of disease has been a side effect of the growth of civilization. *Bioscience* **46:** 115–126.

Dobson, A. P. and Hudson, P. J. (1992). Regulation and stability of a free-living host parasite system—*Trichostrongylus tenuis* in Red Grouse. II: Population models. *Journal of Animal Ecology* **61:** 487–498.

Dobson, A. P. and May, R. M. (1986). Patterns of invasions by pathogens and parasites. In: *Ecology of biological invasions of North America and Hawaii*. (ed. H. A. Mooney and J. A. Drake). Berlin Springer-Verlag: 58–76.

Dobson, A. P. and May, R. M. (1986). *Disease and conservation. Conservation Biology: The Science of Scarcity and Diversity*. M. E. Soulé. Sunderland, MA, Sinauer Associates.

Dobson, A. and Meagher, M. (1996). The population dynamics of brucellosis in the Yellowstone National Park. *Ecology* **77:** 1026–1036.

Dobson, A. P. and Pacala, S. W. (1992). The parasites of *Anolis* lizards the northern lesser antilles. 2. The structure of the parasite community. *Oecologia* **91:** 118–125.

Dobson, A. P. and Roberts, M. G. (1994). The population dynamics of parasitic helminth communities. *Parasitology* **109:** S97–S108.

Dobson, A. P., Pacala, S. V., Roughgarden, J. D., Carper, E. R. and Harrics, E. A. (1992). The parasites of *Anolis* lizards in the northern lesser antilles. 1. Patterns of distribution and abundance. *Oecologia* **91:** 110–117.

Dobson, R. J., Waller, P. J. and Donald, A. D. (1990). Population dynamics of *Trichostrongylus colubriformis* in sheep: the effect of host age on the establishment of infective larvae. *International Journal for Parasitology* **20:** 353–357.

Donahue, J. G., Piesman, J. and Spielman, A. (1987). Reservoir competence of white-footed mice for Lyme disease spirochetes. *American Journal of Tropical Medicine and Hygene* **36:** 92–96.

Drazin, P. G. (1992). *Nonlinear systems*. Cambridge, Cambridge University Press.

Drepper, F. R., Engbert, R. and Stollenwerk, N. (1994). Nonlinear time-series analysis of empirical population-dynamics. *Ecological Modelling* **75:** 171–181.

Dunning, J. B. J., Stewart, D. J., Danieloon, B. J., Noon, B. R., Root, T. L., Lamberson, R. H. *et al.* (1995). Spatially explicit population models: current forms and future uses. *Ecological Applications* **5:** 3–11.

Dwyer, G., Levin, S. A. and Buttel, I. (1990). A simulation-model of the population-dynamics and evolution of myxomatosis. *Ecological Monographs* **60:** 423–447.

Dwyer, G., Elkinton, J. S. and Buonaccorsi, J. P. (1997). Host heterogeneity in susceptibility and disease dynamics: tests of a mathematical model. *American Naturalist* **150:** 685–707.

Dybdahl, M. F. and Lively, C. M. (1995). Host-parasite interactions—infection of common clones in natural-populations of a fresh-water snail (*Potamopyrgus antipodarum*). *Proceedings of the Royal Society of London Series B-Biological Sciences* **260:** 99–103.

Dybdahl, M. F. and Lively, C. M. (1996). The geography of coevolution: comparative population structures for a snail and its trematode parasite. *Evolution* **50:** 2264–2275.

Dybdahl, M. F. and C. M. Lively (1998). Host-parasite coevolution: Evidence for rare advantage and time-lagged selection in a natural population. Evolution **52:** 1057–1066.

Dye, C. and Williams, B. G. (1995). Nonlinearities in the dynamics of indirectly transmitted infections (or, does having a vector make a difference?). In: *Ecology of infectious diseases in natural populations*. (ed. B. T. Grenfell and A. P. Dobson). Cambridge, Cambridge University Press: 260–279.

Dye, C., Barlow, N. D., Begon, M. E., Bowers, R. G., Bolker, B. M., Briggs, C. J. *et al.* (1995). Microparasites group report: persistence of microprasites in natural populations. In: *Ecology of infectious diseases in natural populations*. (ed. B. T. Grenfell and A. P. Dobson). Cambridge, Cambridge University Press: 123–183.

Earn, D. J. D., Rohani, P. and Grenfell, B. T. (1998). Persistence, chaos and synchrony in ecology and epidemiology. *Proceedings of the Royal Society of London Series B-Biological Sciences* **265:** 7–10.

Efron, B. and Tibshirani, R. (1986). Bootstrap methods for standard errors. Confidence intervals, and other measures of statistical accuracy. *Statistical Science* **1:** 54–77.

Efron, B. and Tibshirani, R. (1993). *An introduction to the bootstrap*. London, Chapman and Hall.

Elliot, J. M. (1977). *Statistical analysis of samples of benthic invertebrates*. Ambleside, Freshwater Biological Association.

Ellner, S. P., Barley, B. A., Bobashev, G. V., Gallant, A. R., Grenfell, B. T. and Nychka, D. W. *et al.* (1998). Noise and nonlinearity in measles epidemics: Combining mechanistic and statistical approaches to population modeling. *American Naturalist* **151:** 425–440.

Elsen, J. M., Amigues, Y., Schetcher, F., Ducrocq, V., Andrealetti, O., Eychenne, F. *et al.* (1999). Genetic susceptibility and transmission factors in scrapie: detailed analysis of an epidemic in a closed flock of Romanov. *Archives of Virology* **144:** 431–445.

Engen, S. and Lande, R. (1996). Population dynamic models generating species abundance distributions of the gamma type. *Journal of Theoretical Biology* **178:** 325–331.

Entwistle, P. F., Forkner, A. C., Green, B. M. and Cory, J. S. (1993). Avian dispersal of nuclear polyhedrosis virus after induced epizootics in the pine beauty moth *Panolis flammea* (Lepidoptera: Noctuidae). *Biological Control* **3:** 61–69.

Esch, G. W., Shostak, A. W., Marcogliese, D. J. and Goater, T. M. (1990a). Patterns and processes in helminth communities: an overview. In: *Parasite communities: patterns and processes*. (ed. G. W. Esch, A. O. Bush, and J. M. Aho). London, Chapmann and Hall: 1–19.

Esch, G., Bush, A. O. and Ano, J. M. (1990b). *Parasite communities: patterns and processes*. London, Chapmann and Hall.

Estrada-Pena, A. (1997). Epidemiological surveillance of tick populations: a model to predict the colonization success of *Ixodes ricinus* (Acari: Ixodidae). *European Journal of Epidemiology* **13:** 581–586.

Ewald, P. W. (1983). Host-parasite relations, vectors, and the evolution of disease severity. *Annual Review of Ecology and Systematics* **14:** 465–485.

Ewald, P. W. (1993). The evolution of virulence. *Scientific American* **268:** 86.

Falconer, D. S. and Mackay, T. F. C. (1996). *Introduction to quantitative genetics*. Essex, Longman.

Fellowes, M. D. E., Kraaijeveld, A. R. and Godfray, H. C. J. (1998). Trade-off associated with selection for increased ability to resist parasitoid attack in *Drosophila melanogaster*. *Proceedings of the Royal Society of London Series B-Biological Sciences* **265:** 1553–1558.

Fellowes, M. D. E., Kraaijeveld, A. R. and Godfray, H. C. J. (1999). Cross-resistance following artificial selection for increased defense against parasitoids in *Drosophila melanogaster*. *Evolution* **53:** 966–972.

Fenner, F. (1983). The florey lecture, 1983—biological-control, as exemplified by smallpox. Eradication and myxomatosis. *Proceedings of the Royal Society of London Series B-Biological Sciences* **218:** 259–285.

Feore, S. M., Bennett, M., Chantiey, J., Jones, T., Baxby, D. and Begon, M. E. (1997). The effect of cowpox virus infection on fecundity in bank voles and wood mice. *Proceedings of the Royal Society of London Series B-Biological Sciences* **264:** 1457–1461.

Ferguson, N., Anderson, R. and Gupta, S. (1999). The effect of antibody-dependent enhancement on the transmission dynamics and persistence of multiple-strain pathogens. *Proceedings of the National Academy of Sciences of the United States of America* **96:** 790–794.

Ferroglio, E. (1998). *Epidemiologia della tubercolosi e della brucellosi in ambiente silvestre*. Torino, Università di Torino: 108.

Festa-Bianchet, M. (1989). Individual differences, parasites, and the costs of reproduction for bighorn ewes (*Ovis canadensis*). *Journal of Animal Ecology* **58:** 755–795.

Finkenstädt, B. and Grenfell, B. (2000). Time series modelling of childhood diseases: a dynamical systems approach. *Journal of Royal Statistical Society* **49:** 187–205.

Fish, D. and Dowler, R. C. (1989). Host associations of ticks (Acari: Ixodidae) parasitizing medium sized mammals in a Lyme disease endemic area of south New York. *Journal of Medical Entomology* **26:** 200–209.

Fisher, R. A. (1941). The negative binomial distribution. *Annals of Eugenics* **11:** 182–187.

Fisher, R. A., Corbet, A. C. and Williams, C. B. (1943). The relation between the number of species and the number of individuals in a random sample of an animal population. *Journal of Animal Ecology* **12:** 42–58.

Florez-Duquet, M., Peloso, E. D. and Satinoff, E. (1998). Behavioral fever in aged rats. *Faseb Journal* **12:** 4244.

Foggin, C. M. (1988). *Rabies and rabies-related viruses in Zimbabwe: historical, virological and ecological aspects*. Harare, University of Zimbabwe.

Folstad, I., Nilssen, A. C., Haworsen, O. and Andersen, O. (1991). Parasite avoidance—the cause of post-calving migrations in reindeer. *Canadian Journal of Zoology-Revue Canadienne De Zoologie* **69:** 2423–2429.

Folstad, I. and Karter, A. J. (1992). Parasites, bright males and the immunocompetence handicap. *American Naturalist* **139:** 603–622.

Forbes, L. B., Tessaro, S. V. and Lees, W. (1996). Experimental studies on *Brucella abortus* in moose (*Alces alces*). *Journal of Wildlife Diseases* **32:** 94–104.

Foreyt, W. J. (1989). Fatal *Pasteurella haemolytica pneumonia* in bighorn sheep after direct contact with clincally normal domestic sheep. *American Journal of Veterinary Research* **50:** 341–344.

Foreyt, W. J. and Jessup, D. A. (1982). Fatal pneumonia of bighorn sheep following association with domestic sheep. *Journal of Wildlife Diseases* **18:** 163–168.

Forman, R. T. T. (1995). *Land mosaics: the ecology of landscapes and regions*. Cambridge, England, Cambridge University Press.

Forrest, S. C., Biggins, D. E., Richardson, L., Clark, T. W., Cambell, T. M., Flagerstone, K. W. *et al.* (1988). Population attributes for the black-footed ferret (*Mustela nigripes*) at Meeteetse, Wyoming, 1981–1985. *Journal of Mammalogy* **69:** 261–273.

Forsyth, M. A., Kennedy, S., Wilson, S., Eybatov, T. and Barrett, T. (1998). Canine distemper virus in a Caspian seal. *Veterinary Record* **143:** 662–664.

Fromont, E., Pontier, D. and Langlacs, M. (1998a). Dynamics of a feline retrovirus (FeLV) in host populations with variable spatial structure. *Proceedings of the Royal Society of London Series B-Biological Sciences* **265:** 1097–1104.

Fromont, E., Artois, M. and Pontier, D. (1998b). Epidemiology of feline leukemia virus (FeLV) and structure of domestic cat populations. *Journal of Wildlife Management* **62:** 978–988.

Fulford, A. J. C., Butterworth, A. E., Sturrock, R. F. and Ouma, J. H. (1992). On the use of age intensity data to detect immunity to parasitic infections, with special reference to *Schistosoma mansoni* in Kenya. *Parasitology* **105:** 219–227.

Furlanello, C. and Merler, S. (2000). Boosting of Tree-based Classifiers for Predictive Risk Modeling in GIS. *First International Workshop on Multiple Classifier Systems, Lecture Notes in Computer Sciences*. F. Roli. London, Springer Verlag: 220–229.

Furlanello, C., Merler, S. and Chemini, C. (1997). Tree-based classifiers and GIS for biological risk forecasting. In: *Advances in intelligent systems*. (ed. F. C. Morabito). Amsterdam, IOS Press: 316–323.

Fuxa, J. R. (1989). Importance of epizootiology to biological-control of insects with viruses. *Memorias Do Instituto Oswaldo Cruz* **84:** 81–88.

Fuxa, J. A., Fuxa, J. R. and Ritcher, A. R. (1998). Host-insect survival time and disintegration in relation to population density and dispersion of recombinant and wild-type nucleopolyhedroviruses. *Biological Control* **12:** 143–150.

Garin-Bastuji, B. (1993). Brucelloses bovine, ovine et caprine: controle et prèvention. *Pratique Veterinaire* **25:** 15–22.

Garin-Bastuji, B., Oudar, J., Richard, Y. and Gastello, J. (1990). Isolation of *Brucella melitensis* Biovar 3 from a Chamois (*Rupicapra rupicapra*) in the Southern French Alps. *Journal of Wildlife Diseases* **26:** 116–118.

Gascoyne, S. C., King, A. A., Laurenson, M. K., Borner, M., Schildger, B. and Banat, J. (1993). Aspects of rabies infection and control in the conservation of the african wild dog (*Lycaon pictus*) in the Serengeti region, Tanzania. *Onderstepoort Journal Of Veterinary Research* **60:** 415–420.

Gatto, M. and de Leo, G. A. (1998). Interspecific competition among macroparasites in a density-dependent host population. *Journal of Mathematical Biology* **37:** 467–490.

Gauthier, D. (1991). Sanitary consequences of mountain cattle breeding on wild ungulates. *Ongules/Ungulates* **91**. Proceeding of the International Symposium, SFEPM-I-RGM, Paris-Toulouse France.

Geman, S., Bienenstock, E. and Doursat, R. (1992). Neural networks and the bias variance dilemma. *Neural Computation* **4:** 1–58.

Gemmell, M. A., Lawson, J. R., Roberts, M. G., Kerrin, B. R. and Mason, C. J. (1986). Population dynamics in echinococcosis and cysticercosis: comparison of the response of *Echinococcus granulosus* and *Taenia hydatigena* and *T.ovis* to control. *Parasitology* **93:** 357–369.

Gemmell, M. A., Lawson, J. R. and Roberts, M. G. (1987). Population dynamics in echinococcosis and cysticercosis: evaluation of the biological parameters of *Taenia hydatigena* and *T.ovis* and comparison with those of *Echinococcus granulosus*. *Parasitology* **94:** 161–180.

Gemmill, A. W., Viney, M. E. and Read, A. F. (1997). Host immune status determines sexuality in a parasitic nematode. *Evolution* **51:** 393–401.

Genchi, C., Rizzoli, A., Fabbi, M., Manfredi, M. T., Sambri, V., Magnino, S. *et al.* (1994). *Ecology of* Borrelia burgdorferi *in some areas of northern Italy*. VI International Conference on Lyme Borreliosis, Bologna, Italy.

Genchi, C., Manfredi, M. T., Basono-Solasi, F., Rizzoli, A. P., Chemini, C., Faggi, M. *et al.* (1997). *Ecologia di Ixodes ricinus e Borrelia burgdorferi in Nord Italia*. IV Convegno Internazionale sulle Malattie Infettive dell'Arco Alpino, Siusi allo Sicilar, Bolzano (Italy).

Gennero, S. (1993). Indagini sierologiche su ruminanti selvatici in Piemonte. *Atti Societa Italiana Scienze Veterinarie* **47:** 979–983.

Gern, L. and Rais, O. (1996). Efficient transmission of *Borrelia burgdorferi* between cofeeding *Ixodes ricinus* ticks (Acari: Ixodidae). *Journal of Medical Entomology* **33:** 189–192.

Gern, L., Rouvinez, E., Toutoungi, L. N. and Godfroid, E. (1997). Transmission cycles of *Borrelia burgdorferi sensu lato* involving *Ixodes ricinus* and/or *I. hexagonus* ticks and the European hedgehog, *Erinaceus europaeus*, in suburban and urban areas in Switzerland. *Folia Parasitologica* **44:** 309–314.

Gern, L., Estrada-Peña, A., Fiandsen, F., Gray, J. S., Jaenson, T. G. T., Jogejan, F. *et al.* (1998). European reservoir hosts of *Borrelia burgdorferi sensu lato. Zentralblatt für Bakteriologie* **287:** 196–204.

Getz, W. M. and Pickering, J. (1983). Epidemic models: thresholds and population regulation. *The American Naturalist* **121:** 892–898.

Gibbs, H. C. (1986). Hypobiosis in parasitic nematodes—an update. *Advances in Parasitology* **25:** 129–174.

Gilks, W. R., Richardson, S. and Spiegelhaller, D. J. (ed.) (1996). *Markov chain Monte Carlo in practice*. London, Chapman and Hall.

Gipps, J. W. (1991). *Beyond Captive Breeding: re-introducing endangered mammals to the wild*. Symposia of the Zoological Society of London, Clarendon Press, Oxford.

Glinski, Z. and Jarosz, J. (2000). Disease resistance mechanisms of vascular plants. *Medycyna Weterynaryjna* **56:** 567–570.

Goltsman, M., Kruchenkova, E. P. and MacDonald, D. W. (1996). The Mednyi arctic foxes: treating a population imperilled by disease. *Oryx* **30:** 251–258.

Gotelli, N. J. (1991). Metapopulation models: the rescue effect, the propagule rain, and the core-satellite hypothesis. *American Naturalist* **138:** 768–776.

Goulson, D. and Cory, J. S. (1995). Responses of *Mamestra brassicae* (Lepidoptera, Noctuidae) to crowding—interactions with disease resistance, color phase and growth. *Oecologia* **104:** 416–423.

Goyal, P. K. and Wakelin, D. (1993a). Influence of variation in host strain and parasite isolate on inflammatory and antibody-responses to *Trichinella spiralis* in mice. *Parasitology* **106:** 371–378.

Goyal, P. K. and Wakelin, D. (1993b). Vaccination against *Trichinella spiralis* in mice using antigens from different isolates. *Parasitology* **107:** 311–317.

Grachev, M. A., Kumarev, V. P., Mamaev, L. P., Zorin, V. L., Baranova, L. V., Denikma, N. N. *et al.* (1989). Distemper virus in Baikal seals. *Nature* **338:** 209.

Grafen, A. and Woolhouse, M. E. J. (1993). Does the negative binomial distribution add up? *Parasitology Today* **9:** 475–477.

Grant, R. M. and Feinberg, M. B. (1996). HIV replication and pathogenesis. *Current Opinion In Infectious Diseases* **9:** 7–13.

Gravenor, M. B., Mclean, A. R. and Kwiatkowski, D. (1995). The regulation of malaria parasitemia—parameter estimates for a population-model. *Parasitology* **110:** 115–122.

Gray, J. S. (1982). The effects of the piroplasm Babesia bigemina on survival and reproduction of the blue tick, *Boophilus decoloratus. Journal of Invertebrate Pathology* **39:** 413–415.

Gray, J., Kahl, O., Robertson, J. N., Daniel, M., Estiada-Peña, A., Gettinby, G. *et al.* (1998). Lyme borreliosis habitat assessment. *Zentralblatt für Bakteriologie* **287:** 211–228.

Green, R. M. and Randolph, S. E. (1998a). *Establishing relationships between NOAA AVHRR data and microclimatic factors relevant to tick seasonal dynamics*. 9th Australasian Remote Sensing and Photogrammetry Conference, University of New South Wales, Sydney.

Green, R. M. and Randolph, S. E. (1998b). *Remotely sensed surrogates of climatic variables from NOAA AVHRR for mapping tick distributions across Eurasia*. 9th Australasian Remote Sensing and Photogrammetry Conference, University of New South Wales, Sydney.

Greenman, J. V. and Hudson, P. J. (1997). Infected coexistence instability with and without density-dependent regulation. *Journal of Theoretical Biology* **185:** 345–356.

Greenman, J. V. and Hudson, P. J. (2000). Parasite-mediated and direct competition in a two-host shared macroparasite system. *Theoretical Population Biology* **57:** 13–34.

Greenwood, M., Bradford-Hill, A., Topley, W. W. C. and Wilson, J. (1936). Experimental epidemiology. *Medical Research Council Special Report* **209:** 204.

Gregory, R. D. and Woolhouse, M. E. J. (1993). Quantification of parasite aggregation—a simulation study. *Acta Tropica* **54:** 131–139.

Gregory, R. D., Montgomery, S. S. J. and Montgomery, W. L. (1992). Population biology of *Heligmosomoides polygyrus* (nematoda) in the wood mouse. *Journal of Animal Ecology* **61:** 749–757.

Grenfell, B. T. (1988). Gastrointestinal nematode parasites and the stability and productivity of intensive ruminant grazing systems. *Philosophical Transaction of the Royal Society of London Series B-Biological Science* **321:** 541–563.

Grenfell, B. T. (1992). Parasitism and the dynamics of ungulate grazing systems. *American Naturalist* **139:** 907–929.

Grenfell, B. T. and Dobson, A. P. (ed.) (1995). *Ecology of infectious diseases in natural populations*. Cambridge, Cambridge University Press.

Grenfell, B. T. and Harwood, J. (1997). (Meta)population dynamics of infectious diseases. *Trends in Ecology and Evolution* **12:** 395–399.

Grenfell, B. T., Price, O. F., Albon, S. D. and Clutton-Block, T. H. (1992a). Overcompensation and population-cycles in an ungulate. *Nature* **355:** 823–826.

Grenfell, B. T., Lonergan, M. and Harwood, J. (1992b). Quantitative investigations of the epidemiology of phocine distemper virus (PDV) in European common seal populations. *Science of the Total Environment* **115:** 31–44.

Grenfell, B. T., Wilson, K., Isham, U. S., Boyd, H. E. G. and Dietz, K. (1995). Modelling patterns of parasite aggregation in natural populations: Trichostrongylid nematode-ruminant interactions as a case study. *Parasitology* **111:** S135–S151.

Grenfell, B. T., Wilson, K., Finkenstädt, B. F., Coulson, T. N., Murray, S., Albon, S. D. *et al.* (1998). Noise and determinism in synchronized sheep dynamics. *Nature* **394:** 674–677.

Griffith, B., Scott, J. M., Carpenter, J. W. and Reed, C. (1993). Animal translocations and potential disease transmission. *Journal of Zoo and Wildlife Medicine* **24:** 231–236.

Grosholz, E. D. (1992). Interactions of intraspecific, interspecific, and apparent competition with host-pathogen population dynamics. *Ecology* **73:** 507–514.

Grosholz, E. D. (1994). The effects of host genotype and spatial-distribution on trematode parasitism in a bivalve population. *Evolution* **48:** 1514–1524.

Grosholz, E. D. and Ruiz, G. M. (1995a). Spread and potential impact of the recently introduced european green crab, *Carcinus maenas*, in central California. *Marine Biology* **122:** 239–247.

Grosholz, E. D. and Ruiz, G. M. (1995b). Does spatial heterogeneity and genetic-variation in populations of the xanthid crab *Rhithropanopeus harrisii* (gould) influence the prevalence of an introduced parasitic castrator. *Journal of Experimental Marine Biology and Ecology* **187:** 129–145.

Gross, W. G., Siegel, P. B., Hall, W., Dommersmoth, C. H. and DuBoise, R. T. (1980). Production and persistence of antibodies in chickens to sheep erythrocytes. 2. Resistance to infectious disease. *Poultry Science* **59:** 205–210.

Grossman, C. J. (1985). Interactions between the gonadal-steroids and the immune system. *Science* **227:** 257–261.

Gryseels, B. (1994). Human resistance to schistosome infection: age or experience? *Parasitology Today* **10:** 380–384.

Guberti, V., Rutili, D., Ferrari, G., Patta, C. and Oggiano, A. (1998). Estimate of the threshold abundance for the persistence of the Classical Swine Fever in the Wild Boar population of the Eastern Sardinia (Italy). Bologna *European Union Technical Report: Measure to Control Classical Swine Fever in European Wild Boar*: 54–61.

Gulland, F. M. D. (1991a). *The role of parasites in the population dynamics of Soay sheep on St. Kilda*. Cambridge, University of Cambridge, unpublished PhD thesis.

Gulland, F. M. D. (1991b). Nematodirus species on St. Kilda. *Veterinary Record* **128:** 576–576.

Gulland, F. M. D. (1992). The role of nematode parasites in soay sheep (*Ovis aries* L.) mortality during a population crash. *Parasitology* **105:** 493–503.

Gulland, F. M. D. and Fox, M. (1992). Epidemiology of nematode infections of Soay sheep (*Ovis aries* L.) on St. Kilda. *Parasitology* **105:** 481–492.

Gulland, F. M. D., Albon, S. D., Pemberton, J. M., Mooreiojt, P. R. and Cluttonblock, T. H. (1993). Parasite-associated polymorphism in a cyclic ungulate population. *Proceedings of the Royal Society of London Series B-Biological Sciences* **254:** 7–13.

Gupta, S., Swinton, J. and Anderson, R. M. (1994). Theoretical-studies of the effects of heterogeneity in the parasite population on the transmission dynamics of malaria. *Proceedings of the Royal Society of London Series B-Biological Sciences* **256:** 231–238.

Gupta, S., Maiden, M. C. J., Feavers, I. M., Nee, S., May, R. M. and Anderson, R. M. (1996). The maintenance of strain structure in populations of recombining infectious agents. *Nature Medicine* **2:** 437–442.

Haase, A. T. (1999). Population biology of HIV-1 infection: viral and CD4(+) T cell demographics and dynamics in lymphatic tissues. *Annual Review of Immunology* **17:** 625–656.

Haile, D. G. and Mount, G. A. (1987). Computer simulation of population dynamics of the lone star tick, *Amblyomma americanum* (Acari: Ixodidae). *Journal of Medical Entomology* **24:** 356–369.

Haines-Young, Green, R. H., D. R. and Cousins, S. (ed.) (1993). *Landscape ecology and geographical information systems.* London, Taylor and Francis.

Hall, A. J. (1995). Morbilliviruses in marine mammals. *Trends in Microbiology* **3:** 4–9.

Halvorsen, O. and Andersen, K. (1984). The ecological interaction between arctic charr, *Salvelinus alpinus* (L.) and the plerocercoid stage of *Diphyllobothrium ditremum. Journal of Fish Biology* **25:** 305–316.

Hamilton, W. J. and Poulin, R. (1997). The Hamilton and Zuk hypothesis revisited: a meta-analytical approach. *Behaviour* **134:** 299–320.

Hamilton, W. D. and Zuk, M. (1982). Heritable true fitness and bright birds: a role for parasites? *Science* **218:** 384–387.

Hand, D. G. (1997). *Construction and assessment of classification rules.* New York, Jhon Wiley and sons.

Hanley, K. A., Vollmer, D. M. and Case, T. J. (1995). The distribution and prevalence of helminths, coccidia and blood parasites in two competing species of gecko: implications for apparent competition. *Oecologia* **102:** 220–229.

Hanley, K. A., Petren, K. and Case, T. J. (1998). An experimental investigation of the competitive displacement of a native gecko by an invading gecko: no role for parasites. *Oecologia* **115:** 196–205.

Hanski, I. (1983). Coexistence of competitors in patchy environment. *Ecology* **64:** 493–500.

Hanski, I. (1985). Single-species spatial dynamics may contribute to long-term rarity and commonness. *Ecology* **66:** 335–343.

Hanski, I. (1991). Single-species metapopulation dynamics: concepts, models and observations. *Biological Journal of the Linnean Society* **42:** 17–38.

Hanski, I. (1994). Patch-occupancy dynamics in fragmented landscapes. *Trends in Ecology and Evolution* **9:** 131–135.

Hanski, I. and Gilpin, M. E. (1991). Metapopulation dynamics: brief history and conceptual domain. *Biological Journal of the Linnean Society* **42:** 3–16.

Hanski, I. and Gilpin, M. E. (ed.) (1997). *Metapopulation biology: ecology, genetics, and evolution.* San Diego, Academic Press.

Hanski, I. and Gyllenberg, M. (1993). Two general metapopulation models and the core-satellite species hypothesis. *American Naturalist* **142:** 17–41.

Hanski, I. A. and Simberloff, D. (1997). The metapopulation approach, its history, conceptual domain, and application to conservation. In: *Metapopulation biology: ecology, genetics, and evolution.* (ed. I. Hanski and M. E. Giplin). San Diego, Academic Press: 5–26.

Hanski, I., Moilanen, A. and Gyllenberg, M. (1996). Minimum viable metapopulation size. *American Naturalist* **147:** 527–541.

Harder, A., Wunderlich, F. and Merinovski, P. (1992). Effects of testosterone on *Heterakis spumosa* infections in mice. *Parasitology* **105:** 335–342.

Harder, T. C., Willhaus, T., Leibold, W. and Liess, B. (1992). Investigations on course and outcome of phocine distemper virus infection in harbor seals (*Phoca vitulina*) exposed to polychlorinated-biphenyls: virological and serological investigations. *Journal of Veterinary Medicine Series B-Zentralblatt Fur Veterinarmedizin Reihe B-Infectious Diseases and Veterinary Public Health* **39:** 19–31.

Harder, T. C., Stede, M., Wilhaus, T., Schwartz, J., Heidemann, G. and Liess, B. (1993). Morbillivirus antibodies of maternal origin in harbour seal pups (*Phoca vitulina*). *Veterinary Record* **132:** 632–633.

Harrison, S. (1994). Metapopulations and conservation. In: *Large-scale ecology and conservation biology.* (ed. P. J. Edwards, R. M. May, and N. R. Webb). Oxford, Blackwell: 111–128.

Harrison, S. and Taylor, A. D. (1997). Empirical evidence for metapopulation dynamics. In: *Metapopulation biology: ecology, genetics, and evolution.* (ed. I. Hanski and M. E. Giplin). San Diego, Academic Press: 27–42.

Harrison, S., Murphy, D. D. and Ehrlich, P. R. (1988). Distribution of the checkerspot butterfly, *Euphydryas editha bayensis*: evidence for a metapopulation model. *American Naturalist* **132:** 360–382.

Hart, B. L. (1994). Behavioral defense against parasites—interaction with parasite invasiveness. *Parasitology* **109:** S139–S151.

Hart, B. L. (1997). Behavioural defence. In: *Host-parasite evolution: general principles and avian models.* (ed. D. H. Clayton and J. Moore). Oxford, Oxford University Press: 59–77.

Harvey, P. H. and Pagel, M. D. (1991). *The comparative method in evolutionary biology.* Oxford, Oxford University Press.

Harvey, P. H. and Keymer, A. E. (1991). Comparing life histories using phylogenies. *Philosophical Transactions of the Royal Society of London Series B-Biological Sciences* **332:** 31–39.

Harwood, J. (1998). Conservation biology—What killed the monk seals? *Nature* **393:** 17–18.

Harwood, J. and Hall, A. J. (1990). Mass mortality in marine mammals: its implications for population dynamics and genetics. *Trends in Ecology and Evolution* **5:** 254–257.

Hassell, M. P., Comins, H. N. and May, R. M. (1991). Spatial structure and chaos in insect population dynamics. *Nature* **353:** 255–258.

Hassell, M. P., Comins, H. N. and May, R. M. (1994). Species coexistence and self-organising spatial dynamics. *Nature* **370:** 290–292.

Hasselquist, D., Marsh, J. A., Sherman, P. W. and Wingfield, J. C. (1999). Is avian humoral immunocompetence suppressed by testosterone? *Behavioral Ecology and Sociobiology* **45:** 167–175.

Hastings, B. E., Kenny, D., Lowenstine, L. J. and Foster, J. W. (1991). *Mountain gorillas and measles: ontogeny of a wildlife vaccination program*. Proceedings of the Annual Meeting of the American Association of Zoo Veterinarians.

Haukisalmi, V. and Henttonen, H. (1993). Coexistence in helminths of the bank vole *Clethrionomys glareolus*. 1. Patterns of co-occurrence. *Journal of Animal Ecology* **62:** 221–229.

Hay, S. I. and Lennon, J. J. (1999). Deriving meteorological variables across Africa for the study and control of vector-borne disease: a comparison of remote sensing and spatial interpolation of climate. *Tropical Medicine and Hygiene* **4:** 58–71.

Hay, S. I., Tucker, C. J., Rogers, D. J. and Pocker, M. J. (1996). Remotely sensed surrogates of meteorological data for the study of the distribution and abundance of arthropod vectors of disease. *Annals of Tropical Medicine and Hygiene* **90:** 1–19.

Hay, S. I., Packer, M. J. and Rogers, D. J. (1997). The impact of remote sensing on the study and control of invertebrate intermediate hosts and vectors for disease. *International Journal for Remote Sensing* **18:** 2899–2930.

Hay, S. I., Randolph, S. E. and Rogers, D. J. (Eds.) (2000). *Remote sensing and geographical information systems in epidemiology*. London, Academic Press.

He, H. S., Mladenoff, D. J. and Crow, T. R. (1999). Linking an ecosystem model and a landscape model to study forest species response to climate warming. *Ecological Modelling* **114:** 213–233.

Henter, H. J. and Via, S. (1995). The potential for coevolution in a host-parasitoid system. 1. Genetic variation within an aphid population in susceptibility to a parasitic wasp. *Evolution* **49:** 427–438.

Heesterbeek, J. A. P. and Metz, J. A. J. (1993). The saturating contact rate in marriage and epidemic models. *Journal of Mathematical Biology* **31:** 529–539.

Heesterbeek, J. A. P. and Roberts, M. G. (1995a). Mathematical models for microparasites of wildlife. In: *Ecology of infectious diseases in natural populations*. (ed. B. T. Grenfell and A. P. Dobson). Cambridge, Cambridge University Press: 90–122.

Heesterbeek, J. A. P. and Roberts, M. G. (1995b). Threshold quantities for helminth infections. *Journal of Mathematical Biology* **33:** 415–434.

Heide-Jørgensen, M. P. and Härkönen, T. (1992). Epizootiology of the seal disease in the Eastern North Sea. *Journal of Applied Ecology* **29:** 99–107.

Heide-Jørgensen, M. P., Härkönen, T. and Aberg, P. (1992a). Long-term effects of epizootic in harbor seals in the Kattegat–Skagerrak and adjacent areas. *Ambio* **21:** 511–516.

Heide-Jørgensen, M. P., Härkönen, T., Dietz, R. and Thompson, P. M. (1992b). Retrospective of the 1988 European seal epizootic. *Diseases of Aquatic Organisms* **13:** 37–62.

Henien, K. and Merriam, G. (1990). The elements of connectivity where corridor quality is variable. *Landscape Ecology* **4:** 157–170.

Henke, S. E. (1997). Effects of modified live-virus canine distemper vaccines in gray foxes. *Journal of Wildlife Rehabilitation* **20:** 3–7.

Herbst, L. H., Jacobson, E. R. and Klein, P. A. (1995). *Green turtle fibropapillomatosis: evidence for a viral etiology*. Proceedings of the Joint conference of the American Association of Zoo Veterinarians,the Wildlife Disease Association and the American Association of Wildlife Veterinarians.

Hero, J. M. and Gillespie, G. R. (1997). Epidemic disease and amphibian declines in Australia. *Conservation Biology* **11:** 1023–1025.

Herre, E. A. (1985). Sex-ratio adjustment in fig wasps. *Science* **228:** 896–898.

Herre, A. H. (1993). Population structure and the evolution of virulence in nematode parasites of fig wasps. *Science* **259:** 1442–1445.

Herre, E. A. (1995). Factors affecting the evolution of virulence: nematode parasites of fig wasps as a case study. *Parasitology* **111:** S179–S191.

Herriges, J. D., Thorne, E. T. and Anderson, S. L. (1992). Vaccination to control brucellosis in free-ranging elk on Western Wyoming feed grounds. In: *The biology of deer*. (ed. R. D. Brown). New York, Springer-Verlag. 107–112.

Hess, G. R. (1994). Conservation corridors and contagious disease: a cautionary note. *Conservation Biology* **8:** 256–262.

Hess, G. (1996a). Disease in metapopulation models: implications for conservation. *Ecology* **77:** 1617–1632.

Hess, G. R. (1996b). Linking extinction to connectivity and habitat destruction in metapopulation models. *The American Naturalist* **148:** 226–236.

Hickling, G. (1995). Wildlife Reservoirs of Bovine Tuberculosis in New Zealand. Tuberculosis in wildlife and domestic animals. *Proceedings of the second international* Mycobacterium bovis *conference*. (ed. F. Griffin and G. Lisle): 267.

Higgins, K., Hastings, A., Sarvela, J. N. and Botsford, L. W. (1997). Stochastic dynamics and deterministic skeletons: population behaviour of dungeness crab. *Science* **276:** 1431–1435.

Hilborn, R. and Mangel, M. (1997). *The ecological detective: confronting models with data*. Princeton, NewJersey, Princeton University Press.

Hillgarth, N. and Wingfield, J. C. (1997). Parasite-mediated sexual selection: endocrine aspects. In: *Host–parasite evolution: general principles and avian models*. (ed. D. K. Clayton and J. Moore). Oxford, Oxford University Press: 78–104.

Hochberg, M. E. (1989). The potential role of pathogens in biological control. *Nature* **337:** 262–265.

Holmes, J. C. (1982). Impact of infectious disease agents on the population growth and geographical distribution of animals. In: *Population biology of infectious diseases*. (ed. R. M. Anderson and R. M. May). Berlin, Springer-Verlag: 37–51.

Holmes, P. H. (1985). Pathogenesis of trichostrongylosis. *Veterinary Parasitology* **18:** 89–101.

Holmes, J. C. and Price, P. W. (1986). *Communities of parasites. Community ecology: pattern and processes*. Oxford, Blackwell Scientific Publications: 187–213.

Holmes, J. C. (1995). Population regulation: a dynamic complex of interactions. *Wildlife Research* **22:** 11–19.

Holmstad, P. R. and Skorping, A. (1998). Covariation of parasites in willow ptarmigan *Lagopus Lagopus* L. *Canadian Journal of Zoology* **76:** 1581–1588.

Holt, R. D. (1977). Predation, apparent competition, and the structure of prey communities. *Theoretical Population Biology* **12:** 197–229.

Holt, R. D. (1993). Infection diseases of wildlife, in theory and in practice. *Trends in Evolution and Ecology* **8:** 423–425.

Holt, R. D. and Lawton, J. H. (1993). Apparent competition and enemy-free space in insect host-parasitoid communities. *American Naturalist* **142:** 623–645.

Holt, R. D. and Lawton, J. H. (1994). The ecological consequences of shared natural enemies. *Annual Review of Ecology and Systematics* **25:** 495–520.

Holt, R. D. and Pickering, J. (1985). Infectious disease and species coexistence: a model in Lotka-Volterra form. *American Naturalist* **126:** 196–211.

Hoodless, A. N., Kurtenbach, K., Nuttall, P. A. and Randolph, S. E. (2001). The impact of ticks on pheasant territoriality. *Oikos* (in press).

Howe, G. M. (1972). *Man, environment, and disease in Britian*. New York, Barnes and Noble Books.

Hu, C. M., Humair, P. F., Wallich, R. and Gern, L. (1997). Apodemus sp. rodents, reservoir hosts for *Borrelia afzelii* in an endemic area in Switzerland. *Zentralblatt für Bakteriologie* **285:** 558–564.

Hudson, P. J. (1992). *Grouse in space and time*. Fordingbridge, Hants., Game Conservancy Trust.

Hudson, P. J. and Dobson, A. P. (1991). Control of parasites in natural populations: nematodes and virus infections of red grouse. In: *Bird population studies*. (ed. C. M. Perrins, J. D. Lebreton, and G. J. M. Hirons). Oxford, Oxford University Press, 413–432.

Hudson, P. J. and Dobson, A. P. (1995). Macroparasites: observed patterns. In: *Ecology of infectious diseases in natural populations*. (ed. B. T. Grenfell and A. P. Dobson). Cambridge, Cambridge University Press: 144–176.

Hudson, P. J. and Greenman, J. (1998). Competition mediated by parasites: biological and theoretical progress. *Trends in Ecology and Evolution* **13:** 387–390.

Hudson, P. J., Dobson, A. P. and Newborn, D. (1985). Cyclic and non-cyclic populations of red grouse: a role for parasitism? In: *Ecology and genetics of host–parasite interactions*. (ed. D. Rollinson and R. M. Anderson). London, Academic Press: 77–89.

Hudson, P. J., Dobson, A. P. and Newborn, D. (1992). Do parasites make prey vulnerable to predation? Red grouse and parasites. *Journal of Animal Ecology* **61:** 681–692.

Hudson, P. J., Norman, R., Laurenson, M. K., Newborn, D., Gount, M., Jones, L. D. *et al.* (1995). Persistence and transmission of tick-borne viruses: *Ixodes ricinus* and louping-ill virus in red grouse populations. *Parasitology* **111:** S49-S58.

Hudson, P. J., Dobson, A. P. and Newborn, D. (1998). Prevention of population cycles by parasite removal. *Science* **282:** 2256–2258.

Hudson, P. J., Rizzoli, A., Rosà, R., Chemini, C., Jones, L. D. and Gould, E. A. (2001). TBE virus in northern Italy: Molecular analysis, relationships with density and seasonal dynamics of *Ixodes ricinus* (Acari: Ixodidae). *Medical and Veterinary Entomology* **15:** 1–11

Hughes, V. L. (1998). *The effect of testosterone on parasitic infections in rodents, with respect to disease transmission by ticks*. Zoology Department. Oxford, Oxford University.

Hughes, D. E., Carter, S. D., Robinson, I., Clarke, D. D. and Clarke, C. J. (1992). Anti-canine distemper virus antibodies in common and grey seals. *Veterinary Record* **130:** 449–450.

Hughes, D. S., Possee, R. D. and King, L. A. (1993). Activation and detection of a latent baculovirus resembling *Mamestra brassicae* nuclear polyhedrosis virus in *M. brassicae* insects. *Virology* **194:** 608–615.

Hughes, D. S., Possee, R. D. and King, L. A. (1997). Evidence for the presence of a low-level, persistent baculovirus infection of *Mamestra brassicae* insects. *Journal of General Virology* **78:** 1801–1805.

Humair, P. F. and Gern, L. (1998). Relationship between *Borrelia burgdorferi sensu lato* species, red squirrels (*Sciurus vulgaris*) and *Ixodes ricinus* in enzootic areas in Switzerland. *Acta Tropica* **69:** 213–227.

Humair, P. F., Turrian, M. N., Aeschlimann, A. and Gern, L. (1993a). *Borrelia burgdorferi* in a focus of Lyme borreliosis: epizootiologic contribution of small mammals. *Folia Parasitologica* **40:** 65–70.

Humair, P. F., Turrian, M. N., Aeschlimann, A. and Gern, L. (1993b). *Ixodes ricinus* immatures on birds in a focus of Lyme borreliosis. *Folia Parasitologica* **40:** 237–242.

Humair, P. F., Peter, O., Wallichk, R. and Gern, L. (1995). Strain variation of Lyme disease spirochetes isolated from *Ixodes ricinus* ticks and rodents collected in two endemic areas in Switzerland. *Journal of Medical Entomology* **32:** 433–438.

Humair, P. F., Postic, D., Wallich, R. and Gern, L. (1998). An avian reservoir (*Turdus merula*) of the Lyme borreliosis spirochetes. *Zentralblatt für Bakteriologie* **287:** 521–538.

Hutchings, Kyriazakis, M. R., I., Anderson, D. H., Gordon, I. Jr. and Coop, R. L. (1998). Behavioural strategies used by parasitized and non-parasitized sheep to avoid ingestion of gastro-intestinal nematodes associated with faeces. *Animal Science* **67:** 97–106.

Hutchings, Kyriazakis, M. R., I., Anderson, D. H., Gardon, I. J. and Jackson, F. (1999). Grazing trade-offs between forage intake and faecal avoidance in sheep: the effect of sward height, parasitic status and level of feeding motivation. *Proceedings of the Nutrition Society* **58:** 157A–157A.

Hutchinson, M. F., Nix, H. A., McMahon, J. P. and Ord, K. D. (1995). *Africa. A topographc and climatic database.* The Australian National University, CDROM.

Hyde, J. E. (1990). *Molecular parasitlogy.* Milton Keynes, Open University Press.

Ives, A. R. and Murray, D. L. (1997). Can sublethal parasitism destabilize predator–prey population dynamics? A model of snowshoe hares, predators and parasites. *Journal of Animal Ecology* **66:** 265–278.

Jacobson, E. R. (1994). Causes of mortality and diseases in tortoises : a review. *Journal of Zoo and Wildlife Medicine* **25:** 2–17.

Jacobson, E. R., Brown, M. B., Schumercher, I. M., Collins, B. R., Harris, R. K. and Klein, P. A. (1995). Mycoplasmosis and the desert tortoise (*Gopherus agassizii*) in Las Vegas Valley, Nevada. *Chelonian Conservation and Biology* **1:** 279–284.

Jaensen, T. G. T. and Talleklint, L. (1992). Incompetence of roe deer as reservoirs of the Lyme borreliosis spirochete. *Journal of Medical Entomology* **29:** 813–817.

Jarosz, A. M. and Burdon, J. J. (1990). Predominance of a single major gene for resistance to phakopsora-pachyrhizi in a population of glycine-argyrea. *Heredity* **64:** 347–353.

Jenkins, S. R., Perry, B. D. and Winkler, W. G. (1988). Ecology and epidemiology of raccoon rabies. *Reviews of Infectious Diseases* **10:** S620–S636.

Jessup, D. A., De Forge, J. R. and Sandberg, S. (1991). Biobullet vaccination of captive and free-ranging bighorn sheep. In: *Wildlife production: conservation and sustainable development*. (ed. L. A. Renecker and R. J. Hudson). Fairbanks, University of Alaska.

Jessup, D. A., Boyce, W. M. and Torres, S. G. (1995). *Bighorn sheep health management in California: a fifteen year retrospective.* Proceedings of the Joint conference of the American Association of Zoo Veterinarians, the Wildlife Disease Association and the American Association of Wildlife Veterinarians.

Jewell, P. A. (ed.) (1974). *Island survivors: the ecology of the Soay sheep of St Kilda.* London, Athlone Press.

John, J. L. (1997). The Hamilton-Zuk theory and initial test: an examination of some parasitological criticisms. *International Journal for Parasitology* **27:** 1269–1288.

Johnson, M. J., Behnke, J. M. and Cores, G. C. (1996). Detection of gastrointestinal nematodes by a coproantigen capture ELISA. *Research in Veterinary Science* **60:** 7–12.

Jones, L. D., Davies, C. R., Steele, G. M. and Nuttall, P. A. (1987). A novel mode of arbovirus transmission involving a nonviraemic host. *Science* **37:** 775–777.

Jones, L. D., Gaunt, M., Hails, R. S., Laurenson, M. K., Hudson, P. J. and Reid, H. *et al.* (1997). Transmission of louping ill virus between infected and uninfected ticks co-feeding on mountain hares. *Medical and Veterinary Entomology* **11:** 172–176.

Kaiser, M. N., Sutherst, R. W. and Bourne, A. S. (1991). Tick (Acarina: Ixodidae) infestations on zebu cattle in northern Uganda. *Bulletin of Entomological Research* **81:** 257–262.

Kao, R. R. and Roberts, M. G. (1999). A comparison of wildlife control and cattle vaccination as methods for the control of bovine tuberculosis. *Epidemiology and Infection* **122:** 505–519.

Karban, R. (1998). Caterpillar basking behavior and nonlethal parasitism by tachinid flies. *Journal of Insect Behavior* **11:** 713–723.

Kat, P. W., Alexander, K. A., Smith, J. S. and Munson, L. (1995). Rabies and African wild dogs in Kenya. *Proceedings of the Royal Society of London Series B-Biological Sciences* **262:** 229–233.

Keith, L. B., Cary, J. R., Yuil, T. M. and Keith, I. M. (1985). Prevalence of helminths in a cyclic snowshoe hare population. *Journal of Wildlife Diseases* **21:** 233–253.

Kendall, B. E., Briggs, C. J., Murdoch, W. W., Turchin, P., Ellner, S. P., McCauley, E. *et al.* (1999). Why do population cycle: a synthesis of statistical and mechanistic modeling approaches. *Ecology* **80:** 1768–1805.

Kendall, M. G. and Stuart, A. (ed.) (1963). *The advanced theory of statistics. Distribution theory*. London, Griffin.

Kennedy, C. R. (1969). Seasonal incidence and development of the cestode *Caryophyllaeus laticeps* (Pallas) in the River Avon. *Parasitology* **59:** 783–794.

Kennedy, C. R. (1975). *Ecological animal parasitology*. Oxford, Blackwell Scientific Publications.

Kennedy, C. R. (1994a). The ecology of introductions. In: *Parasitic diseases of fish*. (ed. A. W. Pike and J. W. Lewis). Tresaith, Dyfed, Samara Publishing Limited: 189–208.

Kennedy, C. R. (1994b). Introductions, spread and colonization of new localities by fish helminth and crustacean parasites in the British Isles: a perspective and appraisal. *Journal of Fish Biology* **43:** 287–301.

Kennedy, S., Kuiken, T., Jepson, P. D., Deaville, R., Forsyth, M., Barrett, T. *et al.* (2000). Mass die-off of Caspian seals caused by canine distemper virus. *Emerging Infectious Diseases* **6:** 637–639.

Keymer, A. E. and Anderson, R. M. (1979). The dynamics of infection of *Tribolium confusum* by *Hymenolepis diminuta*: the influence of infective-stage density and spatial distribution. *Parasitology* **79:** 195–207.

Keymer, A. E. (1981). Population dynamics of *Hymenolepis diminuta* in the intermediate host. *Journal of Animal Ecology* **50:** 941–950.

Kimura, K., Isogai, E., Isogai, H., Kaniewaka, Y., Nishikava, T., Ishii, N. and Fujii, N. (1995). Detection of Lyme disease spirochaetes in the skin of naturally infected wild Sika deer (*Cervus nippon yesoensis*) by PCR. *Applied and Environmental Microbiology* **61:** 1641–1642.

Kitala, P. M. and McDermot, J. J. (1995). *Population dynamics of dogs in Machakos District, Kenya: implications for vaccination strategy*. Proceedings of the Third International Conference of the Southern and Eastern African Rabies Group, Harare, SEARG, Harare.

Kitron, U. (1998). Landscape ecology and epidemiology of vector-borne diseases: tools for spatial analysis. *Journal of Medical Entomology* **35:** 435–445.

Kitron, U., Otieno, L. H., Hungerford, L., Odulaja, A., Brignam, W. U., Okeuo, O. *et al.* (1996). Spatial analysis of the distribution of tsetse flies in the Lambwe Valley, Kenya, using Landsat TM satellite imagery and GIS. *Journal of Animal Ecology* **65:** 371–380.

Klein, S. L. (2000). The effects of hormones on sex differences in infection: from genes to behavior. *Neuroscience and Biobehavioral Reviews* **24:** 627–638.

Klein, S. L. and Nelson, R. J. (1997). Sex differences in immunocompetence differ between two Peromyscus species. *American Journal of Physiology-Regulatory Integrative and Comparative Physiology* **42:** R655–R660.

Klein, S. L. and Nelson, R. J. (1998). Sex and species differences in cell-mediated immune responses in voles. *Canadian Journal of Zoology-Revue Canadienne De Zoologie* **76:** 1394–1398.

Klein, S. L. and Nelson, R. J. (1999). Influence of social factors on immune function and reproduction. *Reviews of Reproduction* **4:** 168–178.

Klein, S. L., Hairston, J. E., DeVries, A. C. and Nelson, R. J. (1997). Social environment and steroid hormones affect species and sex differences in immune function among voles. *Hormones and Behavior* **32:** 30–39.

Kloosterman, A., Parmentier, H. K. and Ploeger, H. W. (1992). Breeding cattle and sheep for resistance to gastrointestinal nematodes. *Parasitology Today* **8:** 330–335.

Knell, R. J., Begon, M. and Thompson, D. J. (1996). Transmission dynamics of *Bacillus thuringiensis* infecting *Plodia interpunctella*: a test of the mass action assumption with an insect pathogen. *Proceedings of the Royal Society of London Series B-Biological Sciences* **263:** 75–81.

Kock, R. A. and Woodford, M. H. (1988). *Reintroduction of Pere David's deer* (Elaphurus davidianus), *scimitar-horned oryx* (Oryx dammah) *and the Arabian oryx* (Oryx leucoryx) *to their native habitats: a veterinary perspective*. Proceedings of the Joint Conference of American Association of Zoo Veterinarians and American Association of Wildlife Veterinarians, Toronto, Ontario, Canada.

Kock, R., Chalmers, W. S. K., Mioanzia, J. Chillingworth, C., Wan-bura, J., Coremac, P. *et al.* (1998). Canine distemper antibodies in lions of the Masai Mara. *Veterinary Record* **142:** 662–665.

Kodricbrown, A. and Brown, J. H. (1984). Truth in advertising—the kinds of traits favored by sexual selection. *American Naturalist* **124:** 309–323.

Kolmer, J. A. (1996). Genetics of resistance to wheat leaf rust. *Annual Review of Phytopathology* **34:** 435–455.

Korenberg, E. I. (1994). Comparative ecology and epidemiology of Lyme disease and tick-borne encephalitis in the former Soviet Union. *Parasitology Today* **10:** 157–160.

Kozuch, O., Chunikhin, S. P., Glesikova, M., Nosek, J., Kurenkov, V. B. and Lysy, J. (1981). Experimental characteristics of viraemia caused by two strains of tick-borne encephalitis virus in small rodents. *Acta Virologica* **25:** 219–224.

Kraaijeveld, A. R. and Godfray, H. C. J. (1997). Trade-off between parasitoid resistance and larval competitive ability in *Drosophila melanogaster*. *Nature* **389:** 278–280.

Krebs, J. R. (1997). *Bovine tuberculosis in cattle and badgers*. London, Ministry of Agriculture, Fisheries and Food.

Krebs, J. R., Anderson, R. M., Clotton-Burk, T. H., Donnelly, C. A., Frost, S., Morrison, W. I. *et al.* (1998).

Badgers and bovine TB: conflicts between conservation and health. *Science* **279:** 817–818.

Krebs, J. W., Smith, J. S. *et al.* (1999). Rabies surveillance in the United States during 1998. *Journal of the American Veterinary Medical Assocation* **215:** 1786–1798.

Krecek, R. C., Malan, F. S., Rupprecht, C. E. and Childs, J. E. (1987a). Nematode parasites from Burchell's zebras in South Africa. *Journal of Wildlife Diseases* **23:** 404–411.

Krecek, R. C., Sayre, R. M., Els, H. J., Van Niekerk, J. P. and Malan, F. S. (1987b). Fine structure of a bacterial community associated with cyathostomes (Nematoda: Strongylidae) of zebras. *Proceedings of the Helminthological Society of Washington* **54:** 212–219.

Krecek, R. C., Els, H. J., de Wet, S. C. and Henton, M. M. (1992). Studies on ultrastructure and cultivation of microorganisms associated with zebra nematodes. *Microbial Ecology* **23:** 87–95.

Kunimi, Y. and Yamada, E. (1990). Relationship of larval phase and susceptibility of the armyworm, *Pseudaletia separata* walker (Lepidoptera, noctuidae) to a nuclear polyhedrosis-virus and a granulosis-virus. *Applied Entomology and Zoology* **25:** 289–297.

Kunz, C. (1992). Tick-borne encephalitis in Europe. *Acta Leidensia* **66:** 1–14.

Kurtenbach, K., Kampen, H., Dizij, A., Arndt, S., Seitz, H. M., Schaible, U. E. *et al.* (1995). Infestation of rodents with larval *Ixodes ricinus* (Acari: Ixodidae) is an important factor in the transmission cycle of *Borrelia burgdorferi s.l.* In German woodlands. *Journal of Medical Entomology* **32:** 807–817.

Kurtenbach, K., Carey, D., Hoodless, A. N., Nuttall, P. A. and Randolph, S. E. (1998a). Competence of pheasants as reservoirs for Lyme disease spirochetes. *Journal of Medical Entomology* **35:** 77–81.

Kurtenbach, K., Peacey, M. F., Rijpkema, S. G. T., Hoodless, A. N., Nuttall, P. A. and Randolph, S. E. (1998b). Differential transmission of the genospecies of *Borrelia burgdorferi sensu lato* by game birds and small rodents in England. *Applied and Environmental Microbiology* **64:** 1169–1174.

Kurtenbach, K., Sewell, H. S., Ogden, N. H., Randolph, S. E. and Nuttall, P. A. (1998c). Serum complement sensitivity as a key factor in Lyme disease ecology. *Infection and Immunity* **66:** 1248–1251.

Kuznetsov, Y. A. (1998). *Element of applied bifurcation theory.* New York, Springer-Verlag.

Kyriazakis, I., Tolkamp, B. J. and Hutchings, M. R. (1998). Towards a functional explanation for the occurrence of anorexia during parasitic infections. *Animal Behaviour* **56:** 265–274.

Labuda, M. and Randolph, S. E. (1999). Survival strategy of tick-borne encephalitis virus: cellular basis and environmental determinants. *Zentralblatt Für Bakteriologie* **289:** 513–524.

Labuda, M., Jones, L. D., Williams, T., Danielová, V. and Nuttal, P. A. *et al.* (1993a). Efficient transmission of tick-borne encephalitis virus between co-feeding ticks. *Journal of Medical Entomology* **30:** 295–299.

Labuda, M., Nuttall, P. A., Kozuch, O., Elecková, E., Williams, T., Zuffová, E. *et al.* (1993b). Non-viraemic transmission of tick-borne encephalitis virus: a mechanism for arbovirus survival in nature. *Experientia* **49:** 802–805.

Labuda, M., Austyn, J. M., Zuffová, E., Kozuch, O. and Nuttall, P. A. (1996). Importance of localised skin infection in tick-borne encephalitis virus transmission. *Virology* **219:** 357–366.

Lack, D. (1954). *The natural regulation of animal number.* Oxford, Claredon Press.

Lanciani, C. A. (1975). Parasite-induced alterations in host reproduction and survival. *Ecology* **56:** 689–695.

Laurance, W. F., McDonald, K. R. and Speare, R. (1996). Epidemic disease and the catastrophic decline of Australian rain forest frogs. *Conservation Biology* **10:** 406–413.

Laurance, W. F., McDonald, K. R. and Speare, R. (1997). In defense of the epidemic disease hypothesis. *Conservation Biology* **11:** 1030–1034.

Laurenson, K., Esterhuysen, J., Starder, P. and Van Heerden, J. (1997a). Aspects of rabies epidemiology in Tsumkwe district, Namibia. *Onderstepoort Journal of Veterinary Research* **64:** 39–45.

Laurenson, K., Shiferaw, F. and Sillero-Zubiri, C. (1997b). The Ethiopian wolf: status survey and conservation action plan. In: *Disease, domestic dogs and the ethiopian wolf: the current situation.* (ed. C. Sillero-Zubiri and D. W. Macdonald). Gland, IUCN **32:** 42

Laurenson, K., Sillero-Zubiri, C., Thompson, H., Shiefraw, F., Thergood, S. and Malcolm, J. (1998). Disease as a threat to endangered species: Ethiopian wolves, domestic dogs and canine pathogens. *Animal Conservation* **1:** 273–280.

Lavazza, A., Guberti, V., Ferri, M., Zarni, M. L., Poglayen, G. and Capucci, L. (1997). *Epidemiology of EBHS In Modena province (North Italy).* 4th International Congress of Veterinary Virology, Edinburgh, ESVV.

Lawton, J. H., Nee, S., Letcher, A. J. and Harvey, P. H. (1994). Animal distributions: patterns and processes. In: *Large-scale ecology and conservation biology.* (ed. P. J. Edwards, R. M. May, and N. R. Webb). Oxford, Blackwell: 41–58.

Lenghaus, C., Westbury, H., Collins, B., Ratnamoban, N. and Morrissy, C. (1994). Overview of the RHD project in Australia. In: *Rabbit haemorrhagic disease: issues in assessment for biological control.* (ed. E. R. W. R. Munro). Canberra, Australia., Bureau of Resource *Sciences.* 104–129.

Leon-Vizcaino, L. (1991). *Ecopatologia de la Capra Montes (Capra pyrenaica) en la sierras de Cazorla (Espana).* Gruppo stambecco Europa. IV incontro internazionale, Collana Scientifica PNGP.

Levin, M. L. and Fish, D. (1998). Density-dependent factors regulating feeding success of *Ixodes scapularis* larvae (Acari: Ixodidae). *Journal of Parasitology* **84:** 36–43.

Levins, R. (1969). Some demographic and genetic consequences of environmental heterogeneity for biological control. *Bulletin of the Entomological Sciences of America* **15:** 237–240.

Levins, R. (1970). Extinction. Some mathematical questions in biology. In: *Lectures on mathematics in the life sciences.* (ed. M. Grestenhaber). Providence, RI, American Mathematical Society. **2:** 77–107.

Lewin, R. (1986). Supply-side ecology. *Science* **234:** 25–27.

Lindsay, L. R., Barker, I. K., Surgeonary, G. A., McEwen, S. A. and Campbell, G. D. (1997). The duration of *Borrelia burgdorferi* infectivity. *Journal of Wildlife Diseases* **33:** 766–775.

Lively, C. M. and Apanius, V. (1995). Genetic diveristy in host-parasite interaction. In: *Ecology of infectious diseases in natural populations.* (ed. B. T. Grenfell and A. P. Dobson). Cambridge, Cambridge University Press: 421–449.

Lively, C. M. and Dybdahl, M. F. (2000). Parasite adaptation to locally common host genotypes. *Nature* **405:** 679–681.

Lloyd, M. (1967). Mean crowding. *Journal of Animal Ecology* **36**: 1–30.

Lloyd, S. and Soulsby, E. J. L. (1987). Immunobiology of gastrointestinal nematodes of ruminants. In: *Immune responses in parasitic infections: immunology, immunopathology and immunoprophylaxis.* (ed. E. J. L. Soulsby). Boca Raton, Florida, CRC Press. **1:** 1–41.

Lloyd, S. (1995). Environmental influences on host immunity. In: *Ecology of infectious diseases in natural populations.* (ed. B. T. Grenfell and A. P. Dobson). Cambridge, Cambridge University Press: 327–361.

Longcore, J. E., Pessier, A. P. and Nicholus, D. K. (1999). *Batrachochytrium dendrobatidis* gen. et sp. nov., a chytrid pathogenic to amphibians. *Mycologia* **91:** 219–227.

Loreto, V., Paladin, G. and Vulpiani, A. (1996). Concept of complexity in random dynamical systems. *Physical Review* **53:** 2087–2098.

Lovari, S. (1988). Risultati e prospettive di gestione degli ungulati. *Supplemento Ricerche di Biologia della Selvaggina* **14:** 517–523.

Lowe, H. J. (1942). Rinderpest in Tanganyika Territory. *Empire Journal of Experimental Agriculture* **10:** 189–203.

Lowings, P., Ibata, G., De Mia, G. M., Rutili, D. and Paton, D. (1999). Classical swine fever in Sardinia: epidemiology of recent outbreaks. *Epidemiology and Infection* **122:** 553–559.

Lvov, S. D., Gromashevsky, V. L., Marennikava, S. S., Bogoyavlensky, G. J., Barluk, F. N., Butenko, A. M. *et al.* (1988). Isolation of a poxvirus (Poxviridae, Orthopoxvirus, cowpox complex) from root voles (*Microtus oeconomus* Pal 1776) in the Kolsky peninsula forest-tundra. *Voprosy Virusologii* **1:** 92–94.

Lyles, A. M. and Dobson, A. P. (1993). Infectious-disease and intensive management—population-dynamics, threatened hosts, and their parasites. *Journal of Zoo and Wildlife Medicine* **24:** 315–326.

Lythgoe, K. A. and Read, A. F. (1998). Catching the Red Queen? The advice of the rose. *Trends in Ecology & Evolution* **13:** 473–474.

MacDonald, G. (1957). *The epidemiology and control of malaria.* Oxford, Oxford University Press.

Mace, G. and Sillero-Zubiri, C. (1997). A preliminary populatoin viability analysis for the Ethiopian wolf. In: *The Ethiopian wolf: status survey and conservation action plan.* (ed. C. Sillero-Zubiri and D. Macdonald). Gland, IUCN: 51–60.

Mackie, R. I., Krecek, R. C., Els, H. J., Van Nielurh, J. P., Kirschner, L. M. and Baecher, A. A. W. (1989). Characterization of the microbial community colonizing the anal and vulvar pores of helminths from the hindgut of zebras. *Applied and Environmental Microbiology* **55:** 1178–1186.

Magnarelli, L. A., Anderson, J. F., Stufford, K. C. and Dumler, J. S. (1997). Antibodies to multiple tick-borne pathogens of babesiosis, erlichiosis, and Lyme borreliosis in white-footed mice. *Journal of Wildlife Diseases* **33:** 466–473.

Mairtin, D. O., Williams, D. H., Griffin, J. M., Dolan, L. A. and Eves, J. A. (1998). The effect of a badger removal programme on the incidence of tuberculosis in an Irish cattle population. *Preventive Veterinary Medicine* **34:** 47–56.

Malan, F. S., Horak, I. G., de Vos, V. and Var Wyk, J. A. (1997). Wildlife parasites: lessons for parasite control in livestock. *Onderstepoort Journal of Veterinary Research* **71:** 137–153.

Malgor, R., Nonaka, N., Basmadjian, I., Sakai, H., Carabula, B., Oku, Y. *et al.* (1997). Coproantigen detection in dogs experimentally and naturally infected with *Echinococcus granulosus* by a monoclonal antibody-based enzyme-linked immunosorbent assay. *International Journal for Parasitology* **27:** 1605–1612.

Mamaev, L. V., Denikina, N. N., Belikov, S. I., Volichikov, V. E., Visser, I. K. G., Fleming, M. *et al.* (1995). Characteristics of morbilliviruses isolated from Lake Baikal seals (*Phoca sibirica*). *Veterinary Microbiology* **40:** 251–259.

Marennikova, S. S., Shelukhina, E. M. and Efremova, E. V. (1984). New outlook on the biology of cowpox virus. *Acta Virologica* **28:** 437–444.

Mascara, D., Kawano, T., Magnanelli, A. C., Silva, R. P. S., Santanna, O. A. and Morgante, J. S. (1999). *Schistosoma mansoni*: continuous variation in susceptibility of the vector snail of schistosomiasis, *Biomphalaria tenagophila*—I. Self-fertilization-lineage. *Experimental Parasitology* **93:** 133–141.

Mather, T. M., Telford, S. R. and Adler, G. H. (1991). Absence of transplacental transmission of Lyme disease spirochetes from reservoir mice (*Peromyscus leucopus*) to their offspring. *Journal of Infectious Diseases* **164:** 564–568.

Matthews, L., Woolhouse, M. E. J. and Hunter, N. (1999). The basic reproduction number for scrapie. *Proceedings of the Royal Society of London Series B-Biological Sciences* **266:** 1085–1090.

May, R. M. (1983). Parasitic infections as regulators of animal populations. *American Scientist* **71:** 36–45.

May, R. M. (1988). Conservation and disease. *Conservation Biology* **2:** 26–30.

May, R. M. (1999). Crash tests for real. *Nature* **398:** 371–372.

May, R. M. and Anderson, R. M. (1978). Regulation and stability of host-parasite population interactions. II. Destabilising processes. *Journal of Animal Ecology* **47:** 249–267.

May, R. M. and R. M. Anderson (1983). Epidemiology and genetics in the coevolution of parasites and hosts. *Proceedings of the Royal Society of London Series B-Biological Sciences* **219:** 281–313.

May, R. M. and Novak, M. (1994). Superinfection, metapopulation dynamics, and the evolution of diversity. *Journal of Theoretical Biology* **170**: 95–114.

McCallum, H. I. (2000). *Population parameters: estimation for ecological models*. Oxford, Blackwell Science Ltd.

McCallum, H. I. and Dobson, A. P. (1995). Detecting disease and parasite threats to endangered species and ecosystems. *Trends in Ecology and Evolution* **10:** 190–194.

McCallum, H. I. and Scott, M. E. (1994). Quantifying population processes: experimental and theoretical approaches. In: *Parasitic and infectious diseases: epidemiology and ecology*. (ed. H. I. McCallum and G. Smith). San Diego, Academic Press: 29–45.

McCann, K. and Yodzis, P. (1994). Nonlinear dynamics and population disappearances. *American Naturalist* **144:** 873–879.

McCullagh, P. (1983). Quasi-likelihood functions. *Annals of Statistics* **11:** 59–67.

McCurdy, D. G., Shutler, D., Mullie, A. and Forbes, M. R. (1998). Sex-biased parasitism of avian hosts: relations to blood parasite taxon and mating system. *Oikos* **82:** 303–312.

McKay, M. D., Conover, W. J. and Beckman, R. J. (1979). A comparison of three methods for selecting values of input variables in the analysis of output from a computer code. *Technometrics* **21:** 239–245.

McKenzie, A. A. (1993). Biology of the black-backed jackal *Canis mesomelas* with reference to rabies. *Onderstepoort Journal of Veterinary Research* **60:** 367–371.

McLean, A. R. and Nowak, M. A. (1992). Competition between zidovudine-sensitive and zidovudine-resistant strains of HIV. *AIDS* **6:** 71–79.

Mcleod, P. J., Young, S. Y. and Yearian, W. C. (1982). Application of a baculovirus of pseudoplusia-includens to soybean—efficacy and seasonal persistence (Lepidoptera, Noctuidae). *Environmental Entomology* **11:** 412–416.

Meade, M. S., Florin, V. and Gesler, W. M. (1988). *Medical geography*. New York, Guilford Press.

Mech, L. D. and Goyal, S. M. (1995). Effects of canine parvovirus on gray wolves in Minnesota. *Journal of Wildlife Management* **59:** 565–570.

Medley, G. F., Perry, B. D. and Young, A. S. (1993). Preliminary analysis of the transmission dynamics of *Theileria parva* in eastern Africa. *Parasitology* **106:** 251–264.

Meltzer, D. G. A. (1993). Historical survey of disease problems in wildlife populations: Southern Africa mammals. *Journal of Zoo and Wildlife Medicine* **24:** 237–244.

Merler, S. and Furlanello, C. (1997). Selection of tree-based classifiers with the bootstrap 632+ rule. *Biometrical Journal* **39:** 369–382.

Merler, S., Furlanello, C., Chemini, C. and Nicollini, G. (1996). Classification tree methods for analysis of mesoscale distribution of *Ixodes ricinus* (Acari: Ixodidae) in Trentino, Italian Alps. *Journal of Medical Entomology* **33:** 888–893.

Merrell, C. L. and Wright, D. N. (1978). A serological survey of mule deer and elk in Utah. *Journal of Wildlife Diseases* **14:** 471–478.

Meyer, M. E. and Meagher, M. (1997). *Brucella abortus* infection in the free-ranging bison of Yellowstone National Park. In: *Brucellosis, bison, elk, and cattle in the Greater Yellowstone Area: defining the problem, exploring solutions*. (ed. E. T. Thorne, M. S. Boyce, P. Nicoletti, and T. J. Kreeger). Cheyenne, Wyoming, Game and Fish Department: 20–32.

Michel, J. F. (1974). Arrested development of nematodes and some related phenomena. *Advances in Parasitology* **12:** 279–343.

Mills, J. N., Ksiazek, T. G., Peters C. J. and Childs, J. E. (1999). Long-term studies of hantavirus reservoir populations in the southwestern United States: a synthesis. *Emerging Infectious Diseases* **5:** 135–142.

Minchella, D. J. and Scott, M. E. (1991). Parasitism—A cryptic determinant of animal community structure. *Trends in Ecology and Evolution* **6:** 250–254.

Monagas, W. R. and Gatten, R. E. (1983). Behavioral fever in the turtles *Terrapene carolina* and *Chrysemys picta*. *Journal of Thermal Biology* **8:** 285–288.

Moody, K. D., Terwilliger, G. A., Hansen, G. M. and Barthold, S. W. (1994). Experimental *Borrelia burgdorferi* infection in *Peromyscus leucopus*. *Journal of Wildlife Diseases* **30:** 155–161.

Moore, J. (1984). Altered behavioral-responses in intermediate hosts—an acanthocephalan parasite strategy. *American Naturalist* **123:** 572–577.

Moore, J. and Gotelli, N. J. (1990). A phylogenetic perspective on the evolution of altered host behaviours. In: *Parasitism and host behaviour*. (ed. C. J. Barnard and J. M. Behnke). London, Taylor & Francis: 193–233.

Moore, J. and Gotelli, N. J. (1996). Evolutionary patterns of altered behavior and susceptibility in parasitized hosts. *Evolution* **50:** 807–819.

Morand, S. (1996). Life-history traits in parasitic nematodes: a comparative approach for the search of invariants. *Functional Ecology* **10:** 210–218.

Morand, S., Legendre, P., Gardner, S. L. and Hugot, J. P. (1996). Body size evolution of oxyurid (Nematoda) parasites: the role of hosts. *Oecologia* **107:** 274–282.

Morris, R. S. and Pfeiffer, D. U. (1995). Directions and issues in bovine tuberculosis epidemiology and control in New Zealand. *New Zealand Veterinary Journal* **43:** 256–265.

Moutou, F. (1997). Dog vaccination around the Serengeti. *Oryx* **31:** 14.

Møller, A. P. (1990). Parasites and sexual selection—current status of the Hamilton and Zuk hypothesis. *Journal of Evolutionary Biology* **3:** 319–328.

Møller, A. P., Dufva, R. and Erritzoe, J. (1998). Host immune function and sexual selection in birds. *Journal of Evolutionary Biology* **11:** 703–719.

Muller, C. B. and Schmidhempel, P. (1993). Exploitation of cold temperature as defense against parasitoids in bumblebees. *Nature* **363:** 65–67.

Müller-Graf, C. D. M., Collins, D. A., Packer, C. and Woolhouse, M. E. J. (1997). *Schistosoma mansoni* infection in a natural population of olive baboons (*Papio cynocephalus anubis*) in Gombe Stream National Park, Tanzania. *Parasitology* **115:** 621–627.

Mulvey, M., Aho, J. M., Lyoeard, C., Leberg, P. L. and Smith, M. H. (1991). Comparative population genetic-structure of a parasite (*Fascioloides magna*) and its definitive host. *Evolution* **45:** 1628–1640.

Murray, D. L., Cary, J. R. and Keith, L. B. (1997). Interactive effects of sublethal nematodes and nutritional status on snowshoe hare vulnerability to predation. *Journal of Animal Ecology* **66:** 250–264.

Murray, D. L., Keith, L. B. and Cary, J. R. (1998). Do parasitism and nutritional status interact to affect production in snowshoe hares? *Ecology* **79:** 1209–1222.

Mutapi, F., Ndhlovu, P. D., Hagan, P. and Woolhouse, M. E. J. (1997). A comparison of humoral responses to *Schistosoma haematobium* in areas with low and high levels of infection. *Parasite Immunology* **19:** 255–263.

Nåsell, I. (1999a). On the quasi-stationary distribution of the stochastic logistic epidemic. *Mathematical Biosciences* **156:** 21–40.

Nåsell, I. (1999b). On the time to extinction in recurrent epidemics. *Journal of the Royal Statistical Society Series B-Statistical Methodology* **61:** 309–330.

Nee, S. (1994). How populations persist. *Nature* **367:** 123–124.

Nee, S., Read, A. F., Greenwood, J. J. D. and Hamey, P. H. (1991). The relationship between abundance and body size in British birds. *Nature* **351:** 312–313.

Nee, S., May, R. M. and Hassen, M. P. (1997). Two-species metapopulation models. In: *Metapopulation biology: ecology, genetics, and evolution*. (ed. I. Hanski and M. E. Giplin). San Diego, Academic Press: 123–147.

Neilson, R. P., King, G. A. and Koerper, G. (1992). Toward a rule-based biome model. *Landscape Ecology* **7:** 27–43.

Nilssen, A. C., Haugerud, R. E. and Folstad, I. (1998). No interspecific covariation in intensities of macroparasites of reindeer, *Rangifer tarandus* (L.). *Parasitology* **117:** 273–281.

Nolan, P. M., Hill, G. E. and Stoehr, A. M. (1998). Sex, size, and plumage redness predict house finch survival in an epidemic. *Proceedings of the Royal Society of London Series B-Biological Sciences* **265:** 961–965.

Nonaka, N., Tsukada, H., Abe, N., Oku, Y. and Kamiya, M. (1998). Monitoring of *Echinococcus multilocularis* infection in red foxes in Shiretoko, Japan, by coproantigen detection. *Parasitology* **117:** 193–200.

Norris, K. and Evans, M. R. (2000). Ecological immunology: life history trade-offs and immune defense in birds. *Behavioral Ecology* **11:** 19–26.

Norval, R. A. I., Perry, B. D. and Young, A. S. (1992). *The epidemiology of theileria in Africa*. London, Academic Press.

Nuttall, P. A., Jones, L. D., Labuda, M. and Kaufmann, W. R. (1994). Adaptations of Arboviruses to ticks. *Journal of Medical Entomology* **31:** 1–9.

O'Callaghan, C. J., Medley, G. F., Peter, T. F. and Perry, B. D. (1998). Investigating the epidemiology of Heartwater (*Cowdria ruminantium infection*) by means of a transmission dynamics model. *Parasitology* **117:** 49–61.

O'Callaghan, C. J., Medley, G. F., Peter, T. F., Mahan, S. M. and Perry, B. D. (1999). Predicting the effect of vaccination on the transmission dynamics of heartwater (*Cowdria ruminantium*) infection. *Preventive Veterinary Madicine* **42:** 17–38.

O'Connell, S., Granström, M., Giay, J. S. and Stanek, G. (1998). Epidemiology of European Lyme borreliosis. *Zentralblatt für Bakteriologie* **287:** 229–240.

Ochanda, H., Young, A. S., Wells, C., Medley, G. F. and Perry, B. D. (1996). Comparison of transmission of *Theileria parva* between different instars of *Rhipicephalus appendiculatus*. *Parasitology* **113:** 243–253.

Ogden, N. H., Nuttall, P. A. and Randolph, S. E. (1997). Natural Lyme disease cycles maintained via sheep by cofeeding ticks. *Parasitology* **115:** 591–599.

Okimoto, R., Macfarlane, J. L. and Wolstenholme, D. R. (1994). The mitochondrial ribosomal-RNA genes of the nematodes *Caenorhabditis elegans* and *Ascarissuum*—consensus secondary-structure models and conserved nucleotide sets for phylogenetic analysis. *Journal of Molecular Evolution* **39:** 598–613.

Olsen, B., Jaenson, T. G. T., Noppa, L., Bunihis, J. and Bergstiom, S. (1993). A Lyme borreliosis cycle in seabirds and *Ixodes uriae* ticks. *Nature* **362:** 340–342.

Osterhaus, A., van de Bilt, M., Vedder, L., Martina, B., Niesters, H., Kotomatas, S. *et al.* (1998). Monk seal mortality: virus or toxin? *Vaccine* **16:** 979–981.

Ostfeld, R. S., Hazler, K. R. and Cepeda, O. M. (1996). Temporal and spatial dynamics of *Ixodes scapularis* (Acari: Ixodidae) in a rural landscape. *Journal of Medical Entomology* **33:** 90–95.

Owens, I. P. F. and Wilson, K. (1999). Immunocompetence: a neglected life history trait or conspicuous red herring? *Trends in Ecology* & Evolution **14:** 170–172.

Pacala, S. W. and Dobson, A. P. (1988). The relation between the number of parasites host and host age: population dynamic causes and maximum likelihood estimation. *Parasitology* **96:** 197–210.

Paladin, G., Serva, M. and Volpiani, A. (1995). Complexity in dynamical-systems with noise. *Physical Review Letters* **74:** 66–69.

Park, T. (1948). Experimental studies of interspecies competition. 1. Competition between populations of the flour beetles, *Tribolium confusum* Duval and *Tribolium castaneum* Herbst. *Ecological Monographs* **18:** 267–307.

Pascual, M., Rodo, X., Ellnet, S. P., Colwell, R. and Bouma, M. J. (2000). Cholera dynamics and El Niño-southern oscillation. *Science* **289:** 1766–1769.

Pastoret, P. O. and Brochier, B. (1998). Epidemiology and elimination of rabies in Western Europe. *Veterinary Journal* **156:** 83–90.

Paterson, S., Wilson, K. and Pemberton, J. M. (1998). Major histocompatibility complex variation associated with juvenile survival and parasite resistance in a large unmanaged ungulate population (*Ovis aries* L.). *Proceedings of the National Academy of Sciences of the United States of America* **95:** 3714–3719.

Pavlovsky, E. N. (1966). *The natural nadility of transmissible disease*. Urbana, IL, University of Illinois Press.

Pemberton, J. M., Smith, J. A., Coulsan, T. N., Marshall, T. C., Slate, J., Patterson, S. *et al.* (1996). The maintenance of genetic polymorphism in small island populations: large mammals in the Hebrides. *Philosophical Transactions of the Royal Society of London Series B-Biological Sciences* **351:** 745–752.

Pennycuick, L. (1971). Frequency distributions of parasites in a population of three-spined sticklebacks, *Gasterosteus aculeatus* L., with particular reference to the negative binomial distribution. *Parasitology* **63:** 389–406.

Perco, F. (1987). *Ungulati*. Udine, Carso.

Perez-Eid, C. (1990). Les relations tiques—petits mammifères dans le foyer Alsacien d'encéphalite à tiques. *Acarologia* **31:** 131–141.

Perez-Eid, C., Hannoun, C. and Rodhain, F. (1992). The Alsatian tick-borne encephalitis focus: presence of the virus among ticks and small mammals. *European Journal of Epidemiology* **8:** 178–186.

Peters, R. H. (1983). *The Ecological Implications of Body Size*. Cambridge, Cambridge University Press.

Piesman, J. (1989). Transmission of Lyme disease spirochete (*Borrelia burgdorferi*). *Experimental and Applied Acarology* **7:** 71–80.

Piesman, J., Spielman, A., Etkind, P., Ruebush, T. K. and Juranek, D. D. (1979). Role of deer in the epizootiology of *Babesia microti* in Massachusetts, USA. *Journal of Medical Entomology* **15:** 537–540.

Piesman, J., Mather, T. M., Dammin, G. J., Telford, S. R., Lastavica, C. C. and Spielman, A. (1987). Seasonal variation of transmission risk of Lyme disease and human babesiosis. *American Journal of Epidemiology* **126:** 1187–1189.

Plowright, W. (1982). The effects of rinderpest and rinderpest control on wildlife in Africa. *Symposium of the Zoological Society of London* **50:** 1–28.

Poole, B. C., Chadee, K. and Dick, T. A. (1983). Helminth parasites of pine marten, *Martes americana* (Turton), from Manitoba, Canada. *Journal of Wildlife Diseases* **19:** 10–13.

Postic, D., Assous, M., Grimont, P. A. D. and Baranton, D. G. (1994). Diversity of *Borrelia burgdorferi sensu lato* evidenced by restriction fragment length polymorphism of rrf(5S)-rr(23S) intergenic spacer amplicons. *International Journal of Systemic Bacteriology* **44:** 743–752.

Potts, G. R., Tapper, S. C. and Hudson, P. J. (1984). Population fluctuations in red grouse: analysis of bag records and a simulation model. *Journal of Animal Ecology* **53:** 21–36.

Potts, G. R. (1986). *The Partridge: Pesticides, Predation and Conservation*. London, Collins.

Poulin, R. (1993). The disparity between observed and uniform distributions—a new look at parasite aggregation. *International Journal for Parasitology* **23:** 937–944.

Poulin, R. (1994). The evolution of parasite manipulation of host behavior—a theoretical-analysis. *Parasitology* **109:** S109–S118.

Poulin, R. (1995). Phylogeny, ecology, and the richness of parasite communities in vertebrates. *Ecological Monographs* **65:** 283–302.

Poulin, R. (1996). Sexual inequalities in helminth infections: a cost of being male? *American Naturalist* **147:** 287–295.

Poulin, R. (1998). *Evolutionary ecology of parasites: from individuals to communities*. London, Chapman & Hall.

Poulin, R. (2000). Manipulation of host behaviour by parasites: a weakening paradigm? *Proceedings of the Royal Society of London Series B-Biological Sciences* **267:** 787–792.

Pound, J. M., Miller, J. A., George, J. E. and Lemeilleur, C. A. (2000). The 4-poster passive topical treatment device to apply acaricide for controlling ticks (Acari: Ixodidae) feeding on white-tailed deer. *Journal of Medical Entomology* **37:** 588–594.

Price, P. W., Westoby, M., Rice, B., Atsatt, P. R., Fritz, R. S., Thompson, J. N. *et al.* (1986). Parasite mediation in ecological interactions. *Annual Review of Ecology and Systematics* **17:** 487–505.

Promislow, D. E. L. (1992). Costs of sexual selection in natural populations of mammals. *Proceedings of the Royal Society of London, Series B-Biological Science* **247:** 203–210.

Promislow, D. E. L., Montgomerie, R. and Martin, T. E. (1992). Mortality costs of sexual dimorphism in birds. *Proceedings of the Royal Society of London, Series B-Biological Science* **250:** 143–150.

Pugliese, A. and Rosà, R. (1995). A 2-dimensional model for macroparasitic infections in a host with logistic growth. *Journal of Biologycal System* **3:** 833–849.

Pugliese, A., Rosà, R. and Damaggio, M. L. (1998). Analysis of a model for macroparasitic infection with variable aggregation and clumped infections. *Journal of Mathematical Biology* **36:** 419–447.

Quinnell, R. J., Dye, C. and Shaw, J. J. (1992). Host preferences of the phlebotomine sandfly lutzomyia-longipalpis in amazonian brazil. *Medical and Veterinary Entomology* **6:** 195–200.

Quinnell, R. J., Grafen, A. and Woolhouse, M. E. J. (1995). Change in parasite aggregation with age: a discrete infection model. *Parasitology* **111:** 635–644.

Rand, D. A. and Wilson, H. B. (1991). Chaotic stochasticity: a ubiquitous source of unpredictability in epidemics. *Proceedings of the Royal Society of London Series B-Biological Sciences* **246:** 179–184.

Randolph, S. E. (1975). Patterns of distribution of the tick *Ixodes trianguliceps* Birula on its hosts. *Journal of Animal Ecology* **44:** 451–474.

Randolph, S. E. (1977). Changing spatial relationships in a population of *Apodemus sylvaticus* with the onset of breeding. *Journal of Animal Ecology* **46:** 653–676.

Randolph, S. E. (1991). The effect of *Babesia microti* on feeding and survival in its tick vector, *Ixodes trianguliceps*. *Parasitology* **102:** 9–16.

Randolph, S. E. (1994). Population-dynamics and density-dependent seasonal mortality indexes of the tick *Rhipicephalus appendiculatus* in eastern and southern Africa. *Medical and Veterinary Entomology* **8:** 351–368.

Randolph, S. E. (1995). Quantifying parameters in the transmission of *Babesia microti* by the tick *Ixodes trianguliceps* amongst voles (*Clethrionomys glareolus*). *Parasitology* **110:** 287–295.

Randolph, S. E. (1997). Abiotic and biotic determinants of the seasonal dynamics of the tick *Rhipicephalus appendiculatus* in South Africa. *Medical and Veterinary Entomology* **11:** 25–37.

Randolph, S. E. (1998). Ticks are not insects: consequences of contrasting vector biology for transmission potential. *Parasitology Today* **14:** 186–192.

Randolph, S. E. (2000). Ticks and tick-borne disease systems in space and from space. *Advances in Parasitology* **47:** 217–243.

Randolph, S. E. and Craine, N. G. (1995). General framework for comparative quantitative studies on transmission of tick-borne diseases using Lyme borreliosis in Europe as an example. *Journal of Medical Entomology* **32:** 765–777.

Randolph, S. E. and Rogers, D. J. (1997). A generic population model for the African tick *Rhipicephalus appendiculatus*. *Parasitology* **115:** 265–279.

Randolph, S. E. and Rogers, D. J. (2000). Fragile transmission cycles of tick-borne encephalitis virus may be disrupted by predicted climate change. *Proceedings of the Royal Society of London Series B-Biological Sciences* **267:** 1741–1744.

Randolph, S. E., Gern, L. and Nuttall, P. A. (1996). Co-feeding ticks: epidemiological significance for tick-borne pathogen transmission. *Parasitology Today* **12:** 472–479.

Randolph, S. E., Miklisova, D., Lysy, J., Rogers, D. J. and Labuda, M. (1999). Incidence from coincidence: patterns of tick infestations on rodents facilitate transmission of tick-borne encephalitis virus. *Parasitology* **118:** 177–186.

Randolph, S. E., Green, R. M., Peacey, M. F. and Rogers, D. J. (2000). Seasonal synchrony: the key to tick-borne encephalitis foci identified by satellite data. *Parasitology* **121:** 15–23.

Read, A. F. (1988). Sexual selection and the role of parasites. *Trends in Ecology & Evolution* **3:** 97–102.

Read, A. F. (1990). Parasites and the evolution of host sexual behaviour.In: *Parasitism and host behaviour*. (ed. C. J. Barnard and J. M. Behnke). London, Taylor & Francis: 117–157.

Read, A. F. and Harvey, P. H. (1993). Parasitology—the evolution of virulence. *Nature* **362:** 500–501.

Read, A. F. and Skorping, A. (1995). The evolution of tissue migration by parasitic nematode larvae. *Parasitology* **111:** 359–371.

Read, A. F. and Viney, M. E. (1996). Helminth immunogenetics : why bother. *Parasitology Today* **12:** 337–343.

Read, A. F., Albon, S. D., Antonovics, J., Apanius, V., Dwyer, G., Holt, R. D. *et al.* (1995). Genetics and evolution of infectious diseases in natural populations. In: *Ecology of infectious diseases in natural populations*. (ed. B. T. Grenfell and A. P. Dobson). Cambridge, Cambridge University Press: 450–477.

Rechav, Y. (1981). *Ecological factors affecting the seasonal activity of the brown ear tick* Rhipicephalus appendiculatus. International Conference on Tick Biology and Control, Grahamstown. (ed. G. B. Whitehead, E. J. D. Gobson), Rhodes University, Grahamstown, South Africa, 187–191.

Rechav, Y. (1992). Naturally acquired resistance to ticks. *Insect Science and its Application* **13:** 495–504.

Reeson, A. F., Wilson, K., Gunn, A., Hails, R. S. and Goulson, D. (1998). Baculovirus resistance in the noctuid *Spodoptera exempta* is phenotypically plastic and responds to population density. *Proceedings of the Royal Society of London Series B-Biological Sciences* **265:** 1787–1791.

Reid, H. W., Duncan, J. S., Phillips, J. D. P., Moss, R. and Watson, A. (1978). Studies of louping- ill virus (Flavivirus group) in wild red grouse (*Lagopus lagopus scoticus*). *Journal of Hygiene* **81:** 321–329.

Rementzova, M. M. (1964). La Brucellose des animaux sauvages. *Bulletin Office International des Epizooties* **61:** 99–112.

Rengel, J. and Bohnel, H. (1994). Preliminary trials of oral immunization of wildlife against anthrax. *Berliner und Munchener Tierarztliche Wofhenschirft* **107:** 145–149.

Rhodes, C. J., Atkinson, R. P. D., Anderson, R. M. and MacDonald, D. W. (1998). Rabies in Zimbabwe: reservoir dogs and the implications for disease control. *Philosophical Transactions of the Royal Society of London Series B-Biological Sciences* **353:** 999–1010.

Rhyan, J., Aune, K., Ewalt, D. R., Marquardt, J., Martin, J. W., Payeul, J. B. *et al.* (1997). Survey of free-ranging elk from Wyoming and Montana for selected pathogens. *Journal of Wildlife Diseases* **33:** 290–298.

Richter, D., Spielman, A. and Matuschka F. R. (1998). Effect of prior exposure to non-infected ticks of susceptibility of mice to Lyme Disease spirochetes. *Applied and Environmental Microbiology* **64:** 4596–4599.

Rizzoli, A. P. (1995). *Indagini fisiopatologiche su mammiferi selvatici dell'Arco alpino orientale*. Istituto di Anatomia Patologica e Patologia Aviare. Milano, Università di Milano.

Rizzoli, A. P., Manfredi, M. T., Rosso, F., Rosà, R., Cattadori, I. and Hudson, P. J. (1997). A survey to identify the important macroparasites of rock partridge (*Alectoris graeca saxatilis*) in Trentino, Italy. *Parassitologia* **39:** 331–334.

Rizzoli, A. P., Manfredi, M. T., Rosso, F., Rosà, R., Cattadori, I. and Hudson, P. J. (1999). Intensity of nematode infections in cyclic and non-cyclic rock partridge (*Alectoris graeca saxatilis*) populations. *Parassitologia* **41:** 561–565.

Roberts, M. G. (1995). A pocket guide to host-parasite models. *Parasitology Today* **11:** 172–177.

Roberts, M. G. (1996). The dynamics of bovine tuberculosis in possum populations, and its eradication or control by culling or vaccination. *Journal of Animal Ecology* **65:** 451–464.

Roberts, M. G. and Dobson, A. P. (1995). The population-dynamics of communities of parasitic helminths. *Mathematical Biosciences* **126:** 191–214.

Roberts, M. G. and Grenfell, B. T. (1991). The population dynamics of nematode infections of ruminants: periodic perturbations as a model for management. *IMA Journal of Mathematics Applied in Medicine and Biology* **8:** 83–93.

Roberts, M. G. and Heesterbeek, J. A. P. (1995). The dynamics of nematode infections of farmed ruminants. *Parasitology* **110:** 493–502.

Roberts, M. G. and Heesterbeek, J. A. P. (1998). A simple parasite model with complicated dynamics. *Journal of Mathematical Biology* **37:** 272–290.

Roberts, M. G., Lawson, J. R. and Gemmell, M. A. (1986). Population dynamics in echinococcosis and cysticercosis: Mathematical model of the life cycle of *Echinococcus granulosus*. *Parasitology* **92**: 621–641.

Roberts, M. G., Lawson, J. R. and Gemmell, M. A. (1987). Population dynamics in echinococcosis and cysticercosis: mathematical model of the life cycles of *Taenia hydatigena* and *T.ovis*. *Parasitology* **94:** 181–197.

Roberts, M. G., Smith, G. and Grenfell, B. T. (1995). Mathematical models for macroparasites of wildlife. In:

Ecology of infectious diseases in natural populations. (ed. B. T. Grenfell and A. P. Dobson). Cambridge, Cambridge University Press: 177–208.

Robinson , C. (1994). *Dynamical systems: stability, symbolic dynamics and chaos.* Boca Raton Florida, CRC Press.

Roelke, M. E. and Glass, C. M. (1992). *Strategies for the management of the endangered Florida panther* (Felis concolor coryi) *in an ever-shrinking habitat.* New York Proceedings of the Joint Meeting of the American Association of Zoo Veterinarians and the American Association of Wildlife Veterinarians.

Roelke-Parker, M. E., Munson, L., Pacher, C., Kock, R., Cleaveland, S., Carpenter, M. *et al.* (1996). A canine distemper virus epidemic in Serengeti lions (*Panthera leo*). *Nature* **381:** 172–172.

Rogers, D. J. (1985). Trypanosomiasis 'risk' or 'challenge': a review. *Acta Tropica* **42:** 5–23.

Rogers, D. J. (1991). Satellite imagery, tsetse and trypanosomiasis in Africa. *Preventive Veterinary Medicine* **11:** 201–220.

Rogers, D. J. and Randolph, S. E. (1991). Mortality-rates and population-density of tsetse-flies correlated with satellite imagery. *Nature* **351:** 739–741.

Rogers, D. J. and Randolph, S. E. (1993). Distribution of tsetse and ticks in africa—past, present and future. *Parasitology Today* **9:** 266–271.

Rogers, D. J. and Williams, B. G. (1993). Monitoring trypanosomiasis in space and time. *Parasitology* **106:** S77–S92.

Rogers, D. J. and Williams, B. G. (1994). Tsetse distributions in Africa: seeing the wood and the trees. In: *Large-scale Ecology and Conservation Biology.* (ed. P. J. Edwards, R. M. May, and N. R. Webb). Oxford, Blackwell Scientific Publications: 247–271.

Rogers, D. J., Hay, S. I. and Packer, M. J. (1996). Predicting the distribution of tsetse flies in West Africa using temporal Fourier processed meteorological satellite data. *Annals of Tropical Medicine and Hygiene* **90:** 225–241.

Rogers, D. J. and Randolph, S. E. (2000). The global spread of malaria in a future, warmer world. *Science* **289:** 1763–1766.

Rohani, P. and Miramontes, O. (1995). Host-parasitoid metapopulations: the consequences of parasitoid aggregation on spatial dynamics and searching efficiency. *Procedings of the Royal Society London Series B-Biological Sicences* **260:** 335.

Rohani, P., Godfray, H. C. J. and Hassell, M. P. (1994). Aggregation and the dynamics of host-parasitoid systems—a discrete-generation model with within-generation redistribution. *American Naturalist* **144:** 491–509.

Rohani, P., May, R. M. and Hassell, M. P. (1996). Metapopulations and equilibrium stability: the effects of spatial structure. *Journal of Theoretical Biology* **181:** 97–109.

Rohani, P., Earn, D. J., Finkenstadt, B. and Grenfell, B. T. (1998). Population dynamic interference among childhood diseases. *Proceedings of The Royal Society Of London Series B-Biological Sciences* **265:** 2033–2041.

Roitt, I., Brostoff, J. and Male, D. (1998). *Immunology.* London, Mosby.

Rott, A. S., Müller, C. B. and Godfray, H. C. J. (1998). Indirect population interaction between two aphid species. *Ecology Letters* **1:** 99–103.

Roughgaden, J. (1995). Anolis *lizards of the Caribbean.* Oxford, Oxford University Press.

Ruppert, D. (1997). Empirical-bias bandwidths for local polynomial nonparametric regression and density estimation. *Journal of American Statistical Association* **92:** 1049–1062.

Ruxton, G. D. (1994). Low levels of immigration between chaotic populations can reduce system extinctions by inducing asynchronous regular cycles. *Procedings of the Royal Society London, Series B-Biological Sciences* **256:** 189–193.

Ryan, T. J., Livingstone, P. G., Bailey, J. B., Carter, C. B., Crews, K. B. and Timbs, D. V. (1998). *Tuberculosis control in livestock in New Zealand: the current situation and future directions.* Proceedings of a Meeting held at the East County Hotel, Ennis Co. Clare on the 25th, 26th and 27th March 1998, Society for Veterinary Epidemiology and Medicine.

Saino, N., Møller, A. P. and Bolzern, A. M. (1995). Testosterone effects on the immune system and parasite infestations in the barn swallow (*Hirundo rustica*): an experimental test of the immunocompetence hypothesis. *Behavioral Ecology* **6:** 397–404.

Saino, N., Bolzern, A. M. and Møller, A. P. (1997). Immunocompetence, ornamentation, and viability of male barn swallows (*Hirundo rustica*). *Proceedings of the National Academy of Sciences of the United States of America* **94:** 549–552.

Sait, S. M., Liu, W. C., Thompson, D. J., Godfray, H. C. J. and Begon, M. (2000). Invasion sequence affects predator-prey dynamics in a multi-species interaction. *Nature* **405:** 448–450.

Sanchez, D. A. (1968). *Ordinary differential equations and stability theory: an introduction.* San Francisco, Freeman.

Sandberg, S., Awerbuch, T. E. and Spielman, A. (1992). A comprehensive multiple matrix model representing the life cycle of the tick that transmits the agent of Lyme disease. *Journal of Theoretical Biology* **157:** 203–220.

Saunders, D. A., Hobbs, R. J. and Margules, C. R. (1991). Biological consequences of ecosystem fragmentation: a review. *Conservation Biology* **5:** 18–32.

Saunders, D. A. and Hobbs, R. J. (1991). *Nature* Conservation **2:** The Role of Corridors. Chipping Norton, Australia, Surrey Beatty and Sons.

Saunders, G., Choquenot, D., Mellroy, J. and Packwood, R. (1999). Initial effects of rabbit haemorrhagic disease on free-living rabbit (*Oryctolagus cuniculus*) populations in central-western New South Wales. *Wildlife Research* **26:** 69–74.

Sauter, C. M. and Morris, R. S. (1995). Behavioural studies on the potential for direct transmission of tuberculosis from feral ferrets (*Mustela furo*) and possums (*Trichosurus vulpecula*) to farmed livestock. *New Zealand Veterinary Journal* **43:** 294–300.

Schalk, G. and Forbes, M. R. (1997). Male biases in parasitism of mammals: effects of study type, host age, and parasite taxon. *Oikos* **78:** 67–74.

Schall, J. J. (1983). Lizard malaria: cost to vertebrate host's reproductive success. *Parasitology* **87:** 1–6.

Schall, J. J. (1990). The *Ecology* of lizard malaria. *Parasitology Today* **6:** 204–269.

Schall, J. J. (1992). Parasite-mediated competition in *Anolis* lizards. *Oecologia* **92:** 58–64.

Schluter, D. (1984). A variance test for detecting species associations, with some example. *Applications Ecology* **65:** 998–1005.

Schmitz, O. J. and Nudds, T. D. (1994). Parasite-mediated competition in deer and moose: how strong is the effect of the meningeal worm on moose? *Ecological Applications* **4:** 91–103.

Schuurs, A. and Verheul, H. A. M. (1990). Effects of gender and sex steroids on the immune-response. *Journal of Steroid Biochemistry and Molecular Biology* **35**(2): 157–172.

Schwaiger, F. W., Gostomski, D., Steur, M. J., Duncan, J. L., McKellar, Q. A., Epplen, J. T. *et al.* (1995). An ovine major histocompatibility complex drb1 allele is associated with low fecal egg counts following natural, predominantly ostertagia-circumcincta infection. *International Journal for Parasitology* **25:** 815–822.

Scott, G. R. (1981). Rinderpest. Infectious Diseases of Wild Mammals. J. W. Davis, L. H. Karstad and D. O. Trainer. Ames, Iowa, Iowa State University Press: 18–30.

Scott, M. E. (1987a). Temporal changes in aggregation: a laboratory study. *Parasitology* **94:** 583–595.

Scott, M. E. (1987b). Regulation of mouse colony abundance by *Heligmosomoides polygyrus*. *Parasitology* **95:** 111–124.

Scott, M. E. (1988). The impact of infection and disease on animal populations: implications for conservation biology. *Conservation Biology* **2:** 40–56.

Scott, M. E. and Anderson, R. M. (1984). The populations dynamics of *Gyrodactylus bullatarudis* (Monogenea) on guppies (*Poecilia reticulata*). *Parasitology* **89:** 159–194.

Scott, M. E. and Dobson, A. P. (1989). The role of parasites in regulating host abundance. *Parasitology Today* **5:** 176–183.

Scott, M. E. and Smith, G. (ed.) (1994). *Parasitic and infectious diseases: epidemiology and ecology*. San Diego, Academic Press, Inc.

Settle, W. H. and Wilson, L. T. (1990). Invasion by the variagated leafhopper and biotic interactions: parasitism, competition and apparent competition. *Ecology* **71:** 1461–1470.

Shaw, D. J. and Dobson, A. P. (1995). Patterns of macroparasite abundance and aggregation in wildlife populations: a quantitative review. *Parasitology* **111:** S 111–S 133.

Shaw, D. J., Grenfell, B. T. and Dobson, A. P. (1998). Patterns of macroparasite aggregation in wildlife host populations. *Parasitology* **117:** 597–610.

Siegel, P. B. and Gross, W. B. (1980). Production and persistence of antibodies in chickens to sheep erythrocytes. 1. Directional selection. Poultry *Science* **59:** 1–5.

Sillero-Zubiri, C., King, A. A. and MacDonald, D. W. (1996). Rabies and mortality in Ethiopian wolves (*Canis simensis*). *Journal of Wildlife Diseases* **32:** 80–86.

Silva, M. and Downing, J. A. (1995). The allometric scaling of density and body-mass—a nonlinear relationship for terrestrial mammals. *American Naturalist* **145:** 704–727.

Simberloff, D. and Cox, J. (1987). Consequences and costs of conservation corridors. *Conservation Biology* **1:** 63–71.

Sinclair, A. R. E. (1977). *The African buffalo: a study of resource limitation by populations*. Chicago, University of Chicago Press.

Sinclair, A. R. E. and Norton-Griffiths, M. (ed.) (1979). *Serengeti: dynamics of an ecosystem*. Chicago, University of Chicago Press.

Siva-Jothy, M. T. and Skarstein, F. (1998). Towards a functional understanding of good genes. *Ecology Letters* **1:** 178–185.

Skorping, A. and Read, A. F. (1998). Drugs and parasites: global experiments in life history evolution? *Ecology Letters* **1:** 10–12.

Skorping, A., Read, A. F. and Keymer, A. E. (1991). Life history covariation in intestinal nematodes of mammals. *Oikos* **60:** 365–372.

Smith, A. T. (1980). Temporal changes in insular populations of the pika (*Ochotona princeps*). *Ecology* **61:** 8–13.

Smith, G. (1994). Population biology of the parasitic phase of trichostrongylid nematode parasites of cattle and sheep. *International Journal for Parasitology* **24:** 167–178.

Smith, J. A. (1996). *Polymorphism, parasites and fitness in Soay sheep*. Department of Genetics. Cambridge, University of Cambridge.

Smith, O. (1999). Epidemic! The world of infectious disease. *Science* **283:** 1859–1859.

Smith, G. and Dobson, A. P. (1992). Sexually transmitted diseases in animals. *Parasitology Today* **8:** 159–166.

Smith, G. and Grenfell, B. T. (1994). Modelling of parasite populations: gastrointestinal nematode models. *Veterinary Parasitology* **54:** 127–143.

Smith, G. C., Richards, M. S., Clifton-Hadley, R. S. and Cheeseman, C. L. (1995). Modelling bovine tuberculosis in badgers in England: preliminary results. *Mammalia* **59:** 639–650.

Smith, J. A., Wilson, K., Pilkington, J. G. and Pemberton, J. M. (1999). Heritable variation in resistance to gastro-intestinal nematodes in an unmanaged mammal population. *Proceedings of the Royal Society of London Series B-Biological Sciences* **266:** 1283–1290.

Sorci, G., Møller, A. P. and Boulinier, T. (1997). Genetics of host-parasite interactions. *Trends in Ecology & Evolution* **12:** 196–200.

Sousa, W. P. (1994). Patterns and processes in communities of helminth parasites. *Trends in Ecology and Evolution* **9:** 52–57.

Southwood, T. R. E. (1966). *Ecological methods*. London, Chapman & Hall.

Spencer, J. and Burroughs, R. (1992). Antibody responses to canine distemper vaccine in African wild dogs. *Journal of Wildlife Diseases* **28:** 443–444.

Spielman, A. (1988). Lyme disease and human Babesiosis: evidence incriminating vector and reservoir hosts. In: *The biology of parasitism*: a molecular and immunological approach (ed. P. T. England and A. Shur), 147–165. Alan R. Liss, New York.

Spraker, T. R., Miller, M. W., Williams, E. S., Getzy, D. M., Adrian, W. J., Schoonveld, G. G. *et al.* (1997). Spongiform encephalopathy in free-ranging mule deer (*Odocoileus hemionus*), white-tailed deer (*Odocoileus virginianus*) and Rocky Mountain elk (*Cervus elaphus nelsoni*) in northcentral Colorado. *Journal of Wildlife Disease* **33:** 1–6.

Springer, J. T. (1982). Movement patterns of coyotes in south-central Washington. *Journal of Wildlife Management* **46:** 191–200.

Staats, C. M. and Schall, J. J. (1996). Malarial parasites (Plasmodium) of *Anolis* lizards: biogeography in the Lesser Antilles. *Biotropica* **28:** 388–393.

Stankiewicz, M., Jowett, G. H., Roberts, M. G., Heath, D. D., Cowan, P., Clark, J. M. *et al.* (1996). Internal and external parasites of possums (*Trichosurus vulpecula*) from forest and farmland, Wanganui, New Zealand. *New Zealand Journal of Zoology* **23:** 345–353.

Stankiewicz, M., Cowan, P. E. and Heath, D. D. (1997). Endoparasites of brushtail possums (*Trichosurus vulpecula*) from the South Island, New Zealand. *New Zealand Veterinary Journal* **45:** 257–260.

Staprans, S. I., Hamilton, B. L., Follansbee, S. E., Elbeik, T., Barbosa, P., Grant, R. M. *et al.* (1995). Activation of virus-replication after vaccination of HIV-1-infected individuals. *Journal of Experimental Medicine* **182:** 1727–1737.

Stear, M. J. and Murray, M. (1994). Genetic-resistance to parasitic disease—particularly of resistance in ruminants to gastrointestinal nematodes. *Veterinary Parasitlogy* **54:** 161–176.

Stear, M. J. and Wakelin, D. (1998). Genetic resistance to parasitic infection. *Revue Scientifique et Technique de L'Office International Des Epizooties* **17:** 143–153.

Stear, M. J., Bishop, S. C., Duncan, J. L., McKellar, Q. A., and Murray, M. (1995). The repeatability of fecal egg counts, peripheral eosinophil counts, and plasma pepsinogen concentrations during deliberate infections with *Ostertagia circumcincta*. *International Journal for Parasitology* **25:** 375–380.

Stear, M. J., Bairden, K. Bishop, S. C., Gettinby, G., McKellar, Q. A., Park, M. *et al.* (1998). The processes influencing the distribution of parasitic nematodes among naturally infected lambs. *Parasitology* **117:** 165–171.

Stenseth, N. C. and Maynard Smith, J. (1984). Coevolution in ecosystems—red queen evolution or stasis. *Evolution* **38:** 870–880.

Stephens, P. (1997). *A census and analysis of trends in the large mammal populations of Bale Mountains National Park, Ethiopia*. University of Kent.

Stock, T. M. (1985). *Patterns of community ecology and coevolution of intestinal helminths in grebes*. Edmonton, University of Alberta.

Stock, T. M. and Holmes, J. C. (1988). Functional relationships and microhabitat distributions of enteric helminths of grebes (Podicipedidae): the evidence for interactive communities. *Journal of Parasitology* **74:** 214–227.

Swinton, J., Tuyttens, F., MacDonald, D. W., Nokes, D. J., Cheeseman, C. L. and Clifton-Hadley, R. (1997). Comparison of fertility control and lethal control of bovine tuberculosis in badgers: the impact of perturbation induced transmission. *Philosophical Transactions of the Royal Society of London Series B-Biological Sciences* **352:** 619–631.

Swinton, J., Harwood, J., Grenfell, B. T. and Harwood, J. (1998). Persistence thresholds for phocine distemper virus infection in harbour seal *Phoca vitulina* metapopulations. *Journal of Animal Ecology* **67:** 54–68.

Tapper, S. (1992). *Game heritage: an ecological review from shooting and gamekeeping records*. Fordingbridge, Game Conservancy Trust.

Taylor, L. R. (1961). Aggregation, variance and the mean. *Nature* **189:** 732–735.

Taylor, L. R. and Taylor, R. A. J. (1977). Aggregation, migration and population dynamics. *Nature* **265:** 415–421.

Taylor, L. H. and Read, A. F. (1997). Why so few transmission stages? Reproductive restraint by malaria parasites. *Parasitology Today* **13:** 135–140.

Taylor, L. H. and Read, A. F. (1998). Determinants of transmission success of individual clones from mixed-clone infections of the rodent malaria parasite, *Plasmodium chabaudi. International Journal for Parasitology* **28:** 719–725.

Taylor, L. R., Taylor, R. A. J., Woiwod, I. P. and Perry, J. N. (1983). Behavioural dynamics. *Nature* **303:** 801–804.

Taylor, L. H., Mackinnon, M. J. and Read, A. F. (1998). Virulence of mixed-clone and single-clone infections of the rodent malaria *Plasmodium chabaudi. Evolution* **52:** 583–591.

Telford, S. R. I., Dawson, J. E., Katavolos, P., Warner, C. K., Kolbert, C. P. and Persing, D. H. (1996). Perpetuation of the agent of human granulocytic ehrlichiosis in a deer tick-rodent cycle. *Proceedings of the National Academy of Sciences of the United States of America* **93:** 6209–6214.

Tessaro, S. V. and Forbes, L. B. (1986). *Brucella suis* biotype. 4: A case of granulomatous nephritis in a barren ground caribou (*Rangifer tarandus groenlandicus*) with a review of the distribution of rangiferine brucellosis in Canada. *Journal of Wildlife Diseases* **22:** 479–483.

Thomas, C. D. (1994). Extinction, colonization, and metapopulations: environmental tracking by rare species. *Conservation Biology* **8:** 373–378.

Thomas, A. D. and Reid, N. R. (1944). Rinderpest in game. A description of an outbreak and an attempt at limiting its spread by means of a bush fence. *Onderspoort Journal of Veterinary Science and Animal Industry* **20:** 7–20.

Thomas, C. D. and Jones, T. M. (1993). Partial recovery of a skipper butterfly (*Hesperia comma*) from population refuges: lessons for conservation in a fragmented landscape. *Journal of Animl Ecology* **62:** 472–481.

Thompson, J. N. and Burdon, J. J. (1992). Gene-for-gene coevolution between plants and parasites. *Nature* **360:** 121–125.

Thompson, S. N. and Kavaliers, M. (1994). Physiological bases for parasite-induced alterations of host behavior. *Parasitology* **109:** S119–S138.

Thompson, P. M., Cornwell, H. J. C., Ross, H. M. and Miller, D. (1992). Serological study of phocine distemper in a population of harbour seals in Scotland. *Journal of Wildlife diseases* **28:** 21–27.

Thorne, E. T. and Morton, J. K. (1978). Brucellosis in elk II. Clinical effects and means of transmission as determined through artificial infections. *Journal of Wildlife Diseases* **14:** 280–291.

Thorne, E. T. and Williams, E. S. (1988). Disease and endangered species: the black-footed ferret as a recent example. *Conservation Biology* **2:** 66–74.

Thrall, P. H. and Antonovics, J. (1997). Polymorphism in sexual vs nonsexual disease transmission. *Proceedings of the Royal Society of London Series B-Biological Science* **264:** 581–587.

Thrall, P. H., Antonovics, J. and Wilson, W. G. (1998). Allocation to sexual versus non-sexual disease transmission. *American Naturalist* **151:** 29–45.

Thrusfield, M. (1995). *Veterinary epidemiology*. Oxford, Blackwell Science Ltd.

Thul, J. E., Forrester, D. J. and Abercrombie, C. L. *et al.* (1985). *Ecology* of parasitic helminths of wood ducks, *Aix sponsa*, in the atlantic flyway. *Proceedings of the Helminthological Society of Washington* **52:** 297–310.

Tilman, D. (1994). Competition and biodiversity in spatially structured habitats. *Ecology* **75:** 2–16.

Tinsley, R. C. (1989). The effects of host sex on transmission success. *Parasitology Today* **5:** 190–195.

Tokeshi, M. (1995). On the mathematical basis of variance-mean power relationship. *Researches on Population Ecology* **37:** 43–48.

Tolari, F., Meneguz, P. G., DeMeneghi, D., Rossi, L. and Mancianti, F. (1987). *Indagini sieroepidemiologiche su stambecchi, camosci ed ovini presenti nel Parco Naturale Argentera.* Atti del Convegno Internazionale Lo Stambecco delle Alpi; realtà attuale e prospettive, Valdieri (Italia) 17/19 Settembre 1987.

Tompkins, D. M. and Begon, M. (1999). Parasites can regulate wildlife populations. *Parasitology Today* **15:** 311–313.

Tompkins, D. M. and Wilson, K. (1998). Wildlife disease ecology: from theory to policy. *Trends in Ecology and Evolution* **13:** 476–478.

Tompkins, D. M., Dickson, G. and Hudson, P. J. (1999). Parasite-mediated competition between pheasant and grey partridge: a preliminary investigation. *Oecologia* **119:** 378–382.

Tompkins, D. M., Draycott, R. A. H. and Hudson, P. J. (2000a). Field evidence for apparent competition mediated via the shared parasites of two gamebird species. *Ecology Letters* **3:** 10–14.

Tompkins, D. M., Greenman, J. V. *et al.* (2000b). The role of shared parasites in the exclusion of wildlife hosts: *Heterakis gallinarum* in the ring-necked pheasant and the grey partridge. *Journal of Animal Ecology* **69:** 829–840.

Tompkins, D. M., Greenman, J. V. *et al.* (2001). Differential impact of a shared nematode parasite on two gamebird hosts: implications for apparent competition. *Parasitology* **122:** 187–193.

Tong, H. (1990). *Non-linear time series: a dynamical system approach*. Oxford, Clarendon Press.

Tong, H. (1996). Some comments on nonlinear time series analysis. *Fields Institute Communications* **11:** 17–27.

Trout, R. C., Chasey, D. and Sharp, G. (1997). Seroepidemiology of rabbit haemorrhagic disease (RHD) in wild rabbits (*Oryctolagus cuniculus*) in the United Kingdom. *Journal of Zoology* **243:** 846–853.

Trust, T. J. (1986). Pathogenesis of infectious disease of fish. *Annual Review of Microbiology* **40:** 479–502.

Tryland, M., Sandvik, T., Hansen, H., Hawkenes, G., Holtet, L., Bennelt, M. *et al.* (1998). Characteristics of four cowpox virus isolates from Norway and Sweden. *APMIS* **106:** 623–635.

Tsanava, S. A., Sakvarelidze, L. A. and Shelukhina, E. M. (1989). Serologic survey of wild rodents in georgia for antibodies to orthopoxviruses. *Acta Virologica* **33:** 91–91.

Turchin, P. (1995). Population regulation: old arguments and a new synthesis. In: *Population dynamics*. (ed. N. Cappuccino and P. Price). New York, Academic Press: 19–39.

Turell, M. J. (1988). *The arboviruses: epidemiology and ecology*. Boca Raton, FL, CRC Press.

Tyvand, P. A. (1993). An exact algebraic theory of genetic drift in finite diploid populations with random mating. *Journal of Theoretical Biology* **163:** 315–331.

Vail, S. G., Wileyto, E. P., Hopkins, R. and Smith, G. (2000). *Density effects in larval and nymphal stages of the black-legged tick (Ixodes scapularis) life cycle: experiments and mathematical models*. Third International Conference Ticks and Tick-borne Pathogens: Into the 21st Century.

Van Baalen, M. and Sabelis, M. W. (1995). The scope for virulence management—a comment on ewalds view on the evolution of virulence. *Trends in Microbiology* **3:** 414–416.

Van den Bosch, F., Hengeveld, F. R. and Metz, J. A. J. (1992). Analysing the velocity of animal range expansion. *Journal of Biogeography* **19:** 135–150.

Van-Lawick-Goodall, J. (1971). In the Shadow of Man. Glasgow, William Collins Sons.

Van Riper III, C., Van Riper, S. G., Goft, M. L. and Laird, M. (1986). The epizootiology and ecological significance of malaria in Hawaiian land birds. *Ecological Monographs* **56:** 327–344.

Van Valen, L. (1973). A new evolutionary law. *Evolutionary Theory* **1:** 1–30.

Varley, G. C., Gradwell, G. R. and Hassell, M. P. (1973). *Insect population ecology: an analytical approach*. Oxford, Blackwell Scientific Publications.

Vasconcelos, S. D., Cory, J. S., Wilson, K. R., Sait, S. M. and Hails, R. M. (1996). Modified behavior in Baculovirus-infected lepidopteran larvae and its impact on the spatial distribution of inoculum. *Biological Control* **7:** 299–306.

Verhulst, S., Dieleman, S. J. and Parmentier, H. K. (1999). A trade-off between immunocompetence and sexual ornamentation in domestic fowl. *Proceedings of the National Academy of Science of the USA* **96:** 4478–4481.

Verster , A., Imes, G. D. J. and Smit, J. P. J. (1975). Helminths recovered from the Bontbok, *Damaliscus dorcas dorcas* (Pallas, 1766). *Onderstepoort Journal of Veterinary Research* **42:** 29–32.

Viggers, K. L., Lindenmayer, D. B. and Speatt, D. M. (1993). The importance of disease in reintroduction programmes. *Wildlife Research* **20:** 697–698.

Viney, M. E. (1999). Exploiting the life cycle of *Strongyloides ratti*. *Parasitology Today* **15:** 231–235.

Visser, I. K. G., Vedder, E. J., Vos, H. W., Van de Bildt, M. W. G. and Osterhaus, A. D. M. E. (1993). Continued presence of phocine distemper virus in the Dutch Wadden Sea population. *Veterinary Record* **133:** 320–322.

Volkman, L. E. (1997). Nucleopolyhedrovirus interactions with their insect hosts. *Advances in Virus Research* **48:** 313–348.

Wakelin, D. (1996a). Immunology and genetics of zoonotic infections involving parasites. *Comparative Immunology Microbiology and Infectious Diseases* **19:** 255–265.

Wakelin, D. (1996b). *Immunity to parasites : how parasitic infections are controlled*. Cambridge, Cambridge University Press.

Wakelin, D. and Blackwell, J. (1988). *The genetics of resistance to bacterial and parasitic infection*. London, Taylor & Francis.

Walker, D. E. and Dumler, J. S. (1996). Emergence of the ehrlichioses as human health problems. *Emerging Infectious Diseases* **2:** 18–29.

Wakelin, D. and Goyal, P. K. (1996). Trichinella isolates: parasite variability and host responses. *International Journal for Parasitology* **26:** 471–481.

Wandeler, A. I. (1994). Oral immunization of wildlife. In: *The natural history of rabies*. (ed. G. M. Baer). Boca Raton, CRC Press: 485–503.

Washino, R. K. and Wood, B. L. (1994). Application of remote-sensing to arthropod vector surveillance and control. *American Journal of Tropical Medicine and Hygiene* **50:** S134–S144.

Watt, D. M. and Walker, A. R. (2000). Pathological effects and reduced survival in *Rhipicephalus appendiculatus* ticks infected with *Theileria parva* protozoa. *Parasitology Research* **86:** 207–214.

Watts, D. J. and Strogatz, S. H. (1998). Collective dynamics of 'small-world' networks. *Nature* **393:** 440–442.

Webster, J. P. and Woolhouse, M. E. J. (1999). Cost of resistance: relationship between reduced fertility and increased resistance in a snail-schistosome host-parasite system. *Proceedings of the Royal Society of London Series B-Biological Sciences* **266:** 391–396.

Wedekind, C. (1994). Handicaps not obligatory in sexual selection for resistance genes. *Journal of Theoretical Biology* **170:** 57–62.

Wedekind, C. and Jakobsen, P. J. (1998). Male-biased susceptibility to helminth infection: an experimental test with a copepod. *Oikos* **81:** 458–462.

Weigensberg, I. and Roff, D. A. (1996). Natural heritabilities: can they be reliably estimated in the laboratory? *Evolution* **50:** 2149–2157.

White, K. A. J. and Grenfell, B. T. (1997). Regulation of complex host dynamics by a macroparasite. *Journal of Theoretical Biology* **186:** 81–91.

White, K. A. J. and Wilson, K. (1999). Modelling density-dependent resistance in insect-pathogen interactions. *Theoretical Population Biology* **56:** 163–181.

White, P. C. L., Lewis, A. J. G. and Harris, S. (1997). Fertility control as a means of controlling bovine tuberculosis in badger (*Meles meles*) populations in south-west England: predictions from a spatial stochastic simulation model. *Proceedings of the Royal Society of London Series B-Biological Science* **264:** 1737–1747.

White, L. J., Cox, M. J. and Medley, G. F. (1998). Cross immunity and vaccination against multiple microparasite strains. *Ima Journal of Mathematics Applied in Medicine and Biology* **15:** 211–233.

Wiggins, S. (1990). *Introduction to applied nonlinear dynamical systems and chaos.* Berlin, Springer-Verlag.

Wilkinson, T. L. (1998). The elimination of intracellular microorganisms from insects: an analysis of antibiotic-treatment in the pea aphid (Acyrthosiphon pisum). *Comparative Biochemistry and Physiology A-Molecular and Integrative Physiology* **119:** 871–881.

Williams, E. and Miller, M. W. (2000). Chronic wasting disease in cervids. *Brain Pathology* **10:** 608–608.

Williams, E. S., Thorne, E. T., Appel, M. J. G. and Belitsky, D. W. (1988). Canine distemper in black footed ferrets (*Mustela nigripes*) from Wyoming. *Journal of Wildlife Diseases* **24:** 385–398.

Williams, E. S., Thorne, E. T., Kwiatkowski, D. R. and Oakleaf, B. (1992). Overcoming disease problems in the black-footed ferret recovery programme. *Transactions of the North American Wildlife and Natural Resources Conference* **57:** 474–485.

Williams, E. S., Anderson, S. L., Cavender, J., Lynn, C., List, K., Hearn, C. *et al.* (1996). Vaccination of black-footed ferret (*Mustela nigripes*) x Siberian polecat (*Mustela. eversmanni*) hybrids and domestic ferrets (*Mustela. putorius furo*) against canine distemper. *Journal of Wildlife Diseases* **32:** 417–423.

Williams, E. S., Cain, S. L. and Davies, D. S. (1997). Brucellosis: the disease in bison. In: *Brucellosis, bison, elk, and cattle in the Greater Yellowstone area: defining the problem, exploring solutions.* (ed. E. T. Thorne, M. S. Boyce, P. Nicoletti, and T. J. Kreeger). Cheyenne, Wyoming, Game and Fish Department: 7–19.

Williams-Blangero, S. and Vandeberg, J. L. (1998). Finding the genes that determine susceptibility to parasitic diseases: combining anthropology and genetic epidemiology. *American Journal of Human Biology* **10:** 163.

Williams-Blangero, S., Subedi, J., Upadahayay, R. P., Manral, D. B., Rai, D. R., Jha, B. *et al.* (1999). Genetic analysis of susceptibility to infection with *Ascaris lumbricoides. American Journal of Tropical Medicine and Hygiene* **60:** 921–926.

Wilson, K. (1994). *Analysis of worm and egg counts from the 1992 crash.* Cambridge, Report University of Cambridge.

Wilson, K. and Grenfell, B. T. (1997). Generalised linear modelling for parasitologists. *Parasitology Today* **13:** 33–38.

Wilson, H. B. and Hassell, M. P. (1997). Host-parasitoid spatial models: the interplay of demographic stochasticity and dynamics. *Proceedings of the Royal Society of London Series B-Biological Sciences* **264:** 1189–1195.

Wilson, K. and Reeson, A. F. (1998). Density-dependent prophylaxis: evidence from Lepidoptera-baculovirus interactions? *Ecological Entomology* **23:** 100–101.

Wilson, M. H., Kepler, C. B., Snyder, N. F. R., Dickson, S. R., Dein, F. J., Wiley, J. W. *et al.* (1994). Puerto Rican parrots and potential limitations of the metapopulation approach to species conservation. *Conservation Biology* **8:** 114–123.

Wilson, K., Grenfell, B. T. and Shaw, D. J. (1996). Analysis of aggregated parasite distributions: a comparison of methods. *Functional Ecology* **10:** 592–601.

Wirblich, C., Meyers, G. *et al.* (1994). European brown hare syndrome virus—relationship to rabbit hemorrhagic-disease virus and other caliciviruses. *Journal Of Virology* **68:** 5164–5173.

Wobeser, G. (1995). *Involvement of small wild animals in bovine tuberculosis. Tuberculosis in wildlife and domestic animals.* Proceedings of the second international *Mycobacterium bovis* conference.

Wolff, P. L. and Seal, U. S. (1993). Implications of infectious-disease for captive propagation and reintroduction of threatened species. *Journal of Zoo and Wildlife Medicine* **24:** 229–230.

Woodford, M. H. (1993). International disease implications for wildlife translocation. *Journal of Zoo and Wildlife Medicine* **24:** 265–270.

Woodroffe, R. (1997). The conservation implications of immobilizing, radio-collaring and vaccinating free-ranging wild dogs. In: *The African wild dog: status survey and conservation action plan.* (ed. R. Woodroffe, J. R. Ginsberg, and D. W. Macdonald). Gland, IUCN.

Woodroffe, R. (1998). *Managing disease threats to wild mammals.* Animal Conservation.

Woolhouse, M. E. J. (1991). On the application of mathematical-models of schistosome transmission dynamics. 1. Natural transmission. *Acta Tropica* **49:** 241–270.

Woolhouse, M. E. J. (1992a). A theoretical framework for the immunoepidemiology of helminth infection. *Parasite Immunology* **14:** 563–578.

Woolhouse, M. E. J. (1992b). Evidence for genetic-factors for resistance susceptibility to schistosome infection. *American Journal of Human Genetics* **51:** 206–207.

Woolhouse, M. E. J. (1992c). Immunoepidemiology of intestinal helminths—pattern and process. *Parasitology Today* **8:** 111–111.

Woolhouse, M. E. J. (1998). Patterns in parasite epidemiology: the peak shift. *Parasitology Today* **14:** 428–434.

Woolhouse, M. E. J., Dye, C. *et al.* (1997). Heterogeneities in the transmission of infectious agents: implications for the design of control programs. *Proceedings of the National Academy of Sciences of the United States of America* **94:** 338–342.

Woolhouse, M. E. J., Stringer, S. M. *et al.* (1998). Epidemiology and control of scrapie within a sheep flock. *Proceedings of the Royal Society of London Series B-Biological Sciences* **265:** 1205–1210.

Woolhouse, M. E. J., Matthews, L. *et al.* (1999). Population dynamics of scrapie in a sheep flock. *Philosophical Transactions of the Royal Society of London Series B-Biological Sciences* **354:** 751–756.

Wright, V. L., Farris, A. L. *et al.* (1980). *Effects of Heterakis and Histomonas on the survival of juvenile gray partridge.* Proceedings of Perdix II Gray Partridge Workshop, Moscow, University of Idaho.

Yan, G. Y. and Norman, S. (1995). Infection of tribolium beetles with a tapeworm—variation in susceptibility within and between beetle species and among genetic strains. *Journal of Parasitology* **81:** 37–42.

Yan, G. Y. and Stevens, L. (1995). Selection by parasites on components of fitness in tribolium beetle: the effect of intraspecific competition. *American Naturalist* **146:** 795–813.

Yan, G., Severson, D. W. *et al.* (1997). Costs and benefits of mosquito refractoriness to malaria parasites: implications for genetic variability of mosquitoes and genetic control of malaria. *Evolution* **51:** 441–450.

Yao, Q. and Tong, H. (1995). *On initial-condition sensitivity and prediction in nonlinear stochastic systems.* 50th session of ISI, Beijing.

Yeh, T. (1998). *Disease-induced selection in metapopulation models: a computer simulation analysis.* Durham, NC., Duke University, School of the Environment.

Yodzis, P. (1998). Local trophodynamics and the interaction of marine mammals and fisheries in the Benguela ecosystem. *Journal of Animal Ecology* **67:** 635–658.

Young, A. S., Dolan, T. T. *et al.* (1996). Factors affecting infections in *Rhipicephalus appendiculatus* ticks fed on cattle infected with *Theileria parva. Parasitology* **113:** 255–266.

Yu, I. C., Parker, J. *et al.* (1998). Gene-for-gene disease resistance without the hypersensitive response in Arabidopsis dnd1 mutant. *Proceedings of the National Academy of Sciences of the United States of America* **95:** 7819–7824.

Zaffaroni, E., Fraquelli, C. *et al.* (1996). Abomasal helminth communities in eastern alpine sympatric roe deer (*Capreolus capreolus*) and chamois (*Rupicapra rupicapra*) populations. *Supplemento Ricerche di Biologia della Selvaggina* **24:** 53–68.

Zahavi, A. (1975). Mate selection—a selection for handicap. *Journal of Theoretical Biology* **53:** 205–214.

Zhioua, E., Aeschlimann, A. *et al.* (1994). Infection of field-collected ixodes-ricinus (acari, ixodidae) larvae with *Borrelia burgdorferi* in Switzerland. *Journal of Medical Entomology* **31:** 763–766.

Zohar, A. S. and Rau, M. E. (1986). The role of muscle larvae of *Trichinella spiralis* in the behavioral alterations of the mouse host. *Journal of Parasitology* **72:** 464–466.

Zuk, M. (1990). Reproductive strategies and disease susceptibility: an evolutionary viewpoint. *Parasitology Today* **6:** 231–233.

Zuk, M. (1992). The role of parasites in sexual selection—current evidence and future-directions. *Advances in the Study of Behavior* **21:** 39–68.

Zuk, M. and McKean, K. A. (1996). Sex differences in parasite infections: patterns and processes. *International Journal for Parasitology* **26:** 1009–1023.

Zwillinger, D. (1997). *Handbook of differential equations.* San Diego, Academic Press.

Index